Compendium of Plant Genomes

Series Editor

Chittaranjan Kole, Raja Ramanna Fellow, Government of India,
ICAR-National Research Center on Plant Biotechnology, Pusa,
New Delhi, India

Whole-genome sequencing is at the cutting edge of life sciences in the new millennium. Since the first genome sequencing of the model plant Arabidopsis thaliana in 2000, whole genomes of about 100 plant species have been sequenced and genome sequences of several other plants are in the pipeline. Research publications on these genome initiatives are scattered on dedicated web sites and in journals with all too brief descriptions. The individual volumes elucidate the background history of the national and international genome initiatives; public and private partners involved; strategies and genomic resources and tools utilized; enumeration on the sequences and their assembly; repetitive sequences; gene annotation and genome duplication. In addition, synteny with other sequences, comparison of gene families and most importantly potential of the genome sequence information for gene pool characterization and genetic improvement of crop plants are described.

Interested in editing a volume on a crop or model plant? Please contact Prof. C. Kole, Series Editor, at ckoleorg@gmail.com

More information about this series at http://www.springer.com/series/11805

Ilga M. Porth • Amanda R. De la Torre
Editors

The Spruce Genome

Springer

Editors
Ilga M. Porth
Faculté de foresterie, de géographie
et de géomatique
Université Laval
Quebec, QC, Canada

Amanda R. De la Torre
School of Forestry
Northern Arizona University
Flagstaff, AZ, USA

ISSN 2199-4781 ISSN 2199-479X (electronic)
Compendium of Plant Genomes
ISBN 978-3-030-21000-7 ISBN 978-3-030-21001-4 (eBook)
https://doi.org/10.1007/978-3-030-21001-4

This Springer imprint is published by the registered company Springer Nature Switzerland AG
The registered company address is: Gewerbestrasse 11, 6330 Cham, Switzerland

Preface to the Series

Genome sequencing has emerged as the leading discipline in the plant sciences coinciding with the start of the new century. For much of the twentieth century, plant geneticists were only successful in delineating putative chromosomal location, function, and changes in genes indirectly through the use of a number of "markers" physically linked to them. These included visible or morphological, cytological, protein, and molecular or DNA markers. Among them, the first DNA marker, the RFLPs, introduced a revolutionary change in plant genetics and breeding in the mid-1980s, mainly because of their infinite number and thus potential to cover maximum chromosomal regions, phenotypic neutrality, absence of epistasis, and codominant nature. An array of other hybridization-based markers, PCR-based markers, and markers based on both facilitated construction of genetic linkage maps, mapping of genes controlling simply inherited traits, and even gene clusters (QTLs) controlling polygenic traits in a large number of model and crop plants. During this period, a number of new mapping populations beyond F_2 were utilized and a number of computer programs were developed for map construction, mapping of genes, and for mapping of polygenic clusters or QTLs. Molecular markers were also used in the studies of evolution and phylogenetic relationship, genetic diversity, DNA fingerprinting, and map-based cloning. Markers tightly linked to the genes were used in crop improvement employing the so-called marker-assisted selection. These strategies of molecular genetic mapping and molecular breeding made a spectacular impact during the last one and a half decades of the twentieth century. But still they remained "indirect" approaches for elucidation and utilization of plant genomes since much of the chromosomes remained unknown and the complete chemical depiction of them was yet to be unraveled.

Physical mapping of genomes was the obvious consequence that facilitated the development of the "genomic resources" including BAC and YAC libraries to develop physical maps in some plant genomes. Subsequently, integrated genetic–physical maps were also developed in many plants. This led to the concept of structural genomics. Later on, emphasis was laid on EST and transcriptome analysis to decipher the function of the active gene sequences leading to another concept defined as functional genomics. The advent of techniques of bacteriophage gene and DNA sequencing in the 1970s was extended to facilitate sequencing of these genomic resources in the last decade of the twentieth century.

As expected, sequencing of chromosomal regions would have led to too much data to store, characterize, and utilize with the-then available computer software could handle. But the development of information technology made the life of biologists easier by leading to a swift and sweet marriage of biology and informatics, and a new subject was born—bioinformatics.

Thus, the evolution of the concepts, strategies, and tools of sequencing and bioinformatics reinforced the subject of genomics—structural and functional. Today, genome sequencing has traveled much beyond biology and involves biophysics, biochemistry, and bioinformatics!

Thanks to the efforts of both public and private agencies, genome sequencing strategies are evolving very fast, leading to cheaper, quicker, and automated techniques right from clone-by-clone and whole-genome shotgun approaches to a succession of second-generation sequencing methods. The development of software of different generations facilitated this genome sequencing. At the same time, newer concepts and strategies were emerging to handle sequencing of the complex genomes, particularly the polyploids.

It became a reality to chemically—and so directly—define plant genomes, popularly called whole-genome sequencing or simply genome sequencing.

The history of plant genome sequencing will always cite the sequencing of the genome of the model plant Arabidopsis thaliana in 2000 that was followed by sequencing the genome of the crop and model plant rice in 2002. Since then, the number of sequenced genomes of higher plants has been increasing exponentially, mainly due to the development of cheaper and quicker genomic techniques and, most importantly, the development of collaborative platforms such as national and international consortia involving partners from public and/or private agencies.

As I write this preface for the first volume of the new series "Compendium of Plant Genomes," a net search tells me that complete or nearly complete whole-genome sequencing of 45 crop plants, eight crop and model plants, eight model plants, 15 crop progenitors and relatives, and three basal plants is accomplished, the majority of which are in the public domain. This means that we nowadays know many of our model and crop plants chemically, i.e., directly, and we may depict them and utilize them precisely better than ever. Genome sequencing has covered all groups of crop plants. Hence, information on the precise depiction of plant genomes and the scope of their utilization are growing rapidly every day. However, the information is scattered in research articles and review papers in journals and dedicated Web pages of the consortia and databases. There is no compilation of plant genomes and the opportunity of using the information in sequence-assisted breeding or further genomic studies. This is the underlying rationale for starting this book series, with each volume dedicated to a particular plant.

Plant genome science has emerged as an important subject in academia, and the present compendium of plant genomes will be highly useful to both students and teaching faculties. Most importantly, research scientists involved in genomics research will have access to systematic deliberations on the plant genomes of their interest. Elucidation of plant genomes is of interest not only for the geneticists and breeders, but also for practitioners of an array of plant science disciplines, such as taxonomy, evolution, cytology, physiology, pathology, entomology, nematology, crop production, biochemistry, and obviously bioinformatics. It must be mentioned that information regarding each plant genome is ever-growing. The contents of the volumes of this compendium are, therefore, focusing on the basic aspects of the genomes and their utility. They include information on the academic and/or economic importance of the plants, description of their genomes from a molecular genetic and cytogenetic point of view, and the genomic resources developed. Detailed deliberations focus on the background history of the national and international genome initiatives, public and private partners involved, strategies and genomic resources and tools utilized, enumeration on the sequences and their assembly, repetitive sequences, gene annotation, and genome duplication. In addition, synteny with other sequences, comparison of gene families, and, most importantly, the potential of the genome sequence information for gene pool characterization through genotyping by sequencing (GBS) and genetic improvement of crop plants have been described. As expected, there is a lot of variation of these topics in the volumes based on the information available on the crop, model, or reference plants.

I must confess that as the series editor, it has been a daunting task for me to work on such a huge and broad knowledge base that spans so many diverse plant species. However, pioneering scientists with lifetime experience and expertise on the particular crops did excellent jobs editing the respective volumes. I myself have been a small science worker on plant genomes since the mid-1980s and that provided me the opportunity to personally know several stalwarts of plant genomics from all over the globe. Most, if not all, of the volume editors are my longtime friends and colleagues. It has been highly comfortable and enriching for me to work with them on this book series. To be honest, while working on this series I have been and will remain a student first, a science worker second, and a series editor last. And I must express my gratitude to the volume editors and the chapter authors for providing me the opportunity to work with them on this compendium.

I also wish to mention here my thanks and gratitude to the Springer staff, particularly Dr. Christina Eckey and Dr. Jutta Lindenborn for the earlier set of volumes and presently Ing. Zuzana Bernhart for all their timely help and support.

I always had to set aside additional hours to edit books beside my professional and personal commitments—hours I could and should have given to my wife, Phullara, and our kids, Sourav and Devleena. I must mention that they not only allowed me the freedom to take away those hours from them but also offered their support in the editing job itself. I am really not sure whether my dedication of this compendium to them will suffice to do justice to their sacrifices for the interest of science and the science community.

New Delhi, India Chittaranjan Kole

Preface

The Spruce Genome, an Important Resource to Fundamental Biological Research and Selective Tree Breeding

Main Text

Spruces (*Picea spp.*) are naturally abundant and widely distributed conifer tree species in the Northern hemisphere. Due to their enormous ecological and economic value, management of this important forest genetic resource has focused on conservation and tree improvement. Recently, with the aid of improved sequencing technologies and bioinformatics advances, a draft genome sequence of the 20 Gigabases Norway spruce (*P. abies*) genome was published (Nature 497:581 (2013)). Canadian white spruce hybrid (*P. glauca* × *engelmannii* × *sitchensis*) genome assembly followed in the same year (Bioinformatics 29:1492 (2013)), establishing spruce as a model species in gymnosperm genomics. Continuous efforts to improve the spruce genome assembly are underway, but are challenged by the inherent characteristics of conifer genomes: high amounts of repetitive sequences (introns and transposable elements) and large gene family expansions related to abiotic stress responses, secondary metabolism, and their defense responses against pathogens and herbivory. Because the assembly is still highly fragmented with millions of scaffolds, the generation of ultra-dense genetic maps allows anchoring these scaffolds onto the 12 haploid spruce chromosomes represented by the 12 linkage groups in a spruce genetic map. The generation of RNA-seq data further aids to improve scaffolds. Such data are also particularly valuable in comparative genomics and can highlight the functional divergence between species. Bacterial artificial chromosomes (BACs) sequencing has also served the spruce genomics research community greatly, by (a) unraveling the substantial presence of pseudogenes, (b) supporting the isolation of entire metabolite-biosynthetic genes, (c) facilitating conifer genome comparisons for microsynteny, and last but not least (d) proving indispensable for spruce genome assembly. Some of these BAC sequencing efforts predated the spruce genome sequencing project.

The post-genomic era has seen a surge in genomic applications not only for species amenable to population genomics using whole genome data. In fact, genomics applications using a reduced representation of the massive spruce genome have become very popular (e.g., exome capture sequencing, genotyping-by-sequencing, restriction site-associated DNA sequencing). Throughout this book, we highlight all areas that have been impacted by the acquisition of a high-quality reference genome for spruce. In brief, this volume aims to provide the latest information on (1) status of the genome assembly, (2) detailed insights into whole genome and gene family structure, (3) comprehensive genomic resources available for research, (4) emerging genomics tools for tree improvement programs, (5) genomics related to genetic conservation programs, and (6) functional genomics to improve gene function annotations.

Chapters 1 and 2 focus on the current state of the nuclear and organelle genome assemblies since the first publication of draft genomes and the newest attempts to use whole genome re-sequencing (WGS) data for variant calling. WGS is unprecedented for conifers' complex genomes, where reduced-representation-sequencing-based genotyping has been the state-of-the-art genomic method. By contrast, confident WGS-based variant calling in a population of 1000 individuals for poplar, an angiosperm forest tree species with a 45× smaller genome size, constitutes no major obstacle nowadays. For spruce, however, current challenges regarding such an approach remain. These challenges are highlighted in the respective book chapter on *Picea abies* and potential solutions are extensively discussed. The following two chapters are on repetitive elements, which represent an important 70% fraction of the genome, and retrotransposons are suspected to actually drive spruce genome expansion. The significant differences overall with angiosperm transposable elements dynamics are highlighted; an example also illustrates how BAC sequencing conclusively helps characterize features of a retroelement family important in explaining spruce genome evolutionary dynamics. The epigenomics chapter focuses on the current state of knowledge about epigenetic variation in spruce. One of the chapters devoted to comparative genomics looks at the comparison of nuclear and organelle genomes among spruces, and among spruces and other gymnosperms, focusing on aspects of comparative mapping, and rates of sequence evolution. Another comparative genomics chapter focuses on the sequencing and annotation of a few randomly selected BACs and provides further insights into whole genome evolution comparisons and genome structural features among conifers (spruce versus pine) with regard to genes versus transposons. Spruce genomic resources also have important implications for modern tree selective breeding. A separate chapter is therefore dedicated to genomic selection in white spruce and the increased ability to capture genetic gain by more accurate phenotype prediction models obtained from improved genomic resources. This became possible with the implementation of the genomic pairwise kinship relationship matrix among individuals. This relationship captures the traditional contemporary pedigree (i.e., half- and full-sib family relationships) as well as the historical pedigree through the identification of common DNA variants (SNPs) passed through generations. The following two chapters deal with local adaptation in spruce, the genetic underpinnings of resistance to drought as well as of cold hardiness. This represents a highlight of the current knowledge in clinal genetic variation in spruces. The last two chapters describe genes and gene families implicated in the formation of terpenes and phenols, the most important secondary compounds in spruce defense. Some of these genes have anti-herbivory and pathogen resistance potential. The book will close with an outlook into emerging fields of research in spruce genomics.

Quebec, Canada Ilga M. Porth
Flagstaff, USA Amanda R. De la Torre
November 2019

Contents

1 Sequencing and Assembling the Nuclear and Organelle Genomes of North American Spruces

Inanc Birol and Amanda R. De la Torre

Abstract

Reference genomes provide valuable information to study the molecular biology and the genomic architecture of species, and constitute a baseline for applied sciences such as molecular breeding and gene editing. The sequencing of conifer genomes still lags behind other plant and animal species, with only a few available conifers having full sequence genomes to date. This chapter aims to describe details on the sequencing and bioinformatics analysis of the nuclear and organelle genome assemblies of the economically important white spruce (*Picea glauca*), and closely related *Picea* species *P. sitchensis* and *P. engelmannii*. The chapter finishes by providing some perspectives for future genome assemblies of North American species.

I. Birol
Department of Medical Genetics, Michael Smith Genome Sciences Centre, University of British Columbia, Vancouver, Canada
e-mail: ibirol@bcgsc.ca

A. R. De la Torre (✉)
School of Forestry, Northern Arizona University, 200 E. Pine Knoll, Flagstaff AZ86001, AZ, USA
e-mail: Amanda.de-la-torre@nau.edu

1.1 Introduction

We are at the dawn of an era for conifer genomics. Several groups around the world have developed reference resources for a variety of conifer species (Birol et al. 2013; Nystedt et al. 2013; Zimin et al. 2014; Warren et al. 2015)—valuable reference material to study the molecular biology of these species. Beyond advancing basic science, reference genome assemblies and their detailed annotations are keys to support the development of marker systems for applications in conifer breeding programs (De La Torre et al. 2014a).

The flurry of activity in the field is a manifestation of the wider availability and reducing costs of high throughput sequencing platforms. The Illumina sequencing technology (Illumina Inc., San Diego, CA), in its various forms, has dominated the field of de novo genome sequencing until very recently. Illumina, offers several instruments built on the concept of sequencing by synthesis, where clusters of amplified DNA fragments are interrogated by a series of fluorescently labeled reversible termination reactions (Bentley et al. 2008). On their high-throughput platform, the approach generates over a billion short and accurate sequences, typically achieving error rates under 1% with up to 250 base pair (bp) reads.

Recently, the proven performance and favourable per nucleotide cost of the Illumina

I. M. Porth and A. R. De la Torre (eds.), *The Spruce Genome*, Compendium of Plant Genomes,
https://doi.org/10.1007/978-3-030-21001-4_1

sequencers have been leveraged to provide co-localization information on short reads. The Chromium sequencing library preparation instrument from 10X Genomics (Pleasanton, CA) uses microfluidics to label fragments from large DNA molecules (reaching beyond 100,000 bp), expanding the utility of the short reads generated (Zheng et al. 2016).

In contrast to Illumina sequencing, instruments from Pacific Biosciences (PacBio, Menlo Park, CA), and Oxford Nanopore Technologies (ONT, Oxford, UK) generate long reads from single molecules, opening up new possibilities. These reads can be three orders of magnitude longer than the reads from Illumina platforms, albeit with higher error rates. The PacBio technology is built on real-time observation of DNA polymerase reactions on single-stranded DNA templates immobilized in zero-mode waveguide arrays (Eid et al. 2009). The ONT technology is the only sequencing approach on major commercial platforms that does not synthesise DNA, but reads electrical signals off single-strand molecules while they transition through nanopores.

These sequencing platforms continue to improve, and new technologies promise longer reads, better data quality, and higher throughput. With each wave of progress in sequencing technologies, new frontiers emerge for life sciences. However, this *emergence* is by no means spontaneous; enabling research on new frontiers invariably requires new analytics capabilities. The rapidly evolving field of conifer genomics is a prime example of how large-scale problems benefit from these developments, while also inspiring the development of cutting-edge bioinformatics technologies.

This chapter offers an account of this interaction, within the specifics of building reference resources for spruce genomes. It primarily focuses on algorithms developed at Canada's Michael Smith Genome Sciences Centre (Vancouver, BC), and how they were used in assembling the nuclear genome of the Canadian white spruce (*Picea glauca*) (Birol et al. 2013; Warren et al. 2015), and the organellar genomes of *P. glauca* and the Sitka spruce (*P. sitchensis*) (Jackman et al. 2016; Coombe et al. 2016; Lin et al. 2019). While the story of the spruce genomes is still being written, where projects are underway to complete their assemblies, we expect the algorithms reported here and their future versions will continue to play a significant role in collating and refining these valuable resources.

1.1.1 De Novo Assembly

The first DNA genome of an organism, the 5,375 base pair (bp) sequence of the bacteriophage phi X174, was published in 1977 by Sanger et al. (1977), inspiring generations of researchers attempt larger and larger genomes. While the de novo assembly problem this pioneering work had tackled was tractable even for manual reconstruction of the genome, the problem quickly becomes intractable for larger targets.

De novo sequence assembly refers to the reconstruction of a sequence (usually DNA or RNA sequence, in the context of genomics research) using redundant random sampling of the underlying sequence (usually, genome or transcriptome), without consulting a similar reference sequence. The sampling redundancy of a genome is called the fold coverage, and represented by x. Although wrong, fold coverage is often simply referred to as coverage. More correctly, the latter refers to the fraction of the genome covered by reads, yet the ambiguity in terminology is compensated by the unit used. For example, 30× coverage indicates that a genome is on average covered by 30 reads, while 30% coverage indicates the percentage of the bases in the genome represented by at least one read.

When two sequences overlap partially and unambiguously (meaning, the overlapping sequence is unique in the underlying genome) they can be merged to obtain an extended contiguous sequence, or contig in short. As such, de novo assembly task relies heavily on identifying pairs of overlapping sequence reads in experimental data. When a sequencing experiment generates n reads, in its naïve conception, the problem of identifying pairs of overlapping sequence requires comparison of every sequence to every other sequence, hence is an $O(n^2)$ problem, the notation indicating the order of magnitude of the number of operations needed to perform the task.

Fast-forwarding to date, where the number of reads in a sequencing experiment is in the billions for large genomes, it becomes apparent that performing $O(10^{18})$ operations, even on modern high performance computers, would not be feasible.

Practical de novo sequence assembly algorithms use a variety of approaches to simplify this problem. One common technique in such complex computer science problems is to balance algorithmic complexity with computational memory use. In this problem space, search for read-to-read overlaps can be reformulated as a table look up, and $O(n)$ operation, by limiting the sought after overlaps to be of $k - 1$ bp, and by using the fact that the genomic alphabet is composed of four letters (bases): A for adenine, C for cytosine, G for guanine, and T for thymine.

To accommodate, every read is *shredded* into sub-sequences of length k bp, assuming the reads are at least of this length, and redundant observations of the same k bp sequences, or k-mers for short, are collapsed into single representations. The result of this simplification is a table of k-mers, and all other k-mers overlapping a given k-mer by $k - 1$ bp can be found by consulting this table to interrogate $k - 1$ bp sequences appending one of the four bases. Accounting for overlaps on both ends, this totals to eight lookups for each k-mer. When one conceptualizes k-mers as nodes on graph, and overlaps between k-mers as directed edges, the result is called a de Bruijn graph (Pevzner and Tang 2001).

Although this basic component of the genome assembly is well described, the subsequent stages of assembly pipelines vary from algorithm to algorithm, and are less defined. The most modern sequence assembly algorithms, can use paired end sequencing data to (1) disambiguate contig extensions and (2) build scaffolds. The latter happens when flanking sequences around a difficult-to-assemble sequence is unambiguous. When multiple sequencing data types are available, often bespoke pipelines are implemented to best leverage their information.

1.1.2 The Nuclear Genome

The engine behind reconstructing the 20 billion bp (Gbp) genome of the Canadian white spruce (*Picea glauca*) (Birol et al. 2013; Warren et al. 2015) was several algorithms implemented in the ABySS package (Simpson et al. 2009), including ABySS-Bloom and ABySS release v1.5.2, which made use of memory-efficient Bloom filters for analyzing the sequence of large genomes (Birol et al. 2013). The original draft assembly of the genome in 2013 (Birol et al. 2013) reported a contiguity of NG50 = 22,967 bp, reconstructing 20.8 Gbp in 4.9 million scaffolds. Only two years later, this was surpassed by an NG50 of 83,000 bp (Warren et al. 2015), resulting in the most contiguous spruce assembly to date (Table 1.1). This significant increase in contiguity was achieved by the use of transcriptome re-scaffolding and large fragment mate pair sequences. The coding gene space in white spruce only represents 0.11–0.37% of the genome, therefore the increase in contiguity obtained with transcriptome re-scaffolding was low in comparison to the high increase obtained with the use of large fragment mate pair sequences (Birol et al. 2013). The assembly strategy included three main steps. In the first one, the genome assembly PG29 v2 (Birol et al. 2013) was re-scaffolded using RNA-seq libraries and large-fragment mate pair data. In the second step, a second genome assembly (WS77111, Warren et al. 2015) was created. The draft assembly of genotype WS77111 was sequenced after genomic analyses of PG29 suggested the presence of introgression from other spruces (*P. engelmannii*, *P. sitchensis*) (De La Torre et al. 2014b). The third step included the use of the WS77111 draft assembly for second-stage long-range re-scaffolding of PG29 v3 informed by scaffold alignments to the assembly WS77111 (Birol et al. 2013). Besides increases in contiguity, all spruce (and conifer) reference genomes to date

Table 1.1 Comparison among sequence assemblies published for North American spruce genomes

Species	Individual sequenced	Assembly	Genome size (Gb)	Scaffold NG50 (kb)	References
white spruce	WS77111	v1	22.4	19.9	Warren et al. (2015)
white x Engelmann x Sitka spruce	PG29	v1	20.8	22.9	Birol et al. (2013)
	PG29	v2	20.8	41.9	Birol et al. (2013)
	PG29	v3	20.8	71.5	Warren et al. (2015)
	PG29	v4	20.8	83	Warren et al. (2015)

are still highly fragmented due to the challenges associated with sequencing and assembling highly repetitive, complex, and very large genome sizes (De La Torre et al. 2014a).

1.1.3 Organelle Genomes

Following the sequencing of the nuclear genomes, efforts were dedicated into sequencing the organelle genomes of white spruce and other North American spruces. Chloroplast and mitochondrial genomes were sequenced from clone PG29, the individual used in the PG29 v1-4 nuclear genome assemblies of white x Engelmann x Sitka spruce (Jackman et al. 2015). The 123-kb chloroplast and 5.9 mitochondrial genomes were sequenced with Illumina MiSeq and HiSeq at 80X and 30X of coverage, respectively. Pair-end reads were assembled using ABySS, gaps were closed with additional Illumina sequencing or PacBio long reads, and coding regions were annotated with MAKER (Campbell et al. 2014). Transcript abundance of the annotated mitochondrial genome was quantified using RNA-Seq data from three developmental stages and five tissues. The mitochondrial genome had an N50 of 369 kb and contained 106 protein-coding genes (51 distinct genes), 29 tRNAs, and 8rRNAs, whereas the chloroplast genome was composed by 74 protein-coding genes, 36 tRNAs, and 4rRNAs. In contrast to the highly variable angiosperm chloroplast genomes, the chloroplast genome of white spruce had high levels of synteny and co-linearity in relation to Norway spruce even though the species diverged more than 10 Mya (Jackman et al. 2015). The chloroplast of non-admixed white spruce genotype WS77111 was also sequenced a few years later (Lin et al. 2019). In comparison with the PG29 assembly, the WS77111 genotype produced a higher-quality assembly with a N50 lengths of 3692, 1313, and 949 bp (43X, 88X, and 172X) after assembly with ABySS (Lin et al. 2019), even though having a similar size (123,421 bp vs. 123,266 bp).

Sitka spruce (*Picea sitchensis*) and Engelmann spruce (*Picea engelmannii*) are other economically important species growing in western North America. The chloroplast genome of Sitka spruce was sequenced earlier than white spruce, using Solexa sequencing-by-synthesis technology (Cronn et al. 2008). A more recent assembly was sequenced using long-read technologies (Coombe et al. 2016). Sequencing libraries were prepared with 10X Genomics platform that allows the incorporation of long DNA fragments, and libraries were later sequenced using Illumina HiSeq. Assembly followed with ABySS and gaps were filled with Sealer (Paulino et al. 2015). The final assembly had a size of 124,049 bp, about 3873 bp larger than the previous assembly, and a 99% sequence identity with other spruce chloroplast genomes such as white and Norway spruce (Coombe et al. 2016). The Engelmann spruce chloroplast genome was sequenced using whole-genome shotgun sequencing, and later assembled with ABySS to result in a 123,542 bp assembly (Lin et al. 2019). Finally, the mitochondrial genome of Sitka spruce was sequenced using the Oxford Nanopore MinION technology. Reads were assembled using Unicycler (Wick et al. 2017), resulting in a 5.5-Mbp mitochondrial genome assembly (Jackman et al. 2019).

1.2 Nucleotide Sequence Alignment

Bioinformatics technologies have always been in a race with genomics technologies, especially in the era of high-throughput sequencing, to deliver timely results. For example, as the sequencing throughput on the Illumina platforms increased from millions to tens and hundreds of millions of reads per lane—recently pushing the billion mark on their NovaSeq instrument—sequence alignment methods shifted through several paradigms to offer analytical capability to projects that use these platforms.

At the dawn of the NGS era, short-read instruments were initially marketed as re-sequencing platforms. Their reads ($\sim$30 base pairs (bp) at the time) were meant to be interpreted only with respect to a reference genome. To fill the void of bioinformatics tools capable of handling large volumes of data, Illumina provided Eland (Cox, unpublished), a sequence alignment algorithm based on the pigeon-holing principle—where a sequence is partitioned into n non-overlapping sections, requiring at least one of them to have a perfect match to the reference, allowing guaranteed performance for $n - 1$ mismatches.

For faster searches through the reference genome, MAQ algorithm (Li et al. 2008) introduced the concept of hash indexing with multiple spaced seeds (called noncontiguous seed templates in the paper), essentially allowing for missed hits, as long as there is enough statistical evidence from matching seeds. When the hash indexing paradigm also buckled under the increasing throughputs, the community moved on to using concepts from data compression algorithms, such as FM indexing (Ferragina et al. 2000). Algorithms that implemented this paradigm, such as BWA (Li et al. 2008) and Bowtie (Langmead and Salzberg 2012), offered an order of magnitude faster run times (Hatem et al. 2013), briefly catching up with the increase in sequencing throughputs.

While precise sequence alignment may be needed in many established sequence analysis pipelines, for instance, for variant analysis (Robertson et al. 2010; Van der Auwera et al. 2013), in certain other applications alignment-free methods may be more suitable, such as in quantifying gene expression levels. Methods like Salmon and Kallisto (Patro et al. 2014; Bray et al. 2016) employ a light mapping strategy based on databases of k-mers, sequences of uniform length k. To distinguish the two paradigms, we will call the conventional methods as sequence *alignment* methods, and methods like the ones implemented in Salmon and Kallisto as sequence *mapping* methods.

1.3 Gene Annotation

Information is not knowledge. While the former is about the question "what," the latter is about exploring the "how" and "why." While the former is about foraging and collating data, the latter is about interpreting the data, building hypotheses, and models. The turn of this century witnessed a rapid transformation in biology, which arguably was mostly an observational and descriptive (i.e., information) science, to a more interactive and predictive (i.e., knowledge) science. This transformation has been the starkest in genomics.

The DNA molecule is essentially an information storage medium. It harbours all the information needed for life, and enables the transfer of that information across generations. Since the discovery of the double helix structure of DNA, and the realization that it was the *sequence* of the constituting nucleotides that encode that information, researchers looked for ways to read those sequences. As discussed in previous sections, there are widely available commercial platforms for DNA sequencing, using various experimental approaches, and a range of bioinformatics methods tightly coupled with their strengths and weaknesses.

An assembled genome, in and of itself, is just raw information, even when it is highly contiguous and correct. Even though that information harbours all the essential biology about the species it describes, inferring that link is what makes the assembled genome valuable. The exercise of cataloguing that link is called genome annotation.

Though the wide selection of methods we see in the de novo assembly field is not paralleled in the annotation of assembled genomes, several tools have served the community well. The genome annotation engines, such as Maker and its derivatives (Holt and Yandell 2011; Campbell et al. 2014; Liu et al. 2014), ab initio gene finders, such as SNAP (Korf et al. 2004) and Augustus (Stanke et al. 2006), and specialized tools, such as DOGMA for organellar genomes (Wyman et al. 2004), Prokka for prokaryotic genomes (Seemann 2014), and MaGe for microbial genomes (Vallenet et al. 2006) use a range of different paradigms. But, broadly speaking, they are based on the concept of sequence homology, carrying over what is known in previously annotated genomes of evolutionarily related species to the genome under study. Some annotation tools, such as Maker and Augustus, do this indirectly using machine learning approaches, such as hidden Markov models (HMMs). As recent exciting results from fundamental research on machine learning are fueling various genomics applications (Libbrecht and Noble 2015), it is worth revisiting the problem of automated genome annotation and interpretation.

In the annotation of the spruce genomes, we used one of the most widely used and sophisticated annotation tools, the Maker pipeline (Holt and Yandell 2011). Now in its second version, Maker2, the pipeline uses evidence from data sources, such as RNA-seq, for de novo gene annotations, and integrates predictions from established ab initio gene finders, including SNAP (Korf et al. 2004), Augustus (Stanke et al. 2006), and GeneMark (Besemer and Borodovsky 2005). We have successfully used Maker in a number of genome assembly projects (Quevillon et al. 2005; Diguistini et al. 2009, 2011; Chan et al. 2011; Birol et al. 2013; Feau et al. 2016; Haridas et al. 2013; Keeling et al. 2013; Jackman et al. 2016; Warren et al. 2015; Coombe et al. 2016; Hammond et al. 2017; Jones et al. 2017a, b). To our experience, the pipeline works best for contiguous assemblies, but genes that are split across multiple contigs or fall across scaffold gaps are partially annotated or missed. Also, while the pipeline includes a workflow that integrates InterProScan (Quevillon et al. 2005) to predict protein family (PFam) domains (Finn et al. 2006) for annotated coding transcripts, it does not provide suggested biological functions for its predicted genes.

1.4 Glossary

Contig—It is a set of overlapping reads (DNA segments) that after multiple sequence alignment result in a consensus sequence representing a region of the draft genome being sequenced.

Unitig—It is a high-confidence contig. Contigs consist of one or more unitigs.

Scaffold—It is an ordered and oriented set of one or more contigs. The scaffold representing a single sequence may also contain gaps.

Superscaffold—Represents a group of scaffolds, usually obtained when the contiguity of an existing draft genome is improved by the use of long reads.

HMM—Stands for Hidden Markov Model, a statistical model used to build sequence analysis algorithms in computational molecular biology.

De novo *assembly*—Refers to the assembly of a novel genome of a species that does not present any previous reference sequences available for alignment.

References

Bentley DR, Balasubramanian S, Swerdlow HP, Smith GP, Milton J et al (2008) *Accurate whole human genome sequencing using reversible terminator chemistry*. Nature 456(7218):53–59

Besemer J, Borodovsky M (2005) *GeneMark: web software for gene finding in prokaryotes, eukaryotes and viruses*. Nucleic Acids Res 33:451–454

Birol I, Raymond A, Jackman SD, Pleasance S, Coope R et al (2013) *Assembling the 20 Gb white spruce (Picea glauca) genome from whole-genome shotgun sequencing data*. Bioinformatics 29(12):1492–1497

Bray NL, Pimentel H, Melsted P, Pachter L (2016) *Near-optimal probabilistic RNA-seq quantification*. Nat Biotechnol 34(5):525–527

Campbell MS, Holt C, Moore B, Yandell M (2014) *Genome Annotation and Curation Using MAKER and MAKER-P*. Curr Protoc Bioinformatics **48**: 4 11 1–39

Chan QW, Cornman RS, Birol I, Liao NY, Chan SK et al (2011) *Updated genome assembly and annotation of Paenibacillus larvae, the agent of American foulbrood disease of honey bees*. BMC Genom 12:450

Coombe L, Warren RL, Jackman SD, Yang C, Vandervalk BP et al (2016) *Assembly of the Complete Sitka Spruce Chloroplast Genome Using 10X Genomics' GemCode Sequencing Data*. PLoS ONE 11(9): e0163059

Cronn R, Liston A, Parks M, Gernandt DS, Shen R, Mockler T (2008) Multiplex sequencing of plant chloroplast genomes using Solexa sequencing-by-synthesis technology. Nucleic Acids Res 36:e122. https://doi.org/10.1093/nar/gkn502 PMID: 18753151

De La Torre AR, Birol I, Bousquet J, Ingvarsson PK, Jansson S, Jones SJM, Keeling CI, MacKay J, Nilsson O, Ritland K, Street N, Yanchuk A, Zerbe P, Bohlmann J (2014a) Insights into Conifer Giga-genomes. Plant Physiol 166:1–9

De La Torre AR, Roberts D, Aitken SN (2014b) Genome-wide admixture and ecological niche modeling reveal the maintenance of species boundaries despite long history of interspecific gene flow. Mol Ecol 23:2046–2059

Diguistini S, Liao NY, Platt D, Robertson G, Seidel M, Chan SK et al (2009) *De novo genome sequence assembly of a filamentous fungus using Sanger, 454 and Illumina sequence data*. Genome Biol 10(9):R94

Diguistini S, Wang Y, Liao NY, Taylor G, Tanguay P, Feau N et al (2011) *Genome and transcriptome analyses of the mountain pine beetle-fungal symbiont Grosmannia clavigera, a lodgepole pine pathogen*. Proc Natl Acad Sci U S A 108(6):2504–2509

Eid J, Fehr A, Gray J, Luong K, Lyle J, Otto G et al (2009) *Real-time DNA sequencing from single polymerase molecules*. Science 323(5910):133–138

Feau N, Taylor G, Dale AL, Dhillon B, Bilodeau GJ, Birol I et al (2016) *Genome sequences of six Phytophthora species threatening forest ecosystems*. Genomics Data 10:85–88

Ferragina P, Manzini G (2000) *Opportunistic data structures with applications*, in *41st Annual Symposium on Foundations of Computer Science, Proceedings* 390–398

Finn RD, Mistry J, Schuster-Bockler B, Griffiths-Jones S, Hollich V, Lassmann T et al (2006) *Pfam: clans, web tools and services*. Nucleic Acids Res 34:D247–D251

Haridas S, Wang Y, Lim L, Alamouti SM, Jackman S, Docking R et al (2013) *The genome and transcriptome of the pine saprophyte Ophiostoma piceae, and a comparison with the bark beetle-associated pine pathogen Grosmannia clavigera*. BMC Genom 14:373

Hammond SA, Warren RL, Vandervalk BP, Kucuk E, Khan H, Gibb EA et al (2017) *The North American bullfrog draft genome provides insight into hormonal regulation of long noncoding RNA*. Nature Communications 8(1):1433

Hatem A, Bozdag D, Toland AE, Catalyurek UV (2013) *Benchmarking short sequence mapping tools*. BMC Bioinformatics 14:184

Holt C, Yandell M (2011) *MAKER2: an annotation pipeline and genome-database management tool for second-generation genome projects*. BMC Bioinformatics 12:491

Jackman SD, Warren RL, Gibb EA, Vandervalk BP, Mohamadi H, Chu J et al (2016) *Organellar Genomes of White Spruce (Picea glauca): Assembly and Annotation*. Genome Biology and Evolution 8(1): 29–41

Jones SJ, Haulena M, Taylor GA, Chan S, Bilobram S, Warren RL (2017) *The Genome of the Northern Sea Otter (Enhydra lutris kenyoni)*. Genes (Basel) **8**(12)

Jones SJM, Taylor GA, Chan S, Warren RL, Hammond SA, Bilobram S, Mordecai G et al (2017) *The Genome of the Beluga Whale (Delphinapterus leucas)*. Genes (Basel) **8**(12)

Keeling CI, Yuen MMS, Liao NY, Docking TR, Chan SK, Taylor GA et al (2013) *Draft genome of the mountain pine beetle, Dendroctonus ponderosae Hopkins, a major forest pest*. Genome Biol 14(3):R27

Korf I (2004) *Gene finding in novel genomes*. BMC Bioinformatics 5:59

Langmead B, Salzberg SL (2012) *Fast gapped-read alignment with Bowtie 2*. Nat Methods 9(4):357–359

Li H, Ruan J, Durbin R (2008) *Mapping short DNA sequencing reads and calling variants using mapping quality scores*. Genome Res 18(11):1851–1858

Libbrecht MW, Noble WS (2015) *Machine learning applications in genetics and genomics*. Nat Rev Genet 16(6):321–332

Lin D, Coombe L, Jackman SD, Gagalova KK, Warren RL, Hammond SA, Kirk H et al (2019) *Complete Chloroplast Genome Sequence of a White Spruce (Picea glauca, Genotype WS77111) from Eastern Canada*. Microbiology Resource Announcements **8** (23)

Liu J, Xiao H, Huang S, Li F (2014) *OMIGA: Optimized Maker-Based Insect Genome Annotation*. Mol Genet Genomics 289(4):567–573

Nystedt B, Street NR, Wetterbom A, Zuccolo A, LIn YC, Scofield DG (2013) *The Norway spruce genome sequence and conifer genome evolution*. Nature **497** (7451): 579–84

Patro R, Mount SM, Kingsford C (2014) *Sailfish enables alignment-free isoform quantification from RNA-seq reads using lightweight algorithms*. Nat Biotechnol 32 (5):462–464

Paulino D, Warren RL, Vandervalk BP, Raymond A, Jackman SD, Birol I (2015) Sealer: a scalable gap-closing application for finishing draft genomes. BMC Bioinformatics 16:230. https://doi.org/10.1186/s12859-015-0663-4 PMID: 26209068

Pevzner PA, Tang H (2001) *Fragment assembly with double-barreled data*. Bioinformatics 17(Suppl 1): S225–S233

Quevillon E, Silventoinen V, Pillai S, Harte N, Mulder N, Apweiler R, Lopez R (2005) *InterProScan: protein domains identifier*. Nucleic Acids Res 33:W116–W120

Robertson G, Schein J, Chiu R, Corbett R, Field M, Jackman SD et al (2010) *De novo assembly and analysis of RNA-seq data*. Nat Methods 7(11):909–912

Sanger F, Air GM, Barrell BG, Brown NL, Coulson AR, Fiddes JC (1977) *Nucleotide sequence of bacteriophage phi X174 DNA*. Nature 265(5596):687–695

Seemann T (2014) *Prokka: rapid prokaryotic genome annotation*. Bioinformatics 30(14):2068–2069

Simpson JT, Wong K, Jackman SD, Schein JD, Jones SJM, Birol I (2009) *ABySS: a parallel assembler for short read sequence data*. Genome Res 19 (6):1117–1123

Stanke M, Keller O, Gunduz I, Hayes A, Waack S, Morgenstern B (2006) *AUGUSTUS: ab initio prediction of alternative transcripts*. Nucleic Acids Research **34**(Web Server issue): W435–9

Vallenet D, Labarre L, Rouy Z, Barbe V, Bocs S, Cruvellier S et al (2006) *MaGe: a microbial genome annotation system supported by synteny results*. Nucleic Acids Res 34(1):53–65

Van der Auwera GA, Carneiro MO, Hartl C, Poplin R, del Angel G, Levy-Moonshine A et al (2013) *From FastQ data to high confidence variant calls: the Genome Analysis Toolkit best practices pipeline*. Curr Protoc Bioinformatics **43**: 11 10 1–33

Warren RL, Keeling CI, Yuen MMS, Raymond A, Taylor GA, Vandervalk BP et al (2015) *Improved white spruce (Picea glauca) genome assemblies and annotation of large gene families of conifer terpenoid and phenolic defense metabolism*. Plant J 83(2):189–212

Wick RR, Judd LM, Gorrie CL, Holt KE (2017) Unicycler: Resolving bacterial genome assemblies from short and long sequencing reads Phillippy, AM, editor. PLoS Comput Biol 13:e1005595. https://doi.org/10.1371/journal.pcbi.1005595

Wyman SK, Jansen RK, Boore JL (2004) *Automatic annotation of organellar genomes with DOGMA*. Bioinformatics 20(17):3252–3255

Zheng GX, Lau BT, Schnall-Levin M, Jarosz M, Bell JM, Hindson CM et al (2016) *Haplotyping germline and cancer genomes with high-throughput linked-read sequencing*. Nat Biotechnol 34(3):303–311

Zimin A, Stevens KA, Crepeau MW, Holtz-Morris A, Koriabine M, Marcais G et al (2014) *Sequencing and assembly of the 22-gb loblolly pine genome*. Genetics 196(3):875–890

Variant Calling Using Whole Genome Resequencing and Sequence Capture for Population and Evolutionary Genomic Inferences in Norway Spruce (*Picea Abies*)

2

Carolina Bernhardsson, Xi Wang, Helena Eklöf, and Pär K. Ingvarsson

Abstract

Advances in next-generation sequencing methods and the development of new statistical and computational methods have opened up possibilities for large-scale, high-quality genotyping in most organisms. Conifer genomes are large and are known to contain a high fraction of repetitive elements and this complex genome structure has bearings for approaches that aim to use next-generation sequencing methods for genotyping. In this chapter, we provide a detailed description of a workflow for variant calling using next-generation sequencing in Norway spruce (*Picea abies*). The workflow starts with raw sequencing reads and proceeds through read mapping to variant calling and variant filtering. We illustrate the pipeline using data derived from both whole-genome resequencing data and reduced representation sequencing. We highlight possible problems and pitfalls of using next-generation sequencing data for genotyping stemming from the complex genome structure of conifers and how those issues can be mitigated or eliminated.

2.1 Introduction

Conifers were one of the last plant groups lacking genome assemblies; but recently, several draft genomes have become available for a number of conifers such as Norway spruce (*Picea abies*, Nystedt et al. 2013), Loblolly pine (*Pinus taeda*, Zimin et al. 2014; 2017), Sugar pine (*Pinus lambertiana*, Stevens et al. 2016), and Douglas fir (*Pseudotsuga menziesii*, Neale et al. 2017). This has opened up new possibilities to assess genome-wide levels of genetic diversity in conifers. Earlier studies of genetic diversity in Norway spruce has either been limited to coding regions (e.g., Heuertz et al. 2006; Chen et al.. 2012) or have used various complexity reduction methods, such as genotyping-by-sequencing,

Carolina Bernhardsson and Xi Wang—These authors contributed equally.

C. Bernhardsson · X. Wang · H. Eklöf · P. K. Ingvarsson (✉)
Department of Plant Biology, Linnean Centre for Plant Biology, Swedish University of Agricultural Sciences, 750 05 Uppsala, Sweden
e-mail: par.ingvarsson@slu.se

C. Bernhardsson
e-mail: carolina.bernhardsson@gmail.com

X. Wang
e-mail: xi.wang@umu.se

H. Eklöf
e-mail: helena.dahlberg@umu.se

C. Bernhardsson · X. Wang
Department of Ecology and Environmental Science, Umeå Plant Science Centre, Umeå University, 901 87 Umeå, Sweden

I. M. Porth and A. R. De la Torre (eds.), *The Spruce Genome*, Compendium of Plant Genomes,
https://doi.org/10.1007/978-3-030-21001-4_2

restriction site associated sequencing, or targeted capture sequencing (Baison et al. 2019) to estimate levels of genetic diversity within species. While we have learned a lot about levels of genetic diversity in Norway spruce from such studies, we still lack detailed information on, for instance, levels of nucleotide polymorphism and linkage disequilibrium in non-genic regions. However, with the availability of a reference genome sequence (Nystedt et al. 2013), whole genome resequencing is now also possible in conifers such as Norway spruce.

Conifer genomes are large (20–40 Gb) and have high repetitive content and current draft genome assemblies in conifers are therefore often fragmented into many, relatively short scaffolds. In addition, large fractions of the predicted genome sizes are also missing from reference genomes. The fragmented nature of conifer reference genome assemblies, combined with the high repetitive content make variant calling in conifers difficult. This is true regardless of which techniques have been used to generate sequencing data but perhaps more so for whole-genome resequencing data that can be expected to provide a relatively unbiased coverage of the target genome. In this chapter, we review methods available for variant calling using NGS data and outline some of the issues one may face when performing analyses of data from whole-genome resequencing (WGS) in Norway spruce. In particular, we discuss the performance of variant calling across different genomic contexts, such as coding and non-coding regions and regions known to be composed of repetitive elements. We also compare variant calling using WGS data with data derived from sequence capture probes, designed to target non-repetitive sequences in the *P. abies* genome and discuss how collapsed genomic regions in the assembly complicates the task of filtering for good, reliable variant- and genotype calls. Having access to robust variant calls is important for downstream analyses, such as population genomic analyses or inferences of the demographic history of individuals, populations or the species as a whole. To highlight these issues, we end by assessing how different approaches to variant calling alter the site frequency spectrum of variants and hence possible evolutionary inferences drawn from the data.

2.2 Sample Collection

We sampled 35 individuals of Norway spruce (*Picea abies*) spanning their natural distributions, mainly from Russia, Finland, Sweden, Norway, Belarus, Poland, and Romania for use in whole genome resequencing. Individuals Pab001–Pab015 were all derived from unique populations and no specific measurements were taken when they were collected. Samples were taken from newly emerged needles or dormant buds for each individual and stored at −80 °C until DNA extraction. In contrast, individuals Pab016–Pab035 were sampled from two different areas, one in the eastern and one in the western part of Västerbotten province in northern Sweden. Two different populations were sampled in each area, one old and untouched forest (>100 years old) and one young planted population (<20 years old). For each population, a transect was made and five trees were sampled from each population along the transect. From each tree, a number of fresh shoots were broken off and put into prelabeled ziplock bags before being taken back to the lab for DNA extractions.

Genomic DNA was extracted using Qiagen plant mini kit following manufacturer's instructions. All sequencing was performed at the National Genomics Initiative platform (NGI) at SciLifeLab facilities, Stockholm, Sweden; paired-end libraries with an insert size of 500 bp were run on different Illumina HiSeq platforms (Pab001–Pab006 on HiSeq 2000, while the remaining individuals on HiSeq X). The original location, estimated coverage from raw sequencing reads and coverage of mapped reads for each individual are given in Table 2.1.

Samples analyzed using sequence capturing methods were obtained from Bernhardsson et al. (2019; 1,997 haploid samples) and Baison et al. (2019; 526 diploid samples). For further information on sample collection, DNA extraction

Table 2.1 Sample name, origin, sequencing platform, and setup, estimated sequencing coverage based on raw reads delivered and coverage of reduced and merged BAM files before subdivision into genomic subsets (including possible PCR duplicates) for each of the 35 whole genome resequenced Norway spruce samples

Sample	Origin	Platform	Sequencing setup (bp)	Sequence coverage	BAM coverage
Pab001	Yakutsk, Russia (*P. obovata*)	HiSeq 2000	2 × 100	10.1x	12.6x
Pab002	Gettinge, Sweden	HiSeq 2000	2 × 100	36.9x[b]	12.0x
Pab003	Vitebsk, Belarus	HiSeq 2000	2 × 100	9.5x	12.2x
Pab004	Blizyn, Poland	HiSeq 2000	2 × 100	10.5x	11.8x
Pab005	Toplita, Romania	HiSeq 2000	2 × 100	10.3x	12.2x
Pab006[a]	Köttsjön, Sweden	HiSeq 2000	2 × 100	50.0x[b]	11.1x
Pab007	Hatfjelldal, Norway	HiSeq X	2 × 151	12.6x	14.5x
Pab008	Rovaniemen, Finland	HiSeq X	2 × 151	11.8x	14.5x
Pab009	Kittilä, Finland	HiSeq X	2 × 151	13.2x	14.2x
Pab010	Suomussalmi, Finland	HiSeq X	2 × 151	11.9x	14.5x
Pab011	Hemnes, Norway	HiSeq X	2 × 151	14.6x	14.2x
Pab012	Levanger, Norway	HiSeq X	2 × 151	12.4x	14.5x
Pab013	Grane, Norway	HiSeq X	2 × 151	13.3x	15.1x
Pab014	Tyda, Norway	HiSeq X	2 × 151	12.4x	15.2x
Pab015	Loppi, Finland	HiSeq X	2 × 151	13.6x	14.9x
Pab016	Marsfjället, Sweden	HiSeq X	2 × 151	19.6x	15.5x
Pab017	Marsfjället, Sweden	HiSeq X	2 × 151	21.0x	15.5x
Pab018	Marsfjället, Sweden	HiSeq X	2 × 151	20.4x	15.8x
Pab019	Marsfjället, Sweden	HiSeq X	2 × 151	20.2x	25.5x
Pab020	Marsfjället, Sweden	HiSeq X	2 × 151	21.1x	25.2x
Pab021	Marsfjället, Sweden	HiSeq X	2 × 151	20.9x	25.5x
Pab022	Marsfjället, Sweden	HiSeq X	2 × 151	19.9x	26.0x
Pab023	Marsfjället, Sweden	HiSeq X	2 × 151	19.8x	25.2x
Pab024	Marsfjället, Sweden	HiSeq X	2 × 151	18.9x	24.5x
Pab025	Marsfjället, Sweden	HiSeq X	2 × 151	18.8x	23.8x
Pab026	Långrumpskogen, Sweden	HiSeq X	2 × 151	19.2x	23.8x
Pab027	Långrumpskogen, Sweden	HiSeq X	2 × 151	18.6x	23.5x
Pab028	Långrumpskogen, Sweden	HiSeq X	2 × 151	19.3x	23.5x
Pab029	Långrumpskogen, Sweden	HiSeq X	2 × 151	18.5x	23.2x
Pab030	Långrumpskogen, Sweden	HiSeq X	2 × 151	18.0x	23.2x
Pab031	Långrumpskogen, Sweden	HiSeq X	2 × 151	17.8x	23.2x
Pab032	Långrumpskogen, Sweden	HiSeq X	2 × 151	20.1x	23.5x
Pab033	Långrumpskogen, Sweden	HiSeq X	2 × 151	20.0x	23.6x
Pab034	Långrumpskogen, Sweden	HiSeq X	2 × 151	20.2x	23.1x
Pab035	Långrumpskogen, Sweden	HiSeq X	2 × 151	19.3x	23.3x

[a]reference individual Z4006 (Nystedt et al. 2013); [b]Downsampled to match remaining samples

and sequencing please refer to the original publications.

2.3 SNP and Genotype Calling Pipeline

Over the recent decades, next-generation sequencing (NGS) technologies have developed at an extraordinary rapid pace and have led to ever decreasing costs per mega-base sequence generated. This has in turn led to rapid increase in the number and diversity of sequenced genomes (Goodwin et al. 2016). These new sequencing technologies have already facilitated and improved many research avenues in biology, such as transcriptomics, gene annotation, and RNA splice identification (Bao et al. 2011). Moreover, metagenomic (Schuster 2007) and genome methylation analysis (Morozova and Marra 2008) have also benefited from NGS technologies. However, with NGS platforms the associated error rates are higher and generally read lengths are shorter compared to traditional Sanger sequencing platforms, and therefore, results require much more careful examination, particularly for variant discovery and downstream applications (Liu et al. 2012; Goodwin et al. 2016). Thus, meaningful analysis of NGS data relies crucially on accurate calling of SNPs and genotypes (Nielsen et al. 2011). Here, we first highlight some of the issues that may contribute to ambiguities in SNP and genotype calling, e.g., repetitive DNA, or misalignment of sequence reads, and then review some recent algorithms and tools that are capable of improving the sensitivity and specificity of SNP detection and genotype determination from NGS data. The pipeline for SNP and genotype calling applied in this chapter includes initial reads mapping, post-alignment processing, pre-calling processing, final SNP and genotype calling, and subsequent variant filtering (Fig. 2.1). The steps of this pipeline are exemplified by using whole-genome resequencing data from 35 Norway spruce (*Picea abies*) individuals generated by different Illumina HiSeq platforms.

2.3.1 Initial Reads Mapping (Step 1)

Mapping is the most fundamental step in variant detection using NGS and involves aligning raw or pre-processed sequence reads to a reference genome of either the same or a closely related species to reveal the location(s) of reads within the reference genome (Trapnell and Salzberg 2009; Flicek and Birney 2010; Wang et al. 2015). The accuracy of mapping plays a crucial role in variants calling (Nielsen et al. 2011) as incorrectly aligned reads may lead to errors in SNP and genotype calling, and therefore will restrict the accuracy of all downstream analyses. Although progressively more advanced NGS technologies feature several advantages such as high throughput, lower cost per base and higher accuracy when compared with traditional Sanger-based sequencers (Bao et al. 2011), several obstacles stemming from the inherent characteristics of NGS data need to be considered when performing this step.

2.3.1.1 Short-Read Mapping

The first challenge is that reads from NGS platforms are usually short (most delivered-read lengths fall in the range of 35–400 bp), when compared with the reads from Sanger sequencers (600–800 bp) (Bao et al. 2011; Mardis 2017). In order to use these short reads efficiently, the initial process, before actual alignment, involves the construction of indices for the reference sequence and/or for the raw short-read sequence data known as "indexing," and is intended to speed up subsequent mapping steps. The two most commonly used indexing algorithms have been incorporated in a large number of mapping software tools: (i) indexing algorithms based on hash tables and (ii) indexing algorithms based on Burrows–Wheeler transform (BWT) (Burrows and Wheeler 1994). A "Hash table" is a common data structure which makes it possible to search rapidly through complex and non-sequential data using an index. One option is to scan the hash table of input reads using the reference genome. However, scanning a hash table of input reads usually requires smaller and more variable memory, while scanning the entire reference genome may use more processing

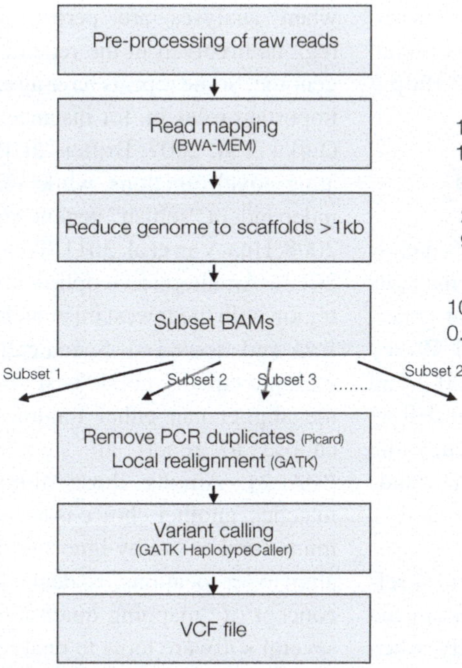

Fig. 2.1 Pipeline of SNP variant and genotype calling based on Norway spruce. As a first step, paired end reads were mapped to the whole genome assembly of Norway spruce (~10 million scaffolds covering 12.6 Gb out of the ~20 Gb estimated genome size) using BWA–MEM with default settings. Sample BAM files were then reduced to include only scaffolds longer than 1 Kb (~2 million scaffolds covering 9.5 Gb of the genome), merged into a single BAM file per sample and then subdivided into 20 genomic subsets with ~100 K scaffolds in each. The genomic subset BAM files were then marked for PCR duplicates using `Picard`, subjected to local realignment around indels using `GATK RealignerTargetCreator` and `IndelRealigner` and genomic intermediate variant calling using `GATK Haplotypecaller` in gvcf mode. A joint genotype call over all 35 samples per genomic subset was then performed using `GATK GenotypeGVCFs` to achieve the final raw vcf files (one file for each of the 20 genomic subsets)

time, especially when there are relatively few reads in the input set (Flicek and Birney 2010). Another option is to scan the hash table of the reference genome by using the set of input reads. Regardless of the size of input reads, the memory of the reference hash table remains constant, but on the other hand, may be large, depending on the size and complexity of the reference genome (Flicek and Birney 2010). Several software tools have been developed based on hash table algorithms, such as `MAQ` (Li et al. 2008a), `SOAP` (Li et al. 2008b), `SHRiMP` (Rumble et al. 2009), and `Stampy` (Lunter and Goodson 2011). BWT-based implementations are much faster compared to hash table-based algorithm at the same sensitivity level (Flicek and Birney 2010), they are also more memory efficient and are particularly useful for aligning repetitive reads (Nielsen et al. 2011). Another advantage of the BWT approach is the ability to store the complete reference genome index on disk and load it completely into memory on almost all standard bioinformatics computing clusters (Flicek 2009; Flicek and Birney 2010). Currently, the most widely used software tools based on the BWT algorithms include `Bowtie` (http://bowtie.cbcb.umd.edu/; Langmead et al.

2009), the `Burrows-Wheeler Aligner` (`BWA`) (http://maq.sourceforge.net/bwa-man.shtml; Li and Durbin 2009), and `SOAP2` (http://soap.genomics.org.cn/; Li et al. 2009a).

2.3.1.2 Extremely Large Amounts of Data

Another challenge for working with NGS data is that the amount produced by most NGS methods are orders of magnitude greater than that generated by earlier techniques (Flicek and Birney 2010). For example, Illumina (HiSeq 2000) and Applied Biosystems (ABI SOLiD 4) can deliver hundreds of millions of sequences per run, while Sanger-style sequencer only produces thousands of reads per run (Hung and Weng 2017). Therefore, in order to map a vastly greater numbers of reads (millions or billions) to a reference genome, any algorithm must be optimized for speed and memory usage, especially when reference genomes are very large. Accordingly, several highly memory efficient short-read alignment programs have been developed. For example, `Bowtie` (http://bowtie-bio.sourceforge.net/index.shtml; Langmead et al. 2009) and `BWA`, as examples of the two most efficient short-read aligners, achieve throughputs of 10–40 million reads per hour on a single computer processor (Treangen and Salzberg 2012).

2.3.1.3 Repetitive DNA Mapping

Repetitive DNA sequences, which are similar or identical to sequences elsewhere in the genome, are abundant across a broad range of organisms, from bacteria to mammals (Treangen and Salzberg 2012). Mapping repetitive DNA sequences to a reference genome creates ambiguities of deciding what to do with reads that map to multiple locations, which, in turn, can produce errors when interpreting results. Mapping of repetitive sequencing reads is therefore one of the most commonly encountered problems in read mapping. One possible solution to this problem is to simply remove all multiple mapped reads. However, this discarding strategy may complicate the calculation of coverage by reducing coverage in an uneven fashion (Li et al. 2008a), and may result in many undetected biologically important variants when analyses are performed across unique regions involved in the repetitive contents of the genome. Some repeats have already proven to play important roles in, for instance, human evolution (Jurka et al. 2007; Britten 2010), sometimes creating novel functions, while sometimes acting as independent "selfish" genetic elements (Kim et al. 2008; Hua-Van et al. 2011; Treangen and Salzberg 2012). An alternative option is to only report the region with the fewest mismatches (e.g., as done in `BWA` and `Bowtie`). Specifically, when there are multiple equally good "best alignment matches", the aligner can either randomly choose one or choose to report all such alignments (e.g., `SOAP2`). Among those aligners reporting all matches, another choice is to report up to a maximum number (`m`) by ignoring multiple reads that align to >`m` locations. To deal with these issues, the concept of "mapping quality" was introduced in several software tools to enable evaluation of the likelihood of correct mapping of reads by considering a number of factors, such as base qualities, the number of base mismatches, and/or the existence and size of gaps in the alignment (Li et al. 2008a; Wang et al. 2015). For example, in order to evaluate the reliability of alignments, `MAQ` assigns a Phred-scaled quality score which measures the probability that the true alignment is not the one found by `MAQ` for each individual alignment (Li et al. 2008a). However, `MAQ`'s formula overestimates the probability of missing the true hit, which results in an underestimation of mapping quality (Li and Durbin 2009). `BWA` was developed with a similar algorithm as `MAQ` but has been modified by assuming the true hit can always be found. Simulation reveals that BWA may therefore overestimate mapping quality, although the deviation is relatively small (Li and Durbin 2009).

In general, the choice of alignment tool and the corresponding parameter settings are very important because the outcome will significantly influence the accuracy of variant calling and further downstream analysis. `BWA`, which is based on the BWT algorithm, is much faster compared to many other programs based on hash table algorithm at the same sensitivity level, has more efficiently memory usage, which is particularly useful for aligning repetitive reads, and has a smaller deviation of mapping quality. This is therefore often the best choice for mapping raw sequence reads to a reference genome. BWA provides three different mapping methods: (1) BWA-MEM that is usually suggested for 70 bp or longer Illumina, 454, Ion Torrent and Sanger reads, and is also generally recommended for high-quality queries as it is faster and more accurate; (2) BWA-backtrack is especially suited for short sequences and (3) BWA-SW which may have better sensitivity when alignment gaps are frequent. Consequently, in our whole genome resequencing project of Norway spruce, we have employed BWA-MEM to align all paired-end reads to the reference genome.

2.3.2 Post-alignment Processing (Step 2)

After mapping reads to the reference, various post-alignment data processing types are usually suggested to facilitate further analytical steps. The most common tasks include output file manipulation (e.g., format converting, indexing) and the creation of summary reports from the alignment process. Appropriate formats of output from mapping not only ensure downstream compatibility with variant callers, such as "Haplotypecaller" from the genome analysis toolkit package (GATK) (McKenna et al. 2010), but also allow for a reduction of the output data set size, for example, transferring from SAM file (a text-based format for storing biological sequences aligned to a reference sequence which was developed by Li et al. 2009c) to the binary version (BAM) file (Altmann et al. 2012; Mielczarek and Szyda 2016). The SAMtools package can be used to manipulate a variety of functions on output files, such as, format changing, sorting, indexing, or merging (Li et al. 2009c). Moreover, a detailed summary statistics of aligned output is necessary for evaluating the overall quality and correctness of alignment, which benefits all following steps such as SNP detection. Some aligners usually generate a simple summary describing the alignment process (e.g., Bowtie2, SOAP2), while for some others such as BWA such information is unavailable. Thus, it is possible to generate a summary report by using other software, e.g., the SAMtools flagstat (Li et al. 2009c) and CollectAlignmentSummaryMetrics module of Picard (http://broadinstitute.github.io/picard/).

After mapping raw sequencing reads to the reference genome using BWA-MEM, we employed SAMtools to manipulate output files, for instance, for sorting and indexing. Besides, we also used flagstat in SAMtools to generate summary statistic reports, including, e.g., the total number of reads that pass or fail QC (quality controls), in order to evaluate the quality of the alignments produced during read mapping. In addition, considering the characteristics of the Norway spruce genome which has an extremely large genome size ($\sim$20 Gb) and very high repetitive content ($\sim$70%) (Nystedt et al. 2013), it is challenging to accurately call variants both from a computational perspective and from the large data volumes that need to be handled and stored in this project. Before SNP and genotype calling, we therefore performed several steps to reduce the computational complexity of the SNP calling process. First, we split both the output file from the mapping step and the reference into smaller subsets so that multiple data sets can be manipulated in parallel. Also, smaller data sets (in this case, the number of scaffolds) are crucial to allow software tools, such as GATK, to function properly.

We first reduced the reference genome by only keeping genomic scaffolds greater than 1 kb in length using bioawk (https://github.com/lh3/bioawk). All BAM files were then "subset-ted" using the reduced reference genome with the view module in SAMtools. All reduced BAM files for each individual were merged into a single BAM file using the merge module in SAMtools (Li et al. 2009c). The final step was to split the merged BAM files of each individual into 20 genomic subsets, with each subset containing roughly 100,000 scaffolds, using SAMtools view. We simultaneously subdivided the reference genome into the corresponding 20 genomic subsets by keeping exactly the same scaffolds using bedtools (Quinlan and Hall 2010); indexing was performed for both, BAM subsets and reference subsets, in order to make them available for subsequent analyses.

2.3.3 Pre-calling Processing (Step 3)

Additional processing steps are usually recommended before proceeding to the actual variant calling, so that variant detection results can be more reliable (Li et al. 2009c; McKenna et al. 2010; Altmann et al. 2012; Mielczarek and Szyda 2016).

2.3.3.1 PCR Duplicates

DNA amplification by polymerase chain reaction (PCR) has been a necessary step in the preparation of libraries for most of the NGS platforms, such as Illumina, 454, IonTorrent, and SOLiD (Mardis 2008; Morozova and Marra 2008; Escalona et al. 2016). The number of duplicate clones in the library will increase if too many PCR cycles were run or when DNA was obtained from a gel slice for uniform fragment lengths (Li et al. 2009c). Consequently, the same read pairs may occur many times in the raw data and this will result in unexpectedly high depths in some regions, while less in others, resulting in a skewed coverage distribution (Mielczarek and Szyda2016). Furthermore, excessively high read depth could potentially lead to large frequency differences between the two alleles of a heterozygous site (Li et al. 2009c), which may bias the number of variants and may substantially influence the accuracy of the variant detection (Wang et al. 2015). Removal of these artifacts is thus an essential pre-calling processing step, in particular for applications based on resequencing data. Picard MarkDuplicates (http://broadinstitute.github.io/picard/) and Samtools Rmdup (Li et al. 2009c) are two commonly used software tools that allow the user to either mark these duplicate reads or completely remove them from the alignment. Both methods rely on similar approaches, however, compared with rmdup, MarkDuplicates is likely a better choice as it also handles inter-chromosomal read pairs, considers the library for each read pair, keeps a read pair from each library, and does not actually remove reads but rather flags the duplicates by setting the SAM flag to 1024 for all but the best read pair (Ebbert et al. 2016).

2.3.3.2 Local Realignment

Another important preprocessing step involves alignment corrections. Artifacts in alignment, usually occurring in regions with insertions and/or deletions (indels) during the mapping step, can result in many mismatching bases relative to the reference in regions of misalignment. Such misaligned regions can easily be mistaken for SNPs during variant calling. Read mapping proceeds by independently mapping each read, and reads covering an indel at the start or end of the read can often be incorrectly mapped even when other reads are correctly mapped across the indel. To alleviate these issues, local realignment is usually performed to realign reads in the vicinity of an identified misalignment in order to minimize the number of mismatching bases (Mielczarek and Szyda 2016). Many software tools have been developed to perform local realignment, such as SRMA (Homer and Nelson 2010), which realign reads only in color space originating from the SOLiD platform. GATK (McKenna et al. 2010) is probably the most commonly used software for performing local realignment and it proceeds in two steps: (1) determining (small) suspicious intervals which are likely in need of realignment (RealignerTargetCreator) and (2) running the actual realigner over those intervals (IndelRealigner).

In this project, we first marked potentially PCR duplicates by using MarkDuplicates in Picard. We then performed local realignment by first detecting suspicious intervals using RealignerTargetCreator, followed by realignment of those intervals using IndelRealigner, both implemented in GATK

(DePristo et al. 2011). Both of these steps of pre-call processing were run separately on the 20 genomic subsets of each individual (700 subsets in total across all 35 individuals).

2.3.4 SNP and Genotype Calling (Step 4)

The development of NGS has made it possible to identify a large number of variants in almost any organism. Sequencing entire genomes potentially allows for the discovery of all existing polymorphisms, and can thus identify not only common but also rare SNPs which have been implicated in controlling many complex phenotypes (Mielczarek and Szyda 2016). This means that it should be possible to, at least in theory, identify the true causal mutations directly from resequencing data rather than relying on using linkage disequilibrium between unknown causal mutations and marker SNPs. However, to achieve this, SNP detection and genotype calling need to be performed across polymorphic sites in the NGS data. SNP calling, also known as variant calling, is aimed at detecting sites which differ from a reference sequence, while genotype calling is the process of determining the actual genotype for each individual based on positions in which an SNP or a variant has already been called (Nielsen et al. 2011; Li et al. 2013). SNP calling and genotype calling are identical when analyzing the genome of a single individual, as the inference of a heterozygous or homozygous non-reference genotype would imply the presence of an SNP (Nielsen et al. 2011; Mielczarek and Szyda 2016). However, when it comes to simultaneously analyzing multiple samples, an SNP is identified if at least one individual is heterozygous or homozygous non-reference at a genome position (Nielsen et al. 2011; Mielczarek and Szyda 2016). Many empirical and statistical methods have been developed to perform SNP and genotype calling to discover genetic variants accurately. Most of them are based on either heuristic or probabilistic methods.

2.3.4.1 Heuristic Methods

For variant detection, a heuristic algorithm determines genotypes based on a filtering step where variants fulfilling pre-set thresholds are kept. Such thresholds include coverage (e.g., minimum = 33), base quality (e.g., minimum = 20) or variant allele frequency (e.g., minimum = 0.08). Then, each allele supported by Fisher's exact test of read counts is applied compared to the expected distribution based on sequencing error alone (Mielczarek and Szyda 2016). Several software tools are developed based on such methods, like `VarScan2` (Koboldt et al. 2012). However, these methods can be improved by using more empirically determined cutoffs (Nielsen et al. 2011).

2.3.4.2 Probabilistic Methods

More recent approaches integrate several sources of information within a probabilistic framework. One of the advantages of a probabilistic framework is that it facilitates SNP calling in regions with medium to low coverage compared with heuristic methods (Altmann et al. 2012). For moderate or low sequencing depths, fixed cutoffs-based genotype calling will lead to under-calling of heterozygous genotypes and some information will inevitably be lost when using static filtering criteria (Nielsen et al. 2011). Another advantage of probabilistic methods is that it can account for uncertainty in genotype call, making it possible to monitor the accuracy of genotype calling (Altmann et al. 2012; Mielczarek and Szyda 2016). Additional information concerning allele frequencies and/or patterns of linkage disequilibrium (LD) can thus be incorporated in downstream analysis (Nielsen et al. 2011; Mielczarek and Szyda 2016).

SNP and genotype calling based on probabilistic methods usually involve the calculation of genotype likelihood that can be used to evaluate the quality scores for each read. To improve the accuracy of calculation of genotype likelihoods, the following parameters can be considered: (1) a weighting scheme should be used that takes correlated errors into account (Li et al.

2008a), (2) recalibrating the per base quality scores by using empirical data will also improve genotype likelihoods (Nielsen et al. 2011) and (3) information from the SNP-calling step should be incorporated into the genotype-calling algorithm, leading to genotype likelihoods that are calculated by conditioning on the site containing a polymorphism (Nielsen et al. 2011). By adopting a Bayesian approach for variant calling, prior genotype probabilities can be combined with the estimated genotype likelihood to calculate the posterior probability of a particular genotype. The genotype with the highest posterior probability will then generally be chosen and this probability or the ratio between the highest and the second highest genotype probabilities will be used as a measure of confidence in the variant call (Nielsen et al. 2011; Li et al. 2013; Mielczarek and Szyda 2016).

A prior probability of genotype must be assumed because it is a prerequisite to be able to calculate the posterior probability for a genotype. When data are analyzed from a single individual, either a uniform prior probability is chosen that assigns equal probability to all genotypes or a non-uniform prior can be based on external information, such as dbSNP (SNP database) entries, the reference sequence or an available population sample (Nielsen et al. 2011; Li et al. 2013). Jointly analyzing multiple individuals will improve the prior probabilities by considering allele or genotype frequencies, using, e.g., maximum likelihood (Li et al. 2009b; Martin et al. 2010) and then, by using a Hardy–Weinberg equilibrium assumption or other assumptions that relate allele frequencies to genotype frequencies to calculate genotype probabilities (Nielsen et al. 2011). Moreover, a significant improvement in genotype-calling accuracy can be achieved by considering linkage disequilibrium (LD) information (1000 Genomes Project Consortium 2010). However, this approach is not very efficient for rare mutations.

2.3.4.3 Commonly Used Software Tools Based on Probabilistic Methods

Several software tools have been developed that combine a probabilistic framework with Bayesian analysis for variant calling, such as `SAMtools` (Li et al. 2009c), `GATK` (DePristo et al. 2011), and `FreeBayes` (Garrison and Martin 2012). All these software also support the use of multiple-sample SNP calling. `Samtools` performs SNP and genotype calling in two steps: (1) `mpileup` in `Samtools` computes the possible genotype likelihood and stores this information in BCF (Binary call formats) and (2) `BCFtools` from the `SAMtools` packages applies the prior and does the actual variant calling based on genotype likelihoods information calculated in the previous step (Li et al. 2009c; Wang et al. 2015). Compared with `Samtools`, `GATK`, based on similar algorithms, features an advantage of automatically applying several filters before processing variant calling or other pre- and post-processing steps, e.g., filtering out reads that fail quality checks and reads with a mapping quality of zero. `HaplotypeCaller`, the most popular SNP- and genotype-calling module in `GATK`, can discard the existing mapping information and completely reassembles reads in the region whenever a region shows signs of variation. It thus allows for more accurate calling in regions that are traditionally difficult to call, such as regions containing different types of variants close to each other. Four steps will be performed by using `HaplotypeCaller` in `GATK`: (1) determining the regions of the genome that it needs to operate on by detecting significant evidence for variation; (2) determining haplotypes by local assembly of the active region by building a De Bruijn-like graph and by identifying potential variant sites through realigning each haplotype against the reference haplotype; (3) determining likelihoods of the haplotypes given the read data by a pairwise alignment of each read against each haplotype using the `PairHMM` algorithm; and

(4) assigning sample genotypes based on Bayes' rule. Liu et al. (2013) compared the performance of four common variant callers, SAMtools, GATK, glftools, and Atlas2, using single- and multi-sample variant-calling strategies, and came to the conclusion that GATK had several advantages over other variant callers for general purpose NGS analyses. Pirooznia et al. (2014) conducted a series of comparisons between single-nucleotide variant calls from NGS data and gold-standard Sanger sequencing in order to evaluate the accuracy of each caller module. They found that GATK provided more accurate calls than SAMtools, and the GATK HaplotypeCaller algorithm outperformed the older UnifiedGenotype algorithm (Pirooznia et al. 2014).

Based on these results, we have performed SNP and genotype calling by using GATK HaplotypeCaller by generating intermediate genomic VCFs (gVCFs) on a per subset and per sample basis (20 gVCFs produced for each individual). These gVCF files were then used for joint calling of multiple samples by using the GenotypeGVCFs module in GATK. SNP and genotype calling by GATK produced 20 VCF files with each file including variants from all 35 individuals (Fig. 2.1). The raw variant calls are likely to contain many false positives arising from errors in the genotyping step or from incorrect alignment of the sequencing data and the called variants therefore need to be subjected to a number of subsequent filtering steps (e.g., missing data, depth, and mapping quality) to produce data that is of sufficient quality for answering the biological questions of interest.

2.4 Variant Filtering

2.4.1 Filtering for Depth and Excessive Heterozygosity

A first filtering step was conducted on each of the 20 genomic subset VCF files separately with vcftools (Danecek et al. 2011) and bcftools (Li et al. 2009a, b, c) to only include biallelic SNPs positioned > 5 bp away from an intron and where the SNP quality parameters fulfilled GATK recommendations for hard filtering (https://gatkforums.broadinstitute.org/gatk/discussion/2806/howto-apply-hard-filters-to-a-call-set).

```
bcftools filter -g 5 -e 'QD < 2.0 || FS > 60.0 || MQ < 40.0 || \
      MQRankSum < -12.5 || ReadPosRankSum < -8.0 || SOR > 3.0' \
      subset_unfiltered | vcftools --vcf - --min-alleles 2 --max-alleles 2 \
      --mac 1 --remove-indels --recode --recode-INFO-all --stdout | bcftools
      \
      +fill-tags | bgzip -c > subset_GATK_biallelic_SNPs.vcf.gz && tabix -p \
      vcf subset_GATK_biallelic_SNPs.vcf.gz
```

In addition to the usual problems of calling variants from data with low coverage, SNP calling in conifers face the extra problem of possible collapsed repetitive regions in the assembly. The spruce genome, as most conifers, is highly repetitive and the v1.0 *P. abies* assembly is missing approximately 30% of the predicted genome size (Nystedt et al. 2013). These genome regions are either completely absent from the reference assembly or present as collapsed regions of high sequence similarity. This introduces problems for read mapping and variant calling as it increases the probability of calling false SNPs in collapsed regions, since reads that in reality derive from different genomic regions are mapped at a single region in the assembly. In order to reduce the impact of these issues on variant calling, we performed a second filtering step so that genotype calls with a depth outside the range 6–30 and a GQ < 15 were recoded to missing data. We then filtered each SNP for being variable with an overall average depth in the range of 8–20 and a "maximum missing" value of 0.8 (max 20% missing data). All info tags were then recalculated with `bcftools fill-tags`.

```
vcftools --gzvcf subset_GATK_biallelic_SNPs.vcf.gz --minDP 6 --maxDP 30 \
     --minGQ 15 --min-meanDP 8 --max-meanDP 20 --max-missing 0.8 \
     --min-alleles 2 --max-alleles 2 --mac 1 --recode --recode-INFO-all \
     --stdout | $bcftool +fill-tags | $bgzip -c \
     subset_GATK_biallelic_SNPs_GT.vcf.gz && tabix -p vcf \
     subset_GATK_biallelic_SNPs_GT.vcf.gz
```

Finally, to remove obviously collapsed SNPs we filtered out all SNPs that displayed a *p*-value for excess of heterozygosity less than 0.05. This was done since SNPs called in collapsed regions should show excess heterozygosities since they are based on reads that are derived from different genomic regions.

```
bcftools filter -e 'ExcHet < 0.05' subset_GATK_biallelic_SNPs_GT.vcf.gz
```

In order to evaluate the filtering parameters, we extracted summary statistics for each of the 20 genomic subsets and performed filtering steps using `bcftools stats` (Li et al. 2009a, b, c). We also wrote a custom python script (v2.7) to extract basic statistics regarding deviations from Hardy–Weinberg equilibrium (`HWE`), excess of heterozygosity (`ExcHet`), number of called samples, allele frequency, number of heterozygous calls, alternative allele ratio for heterozygous calls as well as total depth and total heterozygous depth, from the VCF file of two of the genomic subsets (subsets 5 and 6) using `pysam` (https://github.com/pysam-developers/

pysam, Li et al. 2009c). These data were then analyzed in R (R Core Team 2014).

```
import pysam
import sys

Subset=sys.argv[1]
vcf=pysam.VariantFile("Subset.vcf.gz")
out=open("Subset_Allele_stats.txt","w")

out.write("{}\t{}\t{}\t{}\t{}\t{}\t{}\t{}\t{}\t{}\n".format
("Chrom","Start","End","HWE","ExcHet","Called_samples","Allele_frequenc
y","Heterozygotes","Alt_read_proprotion","Total_Het_DP","Total_DP"))

for record in vcf.fetch():
    Chrom=record.contig
    Start=record.pos -1
    End=record.pos
    HWE=record.info["HWE"][0]
    ExcHet=record.info["ExcHet"][0]
    AF=record.info["AF"][0]
    AC_Het=record.info["AC_Het"][0]
    NS=record.info["NS"]
    Alt_read=0
    Tot_Het_read=0
    Total_DP=0
    for sample in record.samples.values():
        if sample["GT"] == (None,None):
            continue
        else:
            if sample["GT"] == (0,1):
                if sample["DP"] != None:
                    Alt_read=Alt_read + int(sample["AD"][1])
                    Tot_Het_read=Tot_Het_read + int(sample["DP"])
            if sample["DP"] != None:
          Total_DP=Total_DP + int(sample["DP"])
    if Tot_Het_read >0:
        Alt_read_prop= float(Alt_read)/float(Tot_Het_read)
    else:
        Alt_read_prop="NA"
    out.write("{}\t{}\t{}\t{}\t{}\t{}\t{}\t{}\t{}\t{}\n".format (Chrom,
Start, End, HWE, ExcHet, NS, AF, AC_Het, Alt_read_prop, Tot_Het_read,
Total_DP))
```

For analyzing different genomic regions (i.e., repeat regions, outside repeat regions, genic regions, and exonic regions) separately, we subset the allele statistics file (above) to corresponding regions in files `Pabies1.0-all.phase.changed.gff3` and `Pabies1.0_Repeats_2.0_repeatmasker.gff3.gz` (available from ftp://plantgenie.org/Data/ConGenIE/) using a custom made python script (v2.7) and `BEDTools` (Quinlan and Hall 2010; Dale et al. 2011).

```
from pybedtools import BedTool

snps=BedTool("Input_file.txt")
repeats=BedTool("Pabies1.0_Repeats_2.0_repeatmasker.gff3.gz")
genes=BedTool("gene_regions.gff3")
exons=BedTool("exon_regions.gff3")
SNPs_Repeat=snps.intersect(repeats,u=True)
SNPs_NoRepeat=snps.intersect(repeats,v=True)
SNPs_Gene=snps.intersect(genes,u=True)
SNPs_Exon=snps.intersect(exons,u=True)

Repeat_out=open("Repeats.txt","w")
NoRepeat_out=open("NoRepeats.txt","w")
Gene_out=open("Genes.txt","w")
Exon_out=open("Exons.txt","w")

Repeat_out.write("{}\t{}\t{}\t{}\t{}\t{}\t{}\t{}\t{}\t{}\t{}\n".format("Chrom
", "Start", "End", "HWE", "ExcHet", "Called_samples", "Allele_frequency",
"Heterozygotes", "Alt_read_proprotion", "Total_Het_DP", "Total_DP"))
NoRepeat_out.write("{}\t{}\t{}\t{}\t{}\t{}\t{}\t{}\t{}\t{}\t{}\n".format("Chr
om", "Start", "End", "HWE", "ExcHet", "Called_samples", "Allele_frequency",
"Heterozygotes", "Alt_read_proprotion", "Total_Het_DP", "Total_DP"))
Gene_out.write("{}\t{}\t{}\t{}\t{}\t{}\t{}\t{}\t{}\t{}\t{}\n".format("Chrom",
"Start", "End", "HWE", "ExcHet", "Called_samples", "Allele_frequency",
"Heterozygotes", "Alt_read_proprotion", "Total_Het_DP", "Total_DP"))
Exon_out.write("{}\t{}\t{}\t{}\t{}\t{}\t{}\t{}\t{}\t{}\t{}\n".format("Chrom",
"Start", "End", "HWE", "ExcHet", "Called_samples", "Allele_frequency",
"Heterozygotes", "Alt_read_proprotion", "Total_Het_DP", "Total_DP"))

Repeat_out.write("{}\n".format(SNPs_Repeat))
NoRepeat_out.write("{}\n".format(SNPs_NoRepeat))
Gene_out.write("{}\n".format(SNPs_Gene))
Exon_out.write("{}\n".format(SNPs_Exon))
```

2.4.2 Results

The raw unfiltered VCF files contained a total of 750 million records of which 710 million were single nucleotide polymorphisms (SNPs). After hardfiltering using GATK best-practice recommendations (https://gatkforums.broadinstitute.org/gatk/discussion/2806/howto-apply-hard-filters-to-a-call-set), 545 million SNPs (76.8%) remained. These variants were distributed over 94.7% of the 1,970,460 scaffolds in the assembly that are longer than 1 Kb (Table 2.2). When comparing the alternative allele frequency with the proportion of heterozygous calls per SNP, it is obvious that we have a high proportion of SNPs that do not follow Hardy–Weinberg expectations (Fig. 2.2a). SNPs with either too high or too low fraction of heterozygous calls most often also show a greater deviation in allele ratio in their heterozygous calls compared to SNPs following Hardy–Weinberg expectations (Fig. 2.2b–d).

In genotyping-by-sequencing (GBS) data, often applied to species that lack a reference genome, duplicated regions have been shown to behave in a very similar manner (McKinney et al. 2017) to our whole genome sequencing (WGS) data. To overcome the issues relating to false SNPs that stem from collapsed/duplicated genomic regions, McKinney et al. (2017) suggests filtering for sequencing depth, so that all accepted calls have a coverage within the expected span (based on the estimated sequencing depth). We therefore recoded individual genotypes to missing data if they had a depth that was too low to reliably call both alleles (<6 reads covering a site), or if their depth was too high compared to the overall coverage (>30 reads covering a site), which makes it likely that reads were derived from multiple genomic positions. Calls with a genotype quality (GQ) less than 15, which indicates that the difference in likelihood between the best and second-best genotypes is

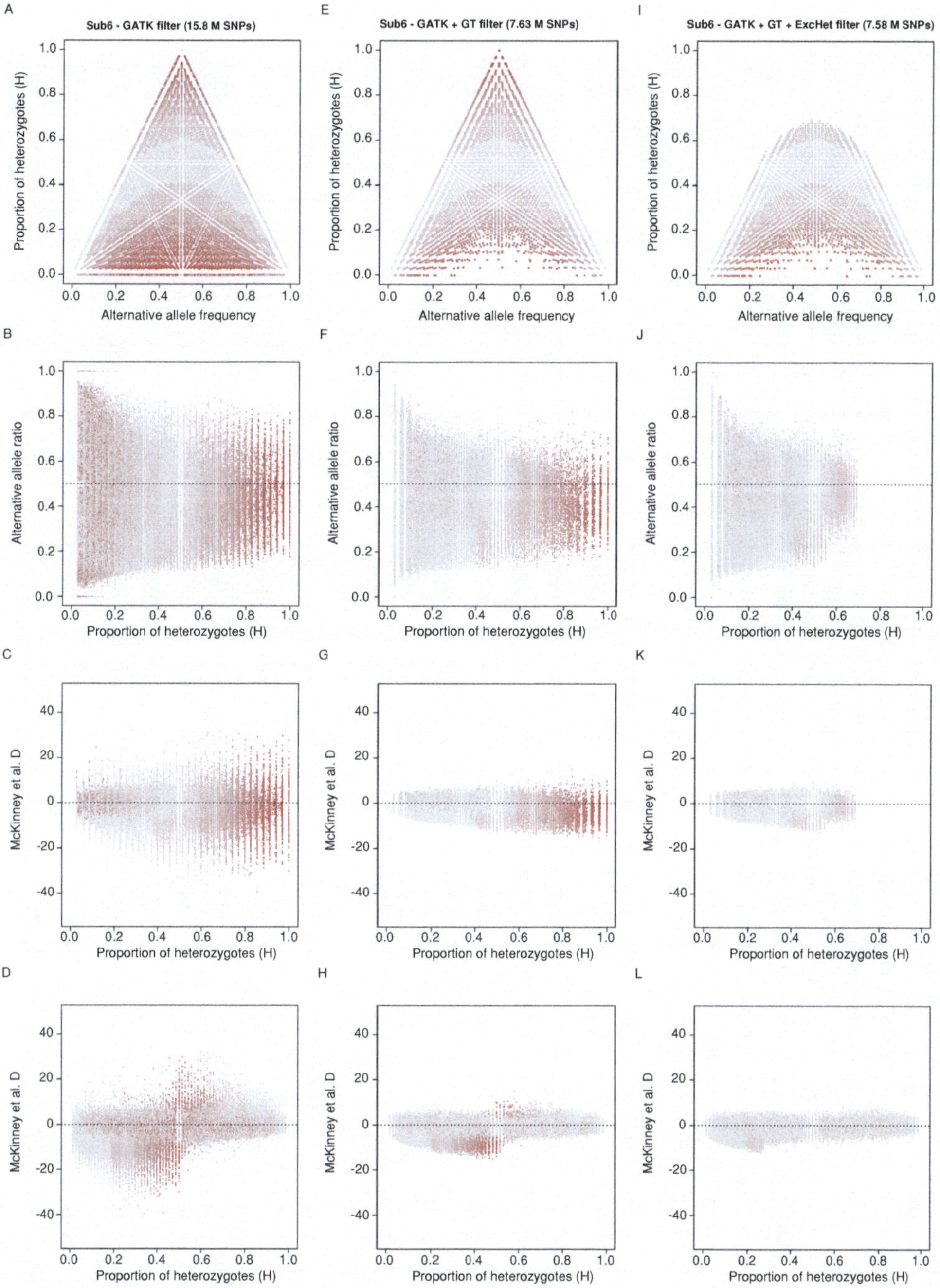

Fig. 2.2 Subset 6 of the WGS data with GATK filtered SNPs (**a–d**, 15.8 M SNPs), GATK+GT filtered SNPs (**e–h**, 7.63 M SNPs), and GATK+GT+ExcessHet filtered SNPs (I-L, 7.58 M SNPs). The first row shows the alternative allele frequency versus proportion of heterozygotes for each SNP. The second row shows the proportion of heterozygotes versus alternative allele ratio of all heterozygotes for each SNP. The third row shows the proportion of heterozygotes versus the standardized deviation of allele ratio (McKinney et al. 2017) for each SNP. The fourth row shows the alternative allele frequency versus the standardized deviation of allele ratio (McKinney et al. 2017) for each SNP. Each SNP is colored according to Hardy–Weinberg deviations, gray for SNPs matching Hardy–Weinberg expectations to bright red for SNPs showing strong deviations

Table 2.2 Summary statistics for each of the 20 genomic subsets of Norway spruce. *Subset*: name of genomic subset; *Genomic length (Mb)*: length (in mega bases) of the combined scaffolds included in the subset; *Average scaffold length (kb)*: Average length (in kilo bases) of each scaffold included in the subset; *Average repeat coverage (%)*: Average percentage of the scaffolds that are covered by known repeats; *Scaffolds with genes (%)*: Percentage of scaffolds included in the subset that has at least 1 annotated gene model; *Unfiltered records (M)*: Number of records (in millions) present in the unfiltered raw VCF file; *Unfiltered SNPs (M)*: Number of SNPs (in millions) present in the unfiltered raw VCF file; *GATK SNPs (M)*: Number of SNPs (in millions) retained after filtering according to GATK best practice recommendations and including only biallelic SNPs > 5 bp away from an indel. *GATK Scaffolds (%)*: percentage of the subset scaffolds that contains retained SNPs after GATK filter; *GATK+GT SNPs (M)*: Number of SNPs (in millions) retained after extra filter for individual depth and genotype quality and overall depth and missing data. *GATK+GT Scaffolds (%)*: percentage of the subset scaffolds that contains retained SNPs after GATK and GT filter; *GATK+GT+ExcHet SNPs (M)*: Number of SNPs (in millions) retained after extra filter for excess of heterozygosity. *GATK+GT+ExcHet Scaffolds (%)*: percentage of the subset scaffolds that contains retained SNPs after GATK+GT+ExcHet filter

Subset	Genomic length (Mb)	Average scaffold length (Kb)	Average repeat coverage (%)	Scaffolds with genes (%)	Unfiltered records (M)	Unfiltered SNPs (M)	GATK SNPs (M)	GATK Scaffolds (%)	GATK +GT SNPs (M)	GATK +GT Scaffolds (%)	GATK+GT +ExcHet SNPs (M)	GATK+GT +ExcHet Scaffolds (%)
1	2,654.8	26.6	53.8	15.7	230.7	217.7	175.7	100.0	114.5	96.1	114.1	96.1
2	1,657.9	16.6	62.4	9.2	113.2	107.8	81.5	96.1	44.4	80.1	44.1	80.1
3	451.9	4.5	60.1	2.0	40.6	38.4	29.6	99.8	15.2	77.6	15.1	77.6
4	394.3	3.9	60.8	1.7	35.4	33.4	25.7	99.8	13.0	76.2	12.9	76.2
5	481.1	4.8	60.6	2.0	43.9	41.4	32.0	99.8	16.6	79.5	16.4	79.4
6	243.3	2.4	62.8	1.1	21.7	20.5	15.8	99.6	7.6	70.4	7.6	70.4
7	245.3	2.5	62.6	1.1	22.0	20.7	16.0	99.6	7.8	70.7	7.7	70.6
8	257.8	2.6	62.5	1.1	23.2	21.9	16.9	99.6	8.5	71.9	8.4	71.9
9	329.2	3.3	66.4	1.3	27.5	25.8	20.4	91.4	11.4	64.7	11.3	64.7
10	256.4	2.8	82.9	0.3	7.8	7.6	4.6	71.9	0.5	16.5	0.5	16.5
11	241.0	2.4	76.0	0.4	13.0	12.6	7.9	93.9	1.5	33.8	1.5	33.8
12	159.1	1.6	57.6	0.7	13.3	12.6	9.6	91.3	3.9	55.5	3.9	55.5
13	179.2	1.8	58.9	0.7	15.0	14.2	10.8	91.7	4.6	56.6	4.5	56.6
14	196.9	2.0	59.6	0.9	16.5	15.7	11.9	92.0	5.3	58.0	5.3	57.9
15	213.0	2.1	59.8	1.0	17.9	17.0	13.0	92.1	6.0	58.6	6.0	58.5
16	229.6	2.3	60.1	1.1	19.2	18.2	14.0	92.3	6.6	59.7	6.6	59.7
17	237.6	2.4	62.8	1.1	17.6	16.8	12.4	95.0	5.3	59.4	5.2	59.3

(continued)

Table 2.2 (continued)

Subset	Genomic length (Mb)	Average scaffold length (Kb)	Average repeat coverage (%)	Scaffolds with genes (%)	Unfiltered records (M)	Unfiltered SNPs (M)	GATK SNPs (M)	GATK Scaffolds (%)	GATK +GT SNPs (M)	GATK +GT Scaffolds (%)	GATK+GT +ExcHet SNPs (M)	GATK+GT +ExcHet Scaffolds (%)
18	262.2	2.6	62.2	1.2	20.1	19.2	14.1	97.9	6.2	66.3	6.1	66.3
19	331.3	3.3	63.9	1.7	25.7	24.4	17.9	97.4	8.5	66.2	8.4	66.1
20	433.6	6.2	74.6	14.0	25.3	23.6	15.3	91.7	8.3	39.7	8.2	39.5
Total	9,455.5	4.8	63.5	2.7	749.6	709.5	545.1	94.7	295.7	63.2	293.9	63.2

small, were also recoded to missing data. We then filtered for an overall average depth per SNP of 8–20 to reduce the proportion of possible erroneous calls further and also added a threshold of maximum 20% missing data in order to keep an SNP for downstream analysis.

There were 296 million SNPs, corresponding to 41.7% of the raw unfiltered SNPs, remaining after these filtering steps and these were distributed over 63.2% of the >1-Kb scaffolds (Table 2.2). When comparing the proportion of heterozygous calls to the alternative allele frequency, a noticeable amount of the SNPs with heterozygous deficiency was filtered out (Fig. 2.2e). The deviation from a balanced allele ratio of heterozygous calls is also visibly lower (Fig. 2.2f–h) even though the data still suffer from SNPs showing heterozygous excess at intermediate alternative allele frequencies (Fig. 2.2h at ~0.5). Such excess of heterozygosity could be explained by real biological phenomena, such as balancing selection (Charlesworth 2006), or arise due to artifacts caused by collapsed/duplicated regions in the assembly as discussed above (McKinney et al. 2017). Regions under balancing selection should be highly localized in the genome, while the risk of collapsed regions in the spruce assembly is a priori expected to be high since we know that the genome contains a very large proportion of retrotransposon-derived sequences (Nystedt et al. 2013; Cossu et al. 2017). To filter out SNPs with a significant excess of heterozygotes therefore seems justifiable for the data at hand. Heterozygous deficiency, on the other hand, can be both explained by errors in variant calling in low-depth regions or by population structure (Hartl and Clark 1989). The latter is a much more likely scenario for our data set, since samples were derived from across the distribution range of *P. abies*, and we also expect this pattern to manifest itself across most of the genome. We therefore chose not to filter for this criterion. By removing all SNPs showing a significant excess of heterozygotes (p-value < 0.05), we ultimately retained 294 million SNPs (corresponding to 41.4% of the unfiltered SNPs) from 63.2% of the > 1 kb scaffolds (Table 2.2, Fig. 2.2i–m).

2.4.3 Comparison of WGS and Reduced Representation Sequencing Data

Even with a number of conifer draft genomes available (Nystedt et al. 2013; Stevens et al. 2016; Zimin et al. 2017; Neale et al. 2017), the methods of choice for analyzing sequence diversity in conifers have been different kinds of reduced representation sequencing methods, such as genotyping-by-sequencing (GBS), restriction site associated DNA sequencing (RADseq), or capture probe sequencing (Syvänen 2005; Andrews et al. 2016). These methods all have in common that they target a small fraction of the target genome, the only thing that differs is how this fraction is selected. In order to compare the behavior of our spruce WGS data with data derived from a reduced representation sequencing technique, we analyzed at set of 526 samples genotyped by a set of 40,018 sequence capture probes that had been designed to target regions within genic regions of the v1.1 *P. abies* genome assembly (Vidalis et al. 2018; Baison et al. 2019). The same filtering steps as described in the preceding section were used for the capture probe data set, although the thresholds for individual depth range (6–40), overall average depth range (10–30), and significance level for excess of heterozygotes (p-value < 1e-10) were altered to fit the size of the data set (the number of samples and estimated sequencing depth). Even though the probes were designed to target unique regions in the assembly (Vidalis et al. 2018), the GATK SNP quality filter show a remarkably high level of SNPs with deviations from Hardy–Weinberg equilibrium (Fig. 2.3a) with large deviations from balanced allele ratios in heterozygous calls (Fig. 2.3b–d).

Adding the depth filters provided a substantial improvement in SNP quality by both reducing the number of SNPs showing excess of heterozygosity as well as improving the balance in allele ratios for heterozygous calls. The sequence capture data set does not show the same level of heterozygote deficiency as seen in the WGS data (Fig. 2.2 vs. Fig. 2.3). However, the

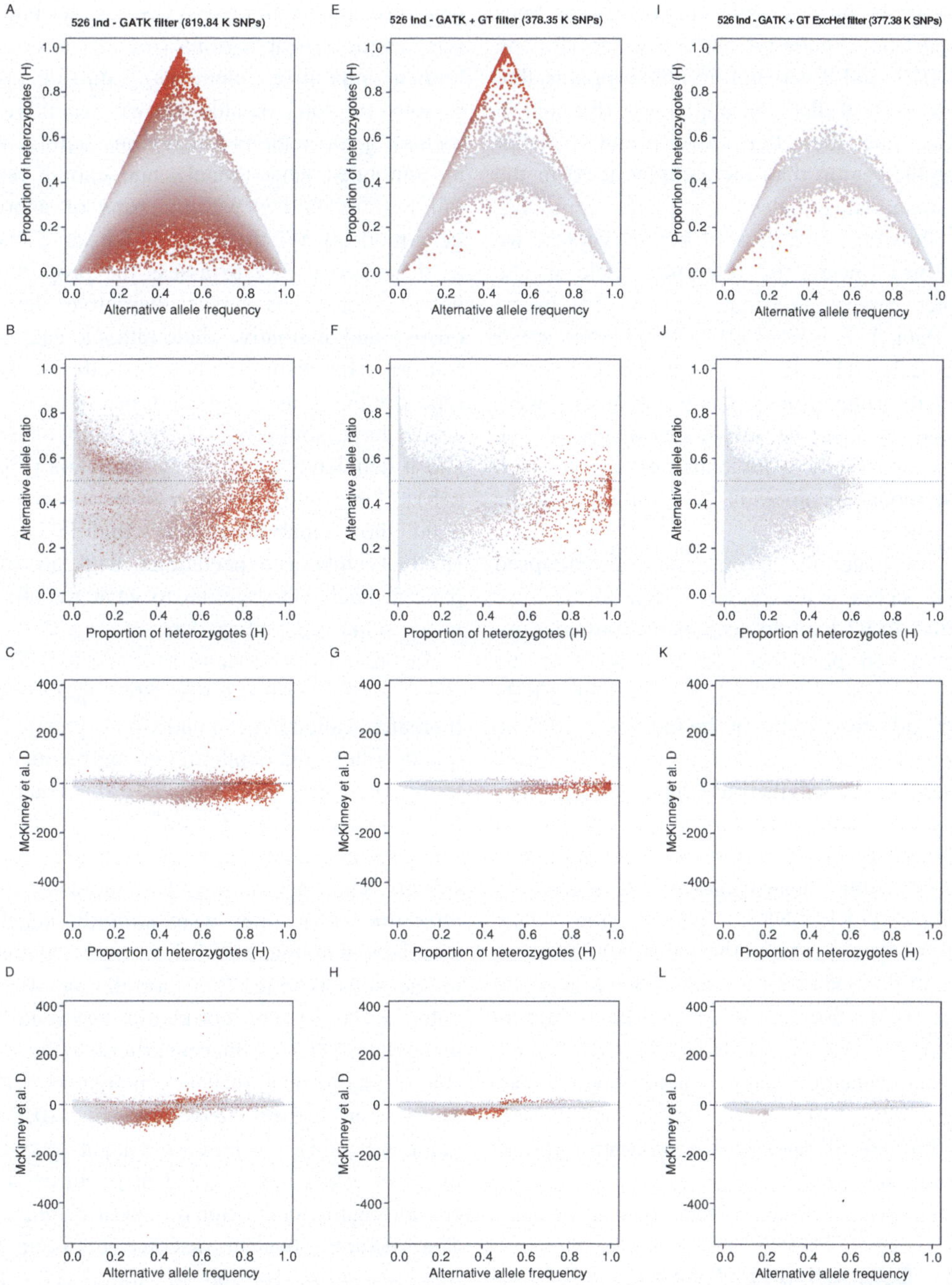

Fig. 2.3 Sequence capture data of 526 individuals with GATK filtered SNPs (**a–d**, 819.8 K SNPs), GATK+GT filtered SNPs (**e–h**, 378.4 K SNPs), and GATK+GT +ExcessHet filtered SNPs (**i–l**, 377.4 K SNPs). The first row shows the alternative allele frequency versus proportion of heterozygotes for each SNP. The second row shows the proportion of heterozygotes versus alternative allele ratio of all heterozygotes for each SNP. The third row shows the proportion of heterozygotes versus the standardized deviation of allele ratio (McKinney et al. 2017) for each SNP. The fourth row shows the alternative allele frequency versus the standardized deviation of allele ratio (McKinney et al. 2017) for each SNP. Each SNP is colorised according to Hardy–Weinberg deviations, gray for SNPs matching Hardy–Weinberg expectations to bright red for SNPs showing strong deviations

capture probe data set only contained plus trees sampled from Southern/Central Sweden (Baison et al. 2019) and it is therefore not surprising that we observe less effects of population structure in this data, since these trees cover a much smaller geographic region than the samples used in the WGS data. Nevertheless, even after removing SNPs showing an excess of heterozygotes, we see a bias toward the reference allele in the sequence capture data that is not visible in the WGS data (bias toward allele frequencies < 0.5 in Fig. 2.3j). This is most probably due to an artifact of using probes, since the probes were designed against the reference alleles of the genome and likely work better in regions with low to moderate amounts of variation (Vidalis et al. 2018).

To evaluate the proportion of collapsed regions within genic regions even further, we analyzed 1,997 haploid sequence captured samples that had previously been mapped to the whole genome, called using a diploid ploidy setting and subset to the probe regions ± 100 bp (Baison et al. 2019; Vidalis et al. 2018). These samples, which should only have homozygous calls, showed an average heterozygosity level of approximately 3.7% (obviously contaminated samples, with heterozygosity levels > 10% removed); 114 k SNPs remained after GATK hard filtering and with no more than 30% missing data, and these showed a median heterozygosity level of 0.01 (first and third quartiles of 0.004 and 0.017, respectively). However, 10% of the SNPs experienced a heterozygosity level > 0.05 (corresponding to 70–98 heterozygous calls depending on call rate), with a maximum of 0.95 (data not shown).

2.4.4 Comparisons of Genic, Intergenic, and Repetitive Regions

In order to analyze how different regions of the spruce genome behave with regards to deviations from Hardy–Weinberg equilibrium, we subdivided the GATK SNP quality + depth and genotype quality filtered WGS data set into four sets: inside repeat regions (regions covered by known repetitive elements), outside repeat regions (regions outside known repetitive elements), genic regions (all regions falling within an annotated gene model), and exonic regions (all regions falling within exons of annotated gene models). We then applied the same analysis as described earlier to assess the proportion of heterozygous calls versus alternative allele frequency and alternative allele ratios to each of the four sets. Interestingly enough, both the "within repeat regions" and "outside repeat regions" data sets behave similarly, showing both an excess and a deficiency of heterozygous calls (Fig. 2.4 a–h). Genic and exonic regions, on the other hand, show much fewer SNPs that deviate from Hardy-Weinberg expectations with an overall pattern much more similar to what we observed in the sequence capture data set (Fig. 2.4i–p).

In order to understand how the two filtering steps, GATK best practice SNP quality filters (hereafter called GATK) and GATK practice SNP quality filters and depth and genotype quality filtering (hereafter called GATK+GT), change the SNPs retained across the different genomic subsets, we analyzed summary statistics from all subsets. The fraction of SNPs retained in subsets after the GATK filter were strongly negatively correlated with fraction of sites covered by repeats in the subsets (Fig. 2.5a, correlation –0.92, p-value = 1.4e-8). This correlation was reduced by using the GATK+GT filtering criteria to the subsets (Fig. 2.5a, correlation between fraction retained SNPs and repeat coverage of –0.77, p-value = 8.1e-5), but remained negative resulting in fewer SNPs being called in genomic subsets containing higher fraction of repetitive sequences. The median physical distance between SNPs increased from 16.5 bp for the GATK filtered subsets (ranging from 15.0 to 55.7 bp per subset, with an overall average of 17.3 bp) to 36.3 bp for the GATK+GT filtered subsets (ranging from 23.2 to 512.8 bp per subset with an overall average of 32.0 bp) (Fig. 2.5c, outliers not shown). Even though this suggest a high level of nucleotide diversity in Norway spruce, most alleles are rare

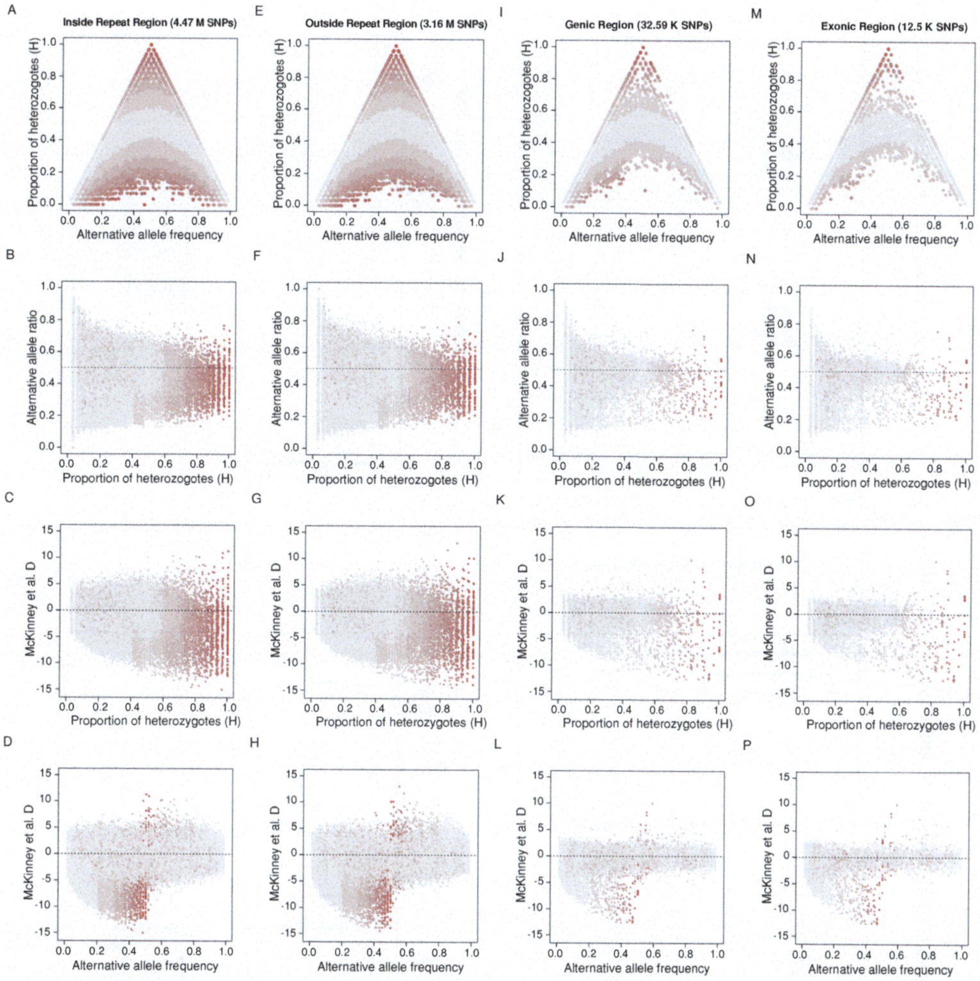

Fig. 2.4 Subset 6 of the WGS data with GATK+GT filtered SNPs divided into four regions. SNPs within repeat regions (**a–d**, 4.47 M SNPs), SNPs outside repeat regions (**e–h**, 3.18 M SNPs), SNPs within genic regions (**i–l**, 32.6 k SNPs), and SNPs within exonic regions (**m–p**, 12.5 k SNPs). The first row shows the alternative allele frequency versus proportion of heterozygotes for each SNP. The second row shows the proportion of heterozygotes versus alternative allele ratio of all heterozygotes for each SNP. The third row shows the proportion of heterozygotes versus the standardized deviation of allele ratio (McKinney et al. 2017) for each SNP. The fourth row shows the alternative allele frequency versus the standardized deviation of allele ratio (McKinney et al. 2017) for each SNP. Each SNP is colored according to Hardy–Weinberg deviations, gray for SNPs matching Hardy–Weinberg expectations to bright red for SNPs showing strong deviations

and as many as 26% of the SNPs appear as singletons in the GATK+GT filtered data set, compared to 21% in the GATK filtered data set. The GATK+GT filter has a large impact on the fraction of scaffolds that retain any SNPs, decreasing from 94.7% for the GATK filtered data set to only 63.2% under GATK+GT (Fig. 2.5b). The transition–transversion ratio similarly decreased with the GATK+GT filter applied (Fig. 2.5d).

At the individual sample level, the GATK+GT filter resulted in an increased proportion of heterozygous calls and a reduced proportion of

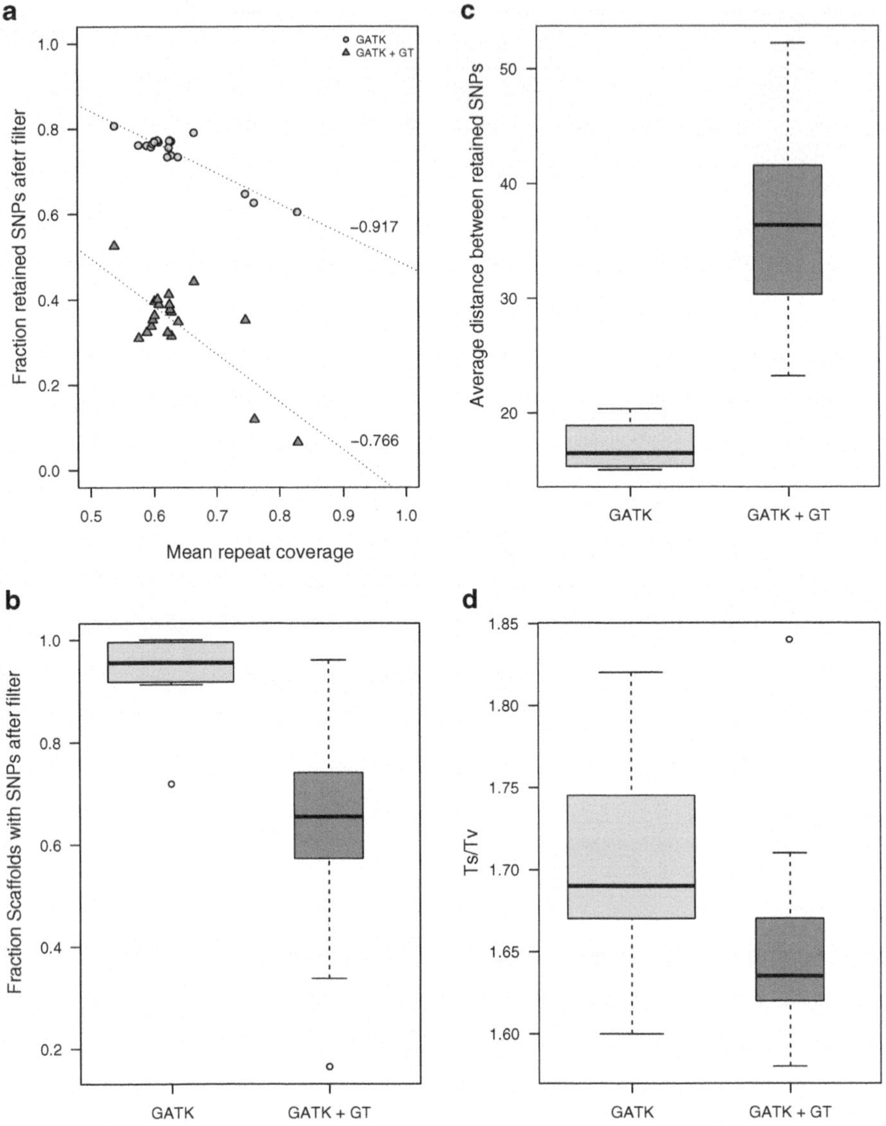

Fig. 2.5 Summary statistics on a per genomic subset basis, over all samples. GATK indicate the GATK SNP quality filtered data set; GATK+GT indicate the GATK SNP quality + depth and genotype quality filtered data set. **a** Correlation between the proportions of retained SNPs after filter (in comparison to the amount of unfiltered raw SNPs) and the average level of repeat coverage. **b** Proportion of scaffolds with retained SNPs after filter. **c** Average physical distance (in bp) between retained SNPs; outliers are not shown. **d** Transition–transversion ratio. The widths of the boxes are proportional to the fraction of retained SNPs after filter

homozygous alternative calls, while the proportion of homozygous reference calls stayed the same in comparison to the `GATK` filter (Fig. 2.6a, b, d). Sample *Pab006*, which is the individual from which the *P. abies* reference assembly was derived (Z4006, Nystedt et al. 2013), has SNPs called as homozygous alternative, a genotype which should be impossible in this individual, but found in 0.07% of the calls using the `GATK` filtering criteria. This fraction was reduced to 0.02% when using the `GATK`

+GT filtering criteria, suggesting that the depth filter does improve the quality of the genotype calls and provides an overall higher average depth in comparison to the GATK filtered data set (Fig. 2.6c). Although the average sequencing depth of called genotypes increases following GATK+GT filtering, the proportion of missing calls also increases, and reaches 30% for samples with the lowest estimated sequencing coverage (Fig. 2.6e). The increase in missingness does not affect the proportion of singletons per sample, which stays roughly constant under both the GATK (average of 0.8%) as well as the GATK+GT (average of 0.9%) filtering criteria (Fig. 2.6f).

2.4.5 Effects of Filtering on Estimates of the Site Frequency Spectrum

To analyze how different summary statistics regarding the site frequency spectrum (SFS) are affected by filtering parameters (GATK and GATK +GT+ExcHet), we analysed Tajima's D (1989) and pairwise nucleotide diversity (π, Hartl and Clark 1989) on a per scaffold basis for genomic subset 6, for inside repeat regions, outside repeat regions, genic regions, and exonic regions separately. The GATK filtered data showed an overall higher estimate of Tajima's D than the GATK+GT +ExcHet filtered data for all four genomic regions (average of –0.35 to –0.13 and –0.57 to –0.53 for GATK and GATK+GT+ExcHet, respectively). These estimates were however highly correlated between the filtering parameters in all four genomic regions (correlation of 0.77–0.82 with p-value < 2.2e-16, Fig. 2.7a–d). The GATK filtered data also showed an overall slightly higher nucleotide diversity level (π) compared to the GATK+GT+ExcHet filtered data (average of 3.7e-4–1.6e-3 and 2.1e-4–1.0e-3 for GATK and GATK+GT+ExcHet, respectively), with lower diversity levels for the same amount of analyzed variants in the fully filtered data set (Fig. 2.7e–h). Smaller differences regarding the SFS between the specified genomic regions could be found in the fully filtered data in comparison to the GATK filtered data, indicating that we managed to remove false SNPs that were due to collapsed genomic regions in the assembly, without altering the SFS by filtering for minor allele frequencies.

2.5 Conclusions and Recommendations

Having high-quality variant calls is essential for downstream analyses such as population genomic studies, inferences of demographic history or complex trait dissection using, for instance, genome-wide association studies (Nielsen et al. 2011). Variant calling in conifers pose a number of challenges arising from the complex highly repetitive nature of conifer genomes. Even among the most complete reference genomes of conifers, large sequence regions of the genome are still lacking and this causes problems for mapping of short (<250 bp) sequencing reads, since reads originating from such missing sequences will not match any region in the reference genome. Such reads may fail to map altogether and will be eliminated from further steps in the variant calling pipeline. However, if they are derived from repetitive regions, highly similar genomic regions may be represented in the assembly allowing read mapping. To a lesser degree, these issues are also true for sequencing reads from repetitive regions that are represented in the assembly as these reads may match equally well multiple regions in the reference genome. The end result in both cases is thus that some regions are being covered by sequencing reads originating from different genomic regions which may ultimately result in the calling of false genetic variants, as nucleotide substitutions that differentiate the paralogous regions are called as true SNPs (Gayral et al. 2013). Such variants can often have high-quality scores, making it hard to filter them out using even the best practice procedures for variant filtering (Nielsen et al. 2011).

These issues are apparent in our data even when using what is considered rather stringent criteria for variant filtering (Fig. 2.2). To

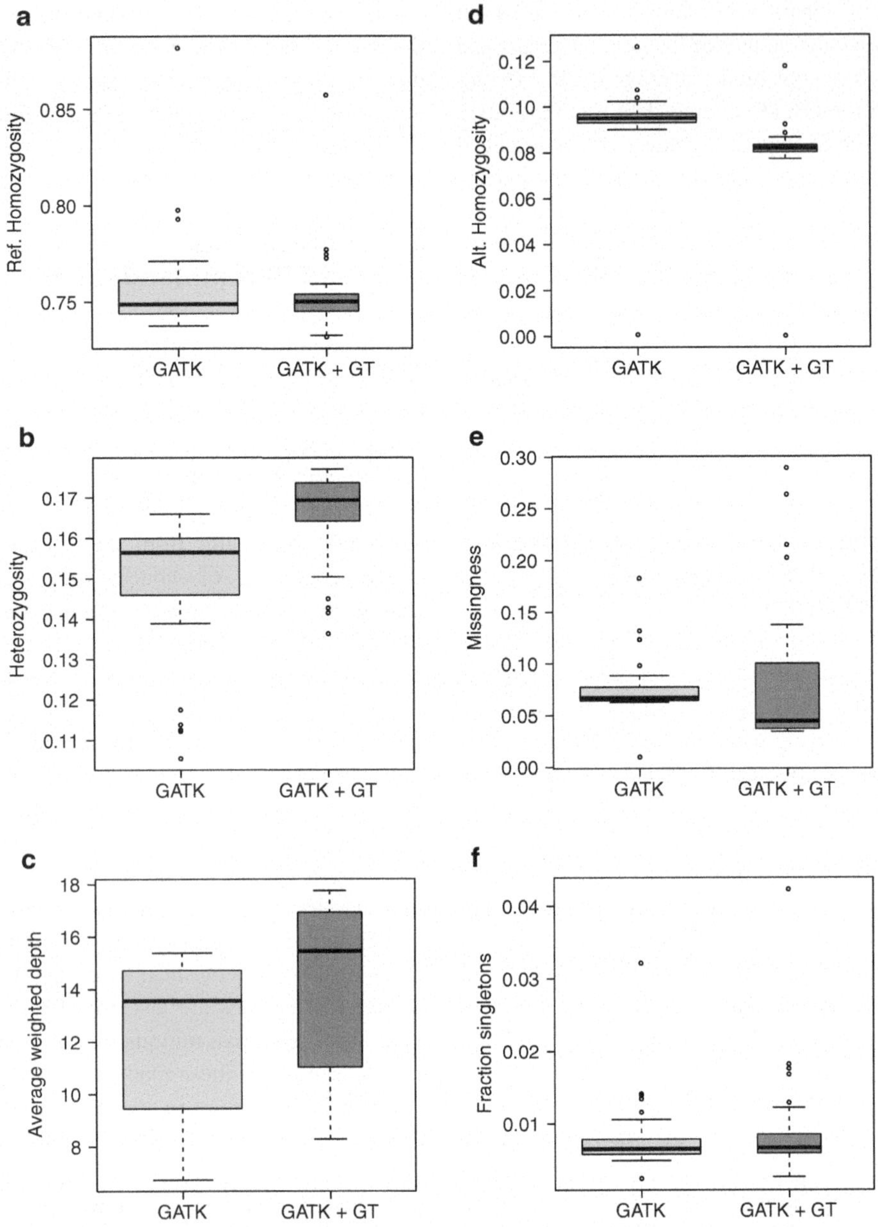

Fig. 2.6 Summary statistics on a per individual basis, over all genomic subsets. GATK indicate the GATK SNP quality filtered data set. GATK+GT indicate the GATK SNP quality + depth and genotype quality filtered data set. **a** Proportion of calls that are homozygous reference. **b** Proportion of calls that are heterozygous. **c** Average depth over all calls. Average depth per genomic subset was weighted toward its number of SNPs before an average was calculated over all genomic subsets. **d** Proportion of calls that are homozygous alternative. **e** Proportion of SNPs with missing data (No calls). **f** Proportion of calls that are singletons. The width of the boxes is proportional to the fraction of retained SNPs after filter

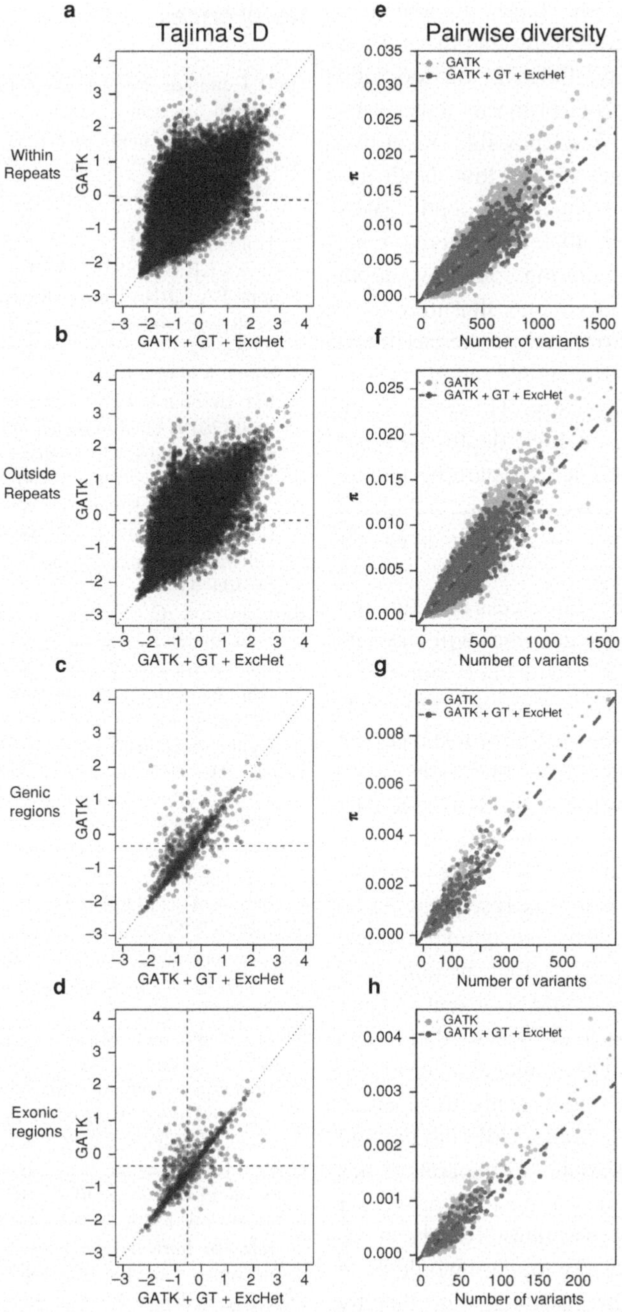

Fig. 2.7 Analyses of the site frequency spectrum (SFS) for genomic subset 6, divided into within repeat regions (row 1), outside repeat regions (row 2), genic regions (row 3), and exonic regions (row 4). **a–d** Correlation between Tajima's D estimated over scaffolds for the GATK and GATK+GT+ExcHet filtered data sets. Horizontal dashed line indicates average Tajima's D for the GATK filtered data. Vertical dashed line indicates average Tajima's D for the GATK+GT+ExcHet filtered data. Diagonal dotted line shows the 1:1 correlation between the data sets. **e–h** Estimates of pairwise nucleotide diversity (π) per scaffold versus number of analyzed variants for GATK filtered data (light gray dots and dotted line) and GATK+GT+ExcHet filtered data (dark gray dots and dashed line). Lines show the linear model (lm) between diversity and number of analyzed variants for each of the data sets

alleviate these issues, we have implemented further filtering on sequencing depth and on excessive heterozygosity. Filtering on too high depth is expected to target primarily false variants that are derived from possible repetitive regions, whereas filtering on too low depth targets variants where low sequencing depth makes confident calling of both alleles in heterozygous individuals unlikely. Filtering on low depth should thus target called variants that may have deflated levels of heterozygosity whereas filtering on high depth primarily should target variants with excessive heterozygosity. However, even very stringent filtering on high depth does not completely eliminate problems regarding excess heterozygosity, and we have therefore included a final filtering step aimed specifically at excess heterozygosity (Fig. 2.2).

To a certain extent, having access to better and more contiguous reference genomes will alleviate some of these issues but it will likely not eliminate them completely. Thus, even as conifer reference genomes improve due to, for instance, the use of long-read sequencing technologies, resequencing, and variant calling from short-read data will still be difficult and fraught with both false positive and false negative SNPs. One possible solution to these issues is to base resequencing on haploid tissue, e.g., from megagametophytes. Using haploid tissue has the benefit that heterozygous genotypes should be absent and any heterozygous genotype call observed is thus likely to arise from either sequencing errors or from mapping of paralogous reads to a single region in the reference genome. Utilizing haploid tissues to rule out issues arising from paralogy has already been practiced in conifers to improve, for instance, transcriptome assembly (Ojeda et al. 2018). Another possible way forward would be to utilize longer reads for resequencing, thereby reducing the probability that a single read may map to multiple regions in the reference genome. Although the cost and throughput of current long-read technologies make genome resequencing of conifers using exclusively long reads infeasible, declining costs and technology development could open up possibilities for resequencing using long reads in the future.

References

1000 Genomes Project Consortium (2010) A map of human genome variation from population-scale sequencing. Nature 467(7319):1061

Altmann A, Weber P, Bader D, Preuß M, Binder EB, Müller-Myhsok B (2012) A beginners guide to SNP calling from high-throughput DNA-sequencing data. Hum Genet 131(10):1541–1554

Andrews KR, Good JM, Miller MR, Luikart G, Hohenlohe PA (2016) Harnessing the power of RADseq for ecological and evolutionary genomics. Nat Rev Genet 17:81–92

Baison J, Vidalis A, Zhou L, Chen Z-Q, Li Z, Sillanpää MJ, Bernhardsson C, Scofield D, Forsberg N, Grahn T et al (2019) Genome-wide association study identified novel candidate loci affecting wood formation in Norway spruce. Plant J 100:83–100

Bao S, Jiang R, Kwan W, Wang B, Ma X, Song YQ (2011) Evaluation of next-generation sequencing software in mapping and assembly. J Hum Genet 56:406–414

Bernhardsson C, Vidalis A, Wang X, Scofield DG, Schiffthaler B, Baison J, Street NR, García-Gil MR, Ingvarsson PK (2019) An ultra-dense haploid genetic map for evaluating the highly fragmented genome assembly of Norway spruce (*Picea abies*). Genes Genomes Genet 9:1623–1632

Britten RJ (2010) Transposable element insertions have strongly affected human evolution. Proc Natl Acad Sci 107(46):19945–19948

Burrows M, Wheeler DJ (1994) A block-sorting lossless data compression algorithm. SRC research report 124. http://www.hpl.hp.com/techreports/Compaq-DEC/SRC-RR-124.pdf

Charlesworth D (2006) Balancing selection and its effects on sequences in nearby genome regions. PLoS Genet 2:e64

Chen J, Källman T, Ma X, Gyllenstrand N, Zaina G et al (2012) Disentangling the roles of history and local selection in shaping clinal variation of allele frequencies and gene expression in Norway spruce (*Picea abies*). Genetics 191:865–881

Cossu RM, Casola C, Giacomello S, Vidalis A, Scofield DG, Zuccolo A (2017) LTR retrotransposons show low levels of unequal recombination and high rates of intraelement gene conversion in large plant genomes. Genome Biol Evol 9:3449–3462

Dale RK, Pedersen BS, Quinlan AR (2011) Pybedtools: a flexible Python library for manipulating genomic datasets and annotations. Bioinformatics 27 (24):3423–3424

Danecek P, Auton A, Abecasis G, Albers CA, Banks E, DePristo MA, Handsaker RE, Lunter G, Marth GT, Sherry ST et al (2011) The variant call format and VCFtools. Bioinformatics 27:2156–2158

DePristo MA, Banks E, Poplin R, Garimella KV, Maguire JR, Hartl C, Philippakis AA, del Angel G, Rivas MA, Hanna M, et al (2011) A framework for variation discovery and genotyping using next-generation DNA sequencing data. Nat Genet 43: 491–498

Escalona M, Rocha S, Posada D (2016) A comparison of tools for the simulation of genomic next-generation sequencing data. Nat Rev Genet 17(8):459

Ebbert MT, Wadsworth ME, Staley LA, Hoyt KL, Pickett B, Miller J, Duce J, Kauwe JS, Ridge PG (2016) Evaluating the necessity of PCR duplicate removal from next-generation sequencing data and a comparison of approaches. BMC Bioinform 17(7):239

Flicek P (2009) The need for speed. Genome Biol 10 (3):212

Flicek P, Birney E (2010) Sense from sequence reads: methods for alignment and assembly. Nat Methods 7 (6):479

Garrison E, Marth G (2012) Haplotype-based variant detection from short-read sequencing. arXiv:1207. 3907 [q-bio.GN]

Gayral P, Melo-Ferreira J, Glémin S, Bierne N, Carneiro M, Nabholz B et al (2013) Reference-free population genomics from next-generation transcriptome data and the vertebrate-invertebrate gap. PLoS Genet 9:e1003457

Goodwin S, McPherson JD, McCombie WR (2016) Coming of age: ten years of next-generation sequencing technologies. Nat Rev Genet 17(6):333

Hartl DL, Clark AG (1989) Principles of population genetic*s*. Sinauer Associates

Homer N, Nelson SF (2010) Improved variant discovery through local re-alignment of short-read next-generation sequencing data using SRMA. Genome Biol 11(10):R99

Hua-Van A, Le Rouzic A, Boutin TS, Filée J, Capy P (2011) The struggle for life of the genome's selfish architects. Biol Direct 6(1):19

Heuertz M, De Paoli E, Källman T, Larsson H, Jurman I et al (2006) Multilocus patterns of nucleotide diversity, linkage dis-equilibrium and demographic history of Norway spruce [*Picea abie*s (L.) Karst]. Genetics 174:2095–2105

Hung JH, Weng Z (2017) Mapping billions of short reads to a reference genome. Cold Spring Harb Protoc 2017 (1):pdb-top093153

Jurka J, Kapitonov VV, Kohany O, Jurka MV (2007) Repetitive sequences in complex genomes: structure and evolution. Annu Rev Genomics Hum Genet 8:241–259

Kim PM, Lam HY, Urban AE, Korbel JO, Affourtit J, Grubert F, Chen X, Weissman S, Snyder M, Gerstein MB (2008) Analysis of copy number variants and segmental duplications in the human genome: evidence for a change in the process of formation in recent evolutionary history. Genome Res 18 (12):1865–1874

Koboldt DC, Zhang Q, Larson DE, Shen D, McLellan MD, Lin L, Miller CA, Mardis ER, Ding L, Wilson RK (2012) VarScan 2: somatic mutation and copy number alteration discovery in cancer by exome sequencing. Genome Res 22(3):568–576

Langmead B, Trapnell C, Pop M, Salzberg SL (2009) Ultrafast and memory-efficient alignment of short DNA sequences to the human genome. Genome Biol 10(3):R25

Li H, Durbin R (2009) Fast and accurate short read alignment with Burrows–Wheeler transform. Bioinformatics 25(14):1754–1760

Li H, Ruan J, Durbin R (2008a) Mapping short DNA sequencing reads and calling variants using mapping quality scores. Genome Res 18(11):1851–1858

Li R, Li Y, Kristiansen K, Wang J (2008b) SOAP: short oligonucleotide alignment program. Bioinformatics 24 (5):713–714

Li R, Yu C, Li Y, Lam TW, Yiu SM, Kristiansen K, Wang J (2009a) SOAP2: an improved ultrafast tool for short read alignment. Bioinformatics 25(15):1966–1967

Li R, Li Y, Fang X, Yang H, Wang J, Kristiansen K, Wang J (2009b) SNP detection for massively parallel whole-genome resequencing. Genome Res 19 (6):1124–1132

Li H, Handsaker B, Wysoker A, Fennell T, Ruan J, Homer N, Marth G, Abecasis G, Durbin R (2009c) The sequence alignment/map format and SAMtools. Bioinformatics 25(16):2078–2079

Li Y, Chen W, Liu EY, Zhou YH (2013) Single nucleotide polymorphism (SNP) detection and genotype calling from massively parallel sequencing (MPS) data. Stat Biosci 5(1):3–25

Liu L, Li Y, Li S, Hu N, He Y, Pong R, Lin D, Lu L, Law M (2012) Comparison of next-generation sequencing systems. J Biomed Biotechnol 2012:251364

Liu X, Han S, Wang Z, Gelernter J, Yang B-Z (2013) Variant callers for next-generation sequencing data: a comparison study. PLoS ONE 8:e75619

Lunter G, Goodson M (2011) Stampy: a statistical algorithm for sensitive and fast mapping of Illumina sequence reads. Genome Res 21(6):936–939

Mardis ER (2008) The impact of next-generation sequencing technology on genetics. Trends Genet 24 (3):133–141

Mardis ER (2017) DNA sequencing technologies: 2006–2016. Nat Protoc 12(2):213

Martin ER, Kinnamon DD, Schmidt MA, Powell EH, Zuchner S, Morris RW (2010) SeqEM: an adaptive genotype-calling approach for next-generation sequencing studies. Bioinformatics 26(22):2803–2810

McKenna A, Hanna M, Banks E, Sivachenko A, Cibulskis K, Kernytsky A, Garimella K, Altshuler D, Gabriel S, Daly M, DePristo MA (2010) The genome analysis toolkit: a MapReduce framework for analyzing next-generation DNA sequencing data. Genome Res 20(9):1297–1303

McKinney GJ, Waples RK, Seeb LW, Seeb JE (2017) Paralogs are revealed by proportion of heterozygotes and deviations in read ratios in genotyping-by-sequencing data from natural populations. Mol Ecol Resour 17:656–669

Morozova O, Marra MA (2008) Applications of next-generation sequencing technologies in functional genomics. Genomics 92(5):255–264

Mielczarek M, Szyda J (2016) Review of alignment and SNP calling algorithms for next-generation sequencing data. J Appl Genet 57(1):71–79

Neale DB, McGuire PE, Wheeler NC, Stevens KA, Crepeau MW, Cardeno C, Zimin AV, Puiu D, Pertea GM, Sezen UU et al (2017) The Douglas-Fir genome sequence reveals specialization of the photosynthetic apparatus in *Pinaceae*. G3 7:3157–3167

Nielsen R, Paul JS, Albrechtsen A, Song YS (2011) Genotype and SNP calling from next-generation sequencing data. Nat Rev Genet 12(6):443

Nystedt B, Street NR, Wetterbom A, Zuccolo A, Lin YC, Scofield DG, Vezzi F, Delhomme N, Giacomello S, Alexeyenko A, Vicedomini R et al (2013) The Norway spruce genome sequence and conifer genome evolution. Nature 497:579

Ojeda D, Mattila T, Ruttink T, Kujala S, Kärkkäinen K, Verta J-P, Pyhajarvi T (2018) Utilization of tissue ploidy level variation in de novo transcriptome assembly of *Pinus sylvestris* 495689

Pirooznia M, Kramer M, Parla J, Goes FS, Potash JB, McCombie WR, Zandi PP (2014) Validation and assessment of variant calling pipelines for next-generation sequencing. Hum Genomics 8:14

Quinlan AR, Hall IM (2010) BEDTools: a flexible suite of utilities for comparing genomic features. Bioinformatics 26:841–842

R Core Team (2014) R: a language and environment for statistical computing. R Foundation for Statistical Computing, Vienna, Austria

Rumble SM, Lacroute P, Dalca AV, Fiume M, Sidow A, Brudno M (2009) SHRiMP: accurate mapping of short color-space reads. PLoS Comput Biol 5:e1000386

Schuster SC (2007) Next-generation sequencing transforms today's biology. Nat Methods 5:16

Stevens KA, Wegrzyn JL, Zimin A, Puiu D, Crepeau M, Cardeno C, Paul R, Gonzalez-Ibeas D, Koriabine M, Holtz-Morris AE et al (2016) Sequence of the sugar pine megagenome. Genetics 204:1613–1626

Syvänen A-C (2005) Toward genome-wide SNP genotyping. Nat Genet 37:S5–S10

Tajima F (1989) Statistical method for testing the neutral mutation hypothesis by DNA polymorphism. Genetics 123:585–595

Trapnell C, Salzberg SL (2009) How to map billions of short reads onto genomes. Nat Biotechnol 27(5):455

Treangen TJ, Salzberg SL (2012) Repetitive DNA and next-generation sequencing: computational challenges and solutions. Nat Rev Genet 13(1):36

Vidalis A, Scofield DG, Neves LG, Bernhardsson C, García-Gil MR, Ingvarsson PK (2018) Design and evaluation of a large sequence-capture probe set and associated SNPs for diploid and haploid samples of Norway spruce (*Picea abies*) 291716

Wang J, Scofield D, Street NR, Ingvarsson PK (2015) Variant calling using NGS data in European aspen (*Populus tremula*). In: Sablo G, Kumar S, Ueno S, Kuo J, Varotto C (eds) Advances in the understanding of biological sciences using next generation sequencing (NGS) approaches, pp 43–61. Springer, NY

Zimin A, Stevens KA, Crepeau MW, Holtz-Morris A, Koriabine M, Marçais G, Puiu D, Roberts M, Wegrzyn JL, de Jong PJ et al (2014) Sequencing and assembly of the 22-gb Loblolly pine genome. Genetics 196:875–890

Zimin AV, Stevens KA, Crepeau MW, Puiu D, Wegrzyn JL, Yorke JA, Langley CH, Neale DB, Salzberg SL (2017) An improved assembly of the Loblolly pine mega-genome using long-read single-molecule sequencing. GigaScience 6:1–4

Transposable Elements in Spruce

3

Giovanni Marturano, Camilla Canovi, Federico Rossi, and Andrea Zuccolo

Abstract

Transposable elements (TEs), along with other repetitive sequences, were dismissed for a long time as junk DNA. Over the years, much evidence accumulated, clarifying how TEs are instead major components of host genomes and have a substantial role in shaping genome structure, functioning, and evolution. In plants, TEs were mostly studied in important model and crop species, in particular, Arabidopsis, rice, and maize. The studies focusing on TEs in gymnosperms lagged behind for different reasons, last but not least, the enormous genome sizes for most species belonging to this group. Recently, the decrease in sequencing cost and advances in assembly algorithms allowed whole-genome sequencing of conifers such as Norway spruce, white spruce, and loblolly pine (Nystedt et al. 2013, Birol et al. 2013, Neale et al. 2014). The availability of such genomic data enabled a more comprehensive and insightful study of TEs starting to depict patterns quite different from the ones already described for angiosperms. In this chapter, we present and discuss the available and newly generated data regarding Norway spruce TEs comparing evidence gathered from genomic and transcriptomic analyses with analogous data of other conifers, crops, and model species, pointing out similarities and differences.

G. Marturano · A. Zuccolo (✉)
Institute of Life Sciences, Scuola Superiore Sant'Anna, P.zza Martiri della libertà 33, 56127 Pisa, Italy
e-mail: andrea.zuccolo@sssup.it; azuccolo@appliedgenomics.org

C. Canovi
Dipartimento di Scienze Agrarie, Alimentari e Agro-Ambientali, Università di Pisa, Via del Borghetto 80, 56124 Pisa, Italy

F. Rossi
Medical and Molecular Genetics Department, King's College, London, UK

3.1 The Repetitive Fraction of Plant Genomes

Many closely related organisms having comparable complexities nevertheless show a remarkable variation of their DNA content per haploid genome (C-value) (Thomas 1971). This poor relationship between biological complexity and DNA content was described as the "C-value paradox." The variability in genome size is striking in plants and it is particularly evident in angiosperms, where the C-value across different species spans several orders of magnitude topping a factor of 2,000 when the 63 Mbp of *Genlisea margaretae* genome is compared to the 150-Gbp genome for *Paris japonica* (Greilhuber et al. 2006; Pellicer et al. 2010). The "C-value paradox" was eventually resolved by

I. M. Porth and A. R. De la Torre (eds.), *The Spruce Genome*, Compendium of Plant Genomes,
https://doi.org/10.1007/978-3-030-21001-4_3

demonstrating that most part of the genome size variation observed could be explained by the repetitive fraction of the genome (Rosbash et al. 1974) constituted by nucleic acid sequences occurring in multiple copies.

The repetitive sequences can be classified according to their spatial arrangement in the host genome in two groups: tandem repeats and dispersed repeats. Tandem repeats are arranged in monomeric units, located one after the other. Notable examples are ribosomal RNA genes, centromeric repeats, telomeric repeats, and satellite DNA. Dispersed repeats are scattered throughout the host genome and include the paralogous copies of gene families' TEs (Richard et al. 2008). TEs are a major component of the genome's repetitive fraction and heavily contribute to the observed genome size variation across different species. Indeed, the TE content positively correlates with genome size (Gaut and Ross-Ibarra 2008). For example, the 125-Mbp Arabidopsis genome contains approximately 27,000 genes and about 10% of its size is represented by TEs (Arabidopsis Genome Initiative 2000); the ~2,500-Mbp maize genome contains about 32,000 genes and TEs altogether constitute more than 84% of the genome (Liu and Bennetzen 2008; Baucom et al. 2009; Schnable et al. 2009).

3.2 Classification of Transposable Elements in Plant Genomes

TEs are DNA sequences, able to change their position by moving into chromosomal sites different from their original one by means of a process called transposition. These "jumping genes" were first identified in maize in the mid-1940s by the biologist Barbara McClintock and were initially considered as a kind of genetic weirdness (Fedoroff 1998). TEs can be arranged into two main classes according to the method they adopt for transposition: retrotransposons (or class I elements) and DNA-transposons (or class II elements).

Retrotransposons exploit for their replication a "copy and paste"-like mechanism, which involves an RNA intermediate molecule (Finnegan 1989) and leads to a massive increase in their copy numbers. Class I TEs include Long Terminal Repeat (LTR) Retroelements (RTs), Long Interspersed Nuclear Elements (LINEs), and Short Interspersed Nuclear Elements (SINEs). Class II elements, on the other hand, usually replicate following a "cut and paste"-like replication mechanism. Class II TEs, because of their type of transposition mechanism, are usually less abundant than retrotransposons. CACTA-like, hAT-like, Mutator-like, Mariner-like, MITEs (Miniature Inverted-Repeat Transposable Elements), and Helitrons are the most common Class II elements inside plant genomes (Grzebelus 2018).

3.3 Transposable Elements: From "Junk" to "Parasite" DNA to Major Players in Shaping Genome Structure, Function, and Evolution

The availability of increasingly comprehensive genomic sequence data made possible by the development of ever sophisticated DNA sequencing technologies starting from the late 1970s has provided substantial knowledge regarding the information content, the structure and the organization of genomes and, consequently, on their repetitive fraction. In the early 1970s, the idea was introduced that most of eukaryote DNA, including TEs, was "junk DNA" without function (Ohno 1972). A few years later, the term "selfish DNA" was proposed, to indicate that DNA sequences are able to proliferate within a genome (such as the TEs) without providing any clear evolutionary advantage to the host organism (Dawkins 1976). About a decade later, these concepts were further elaborated through a comparison of the proliferation of "selfish" DNA sequences within the genome to the expansion of a parasite inside its host (Orgel and Crick 1980). This view overcame the limited assumption that the whole organism's DNA has a specific function "filtered" by natural selection and contributed to explain the C-value paradox (Fedoroff 2012).

Over the years, evidence accumulated demonstrating how TEs not only compose extensive portions of many eukaryotic genomes, but also, that TEs have a pivotal role in controlling genome functioning and in shaping genome structure. TEs activity can initiate a wide array of dramatic changes in gene expression and function such as the rewiring of regulatory networks and the creation of novel genes (Yao et al. 2001). TEs can inactivate genes by inserting into their coding regions or promoters. Alternatively, TEs can provide regulatory motifs recognized by transcriptional enhancers or repressors, thus affecting gene activity. LTR-RTs can also determine the deletion of genes present between LTRs, when an intra-element recombination event takes place, or the duplication of a gene coding sequence harbored in the same region through self-replication and insertion into a new genomic site (Lisch 2013).

However, the majority of TEs activity is likely selectively neutral, since, under certain circumstances, transposition can have catastrophic effects on the functioning and viability of the host genomes (Charlesworth et al. 2004; Brookfield 2005). Because of its potentially detrimental effect, TEs transcription and insertion into new genomic sites is controlled by epigenetic mechanisms. Indeed, TEs in eukaryotic genomes are generally "epigenetically silenced," i.e., they are present in a reversibly inactive form. In plants, epigenetic mechanisms for TEs silencing include DNA methylation, histone modification, and the presence of histone variants (Lisch 2009; Slotkin and Martienssen 2007). DNA methylation is established through the RNA-directed DNA methylation pathway in which 21-, 22-, and 24-nucleotide (nt) small RNAs guide the methylation machinery to loci transcribed by RNA polymerases II, IV, and V. These epigenetic processes evolved to control the destructive power of the "parasitic" TE sequences, and have deep effects on both genome expansion and genome recombination events in higher eukaryotes (Fedoroff 2012).

On the other hand, TEs are by far the most variable portion of the genome (Lisch 2013) and represent an endogenous source of variability that, under certain circumstances, can be beneficial to the evolution of the host genome. Indeed, TEs can sometimes be advantageous at the species level as a source of evolutionary innovation capable of modifying existing genes or creating novel genes through the mechanism of exaptation (Joly-Lopez and Bureau 2018). TEs exaptation has provided a major contribution to the evolution of gene regulation (De Souza et al. 2013); in plant genomes, "exapted transposable element genes" (ETEs) involved in physiological functions such as flowering (Cowan et al. 2005), development (Bundock and Hooykaas 2005), light signaling (Hudson et al. 2003), and stress response (McClintock 1984, Chénais et al. 2012, Makarevitch et al. 2015) were found.

3.4 Transposable Elements in Spruce

3.4.1 Abundance

Conifers have an ancient origin; the major radiation of conifer families occurred between 250 and 65 million years ago (MYA) (Leslie et al. 2012). Their genomes are very large, ranging from 18 to 35 Gbp (Murray et al. 2004) and are characterized by a large fraction of old and highly diverged repeated sequences, mostly consisting of TEs (Morse et al. 2009; Magbanua et al. 2011; Nystedt et al. 2013; Zuccolo et al. 2015).

An extensive assessment of TEs abundance in the Norway spruce genome was carried out within the frame of the Spruce Genome Project (www.congenie.org) that produced a genome assembly for this species (Nystedt et al. 2013). Because the genome assembly was expected to suffer from a significant underrepresentation of TEs, their abundance was estimated separately by analyzing 100,000 random sheared reads, obtained using 454 Roche sequencing technology. Altogether, the fraction of the genome confidently assigned to TEs totaled to around 70% (Nystedt et al. 2013). The most abundant and variable order of TEs in *Picea abies* genome is represented by LTR-RTs contributing for at least 58% of the genome size with the Ty1-*copia* superfamily being less abundant than the Ty3-*gypsy* one (16% and 35%, respectively). Notably, an additional 7% of the genome is represented by LTR-RTs not yet classified as members of any of the two superfamilies. Class I LINEs and Class II DNA-transposons abundance in the spruce genome was 1% for each of the two groups (Nystedt et al. 2013). The amount of unassigned TEs in the genome was close to 10%. This fraction likely includes very old and diverged TEs awaiting their confident assignment to any known TE class. In sum, these data are consistent with what has subsequently been seen in other conifer genomes such as *Pinus taeda* (Wegrzyn et al. 2014; Neale et al. 2014) and *Pinus lambertiana* (Stevens et al. 2016).

In the next approach, a manually curated collection of 2,400 completed conifer TEs was used to estimate TEs abundance on a dataset of 32,632 fosmid-cloned Norway spruce sequences, assumed to exhibit an unbiased representation of the different components of the Norway spruce genome (Cossu et al. 2017). The results obtained for the most abundant TE class, i.e., the LTR-RTs, were mostly consistent with those described by Nystedt et al. (2013). Ty1-*copia* and Ty3-*gypsy* super-families were assessed to represent 17% and 39% of the genome, respectively. LINEs were estimated to represent 1% of the genome and CACTA 1%.

Assuming an average length of 8 kbp for Ty3-*gypsy* and 5 kbp for and Ty1-*copia* elements, we inferred an overall amount of 975,000 Ty3-gypsy and 680,000 Ty1-*copia* elements in the Norway spruce genome. A similar calculation for LINE and CACTA elements, assuming an average length of 6 kbp and 10 kbp, respectively, estimated 20,000 CACTA and 33,300 LINE copies. However, these results are likely underestimates of the true values, since results were obtained assuming that all considered TE copies were complete.

3.4.2 Conservation and Distribution

One hundred reverse transcriptase (RT) amino acidic sequences were randomly collected from each genome of *P. abies* and five other gymnosperms (*Picea glauca, Pinus lambertiana, Pinus taeda, Pseudotsuga menziesii, and Ginkgo biloba*) for both, the Ty1-*copia* and Ty3-*gypsy*, LTR-RT superfamilies. Altogether, 1,200 RT domains were retrieved in these species covering an evolutionary timeframe of about 300 MY. Multiple sequence alignments of the RT paralogs were then used to carry out phylogenetic analysis using the Neighbor-Joining (NJ) method (Fig. 3.1). The analysis showed that LTR-RTs are widespread and conserved in gymnosperm genomes. Several sub-families are present in almost all of the genera considered. Paralogs from the same species usually group together with high bootstrap support within the most internal clades, suggesting that the amplification of these elements mainly occurred after the divergence of the species.

A detailed phylogenetic investigation carried out by Cossu et al. (2017) but limited to Norway spruce mostly confirmed these results and identified 16 Ty3-*gypsy* and 7 Ty1-*copia* groups, respectively, supported by high bootstrap values. The overall heterogeneity of spruce LTR-RTs was also demonstrated by a clustering analysis based on sequence similarity and carried out by Nystedt et al. (2013) using the LTRs collected from 120 complete LTR-RTs. At least 86% of the LTRs did not cluster, when an 80% similarity threshold was employed, but remained as singletons, suggesting a high level of divergence between the LTR-RT families, and furthermore, suggesting the existence of several low abundance families. This result is consistent with what has been seen in other conifers such as *Pinus taeda* (Wegrzyn et al. 2014; Neale et al. 2014).

Many of the typical features of conifer LTR-RTs can be described by the Ty1-copia element

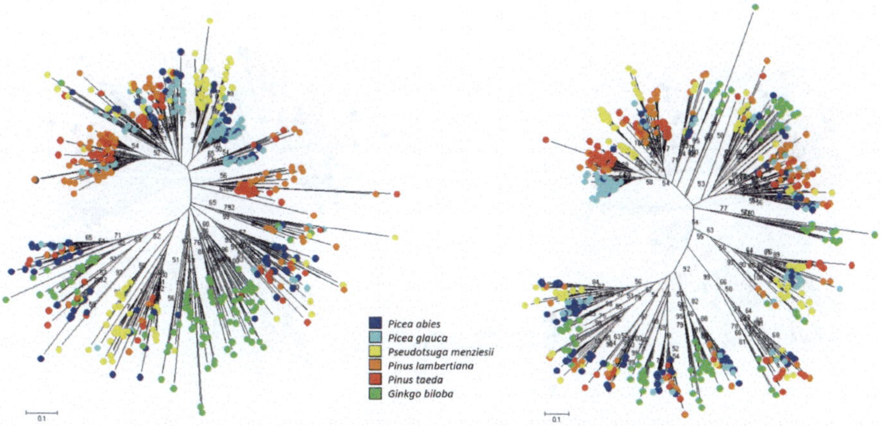

Fig. 3.1 Neighbor-joining trees of Ty3-*gypsy* (left) and Ty1-*copia* (right) elements for six gymnosperm species built using the reverse transcriptase domain sequence. Bootstrap values were calculated for 1,000 replicates. Only values higher than 50 are reported

PARTC (*P. abies* Retrotransposon Copia) isolated in Norway spruce (Zuccolo et al. 2015). By comparing *PARTC* completed elements from *P. abies* against the genomes of six other conifers (*Picea glauca, Pinus sylvestris, Pinus taeda, Abies sibirica, Taxus baccata, and Juniperus communis*), paralogous copies were found for the *Picea, Pinus,* and *Abies* genera, showing a wide distribution of *PARTC* coupled with a high degree of conservation. Indeed, in all these cases, the sequence similarity was well above 80% over the entire completed element length. In the genomes of species belonging to *Taxus* and *Juniperus* genera, however, *PARTC* copies were not conserved enough to be grouped into the same family under the Wicker classification scheme (Wicker et al. 2007) (i.e., they did not share at least 80% sequence similarity over at least 80% of their LTRs lengths), even though sequences were clearly related. Altogether, *PARTC* copies have been highly conserved over 200 MY of conifer evolution. The estimated contribution of *PARTC* to the *P. abies* genome size constitutes roughly 1%, corresponding to more than 39,000 complete copies.

Subsequently, the phylogenetic analysis of TEs was further extended to the class 1 *LINE* and the class 2 *CACTA* superfamilies, often overlooked in previous studies. Coding domains belonging to LINE and CACTA elements were retrieved from the publicly available genome assemblies (www.congenie.org; www.pinegenome.org) of five conifers (*Picea abies, P. glauca, Pinus taeda, P. lambertiana,* and *Pseudotsuga menziesii*) and the only extant Gingkophyta member, *Gingko biloba* (Guan et al. 2016). As already introduced above, paralogs were again used to build NJ trees (Fig. 3.2). In the case of conifers, the analysis revealed for both, LINEs and CACTAs, that several sub-families have paralogs in almost all of the examined conifer genera suggesting their ancient origin and pronounced retention across evolution. The paralogs belonging to the same genus mostly cluster together in internal clades, supported by high bootstrap values, and barely map together with those from other genera, suggesting that the amplification events, creating the extant copies, took place after genera separation. A similar pattern, as expected, was shown by examining the paralogs of the Ginkgophyta *G. biloba* indicating that their amplification took place already after the divergence between *G. biloba* and the other species belonging to the Pynophita division, and therefore, their paralogous sequences diverged from the rest of the examined gymnosperms paralogs more profoundly (Fig. 3.2).

The two TE classes were then analyzed in greater detail in *P. abies* alone; 200 complete LINEs isolated from *P. abies* were arranged into

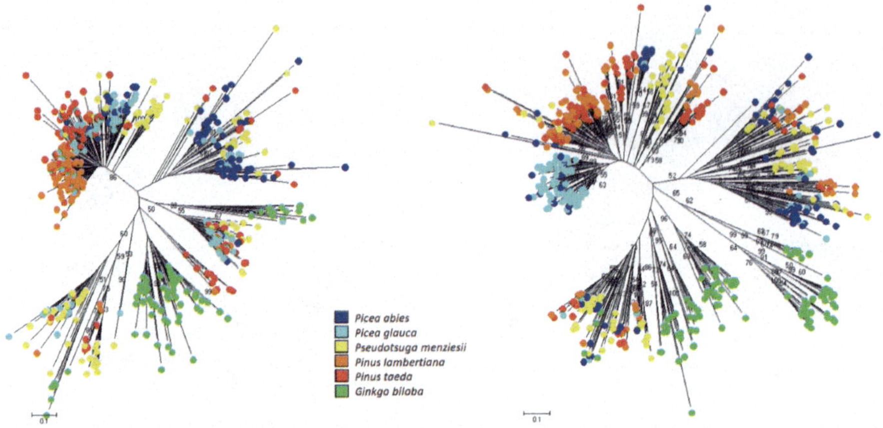

Fig. 3.2 Neighbor-joining trees of CACTA (left) and LINE (right) elements for six gymnosperm species built using transposase and reverse transcriptase domains, respectively. Bootstrap values were calculated for 1,000 replicates. Only values higher than 50 are reported

six groups on the basis of sequence similarity assessed by dot plot comparisons. A further analysis of these 200 elements targeted their coding domains. LINEs coding domains include two Open Reading Frames (ORFs). ORF1 is poorly conserved and its function is not well understood although it is supposed to be related to binding nucleic acids (Kolosha and Martin 2003; Martin et al. 2005). The LINE ORF2 encodes for the enzymes RT and endonuclease (EN) and is highly conserved. Consistently to what has often been seen in plant LINEs, in each of the six *P. abies* LINE groups, ORF2 is well defined and highly conserved, while the identification of ORF1 remains quite elusive because of its great sequence variability and poor conservation. Both ORFs are often fragmented and characterized by several stop codons and frameshift mutations suggesting the ancient origin of the extant LINE elements in *P. abies*.

A phylogenetic analysis was carried out on 200 randomly chosen *P. abies* CACTA elements focusing on their transposase coding domains. The NJ tree built on their alignment showed five major groups (Fig. 3.3). One hundred randomly chosen complete CACTAs underwent a close manual inspection of their transposase coding domains. In the overall majority of the cases (98 out of 100), the transposase coding domains appear to be riddled with stop codons and frameshift mutations indicating, similarly to LINEs, the ancient origin of the extant complement of these elements.

3.4.3 LTR-RTs Insertion Time

At the moment of insertion into the host genome, LTR, flanking intact LTR-RTs, are identical; over time each of the two LTRs accumulates mutations independently. Starting from this assumption and provided that a suitable substitution rate is available, the insertion times of LTR-RT can be calculated comparing the pairs of LTRs flanking intact elements (SanMiguel and Bennetzen 1998). The substitution rate for LTRs in Norway spruce was estimated by analyzing orthologous LTR-RT insertions in *P. abies* and *P. glauca* (Nystedt et al. 2013). The rationale for the analysis was that the nucleotide differences between any pair of orthologs must correspond to the mutations accumulated following post-speciation. Using this approach, the substitution rate was estimated being within the interval of 1.43×10^{-9}–2.2×10^{-9} mutations per synonymous site per year according to the different estimates for *P. abies* and *P. glauca* divergence time used. The divergence times were inferred from fossil records analysis and ranged from 13 to 20 MYA (Bouillé and Bousquet 2005).

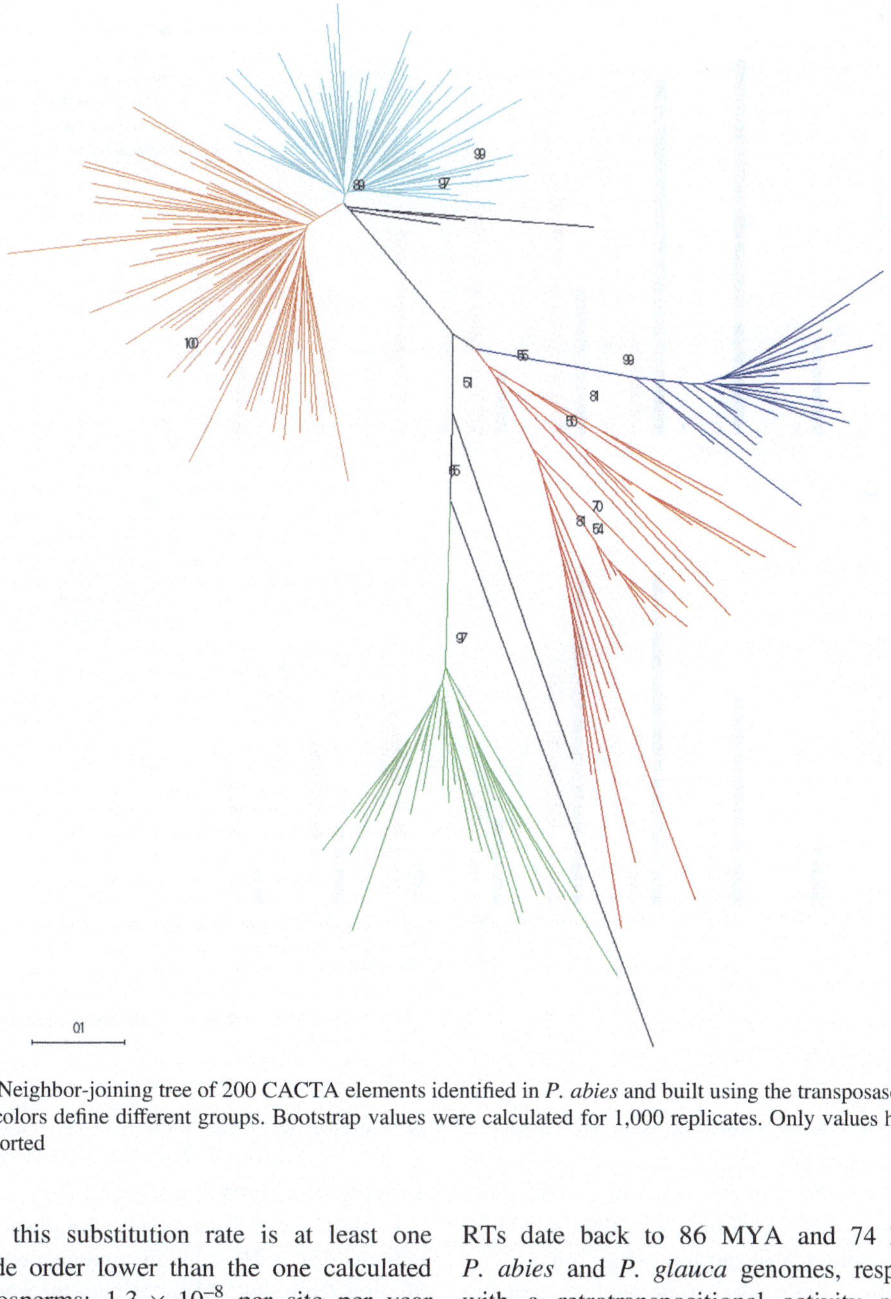

Fig. 3.3 Neighbor-joining tree of 200 CACTA elements identified in *P. abies* and built using the transposase domains. Different colors define different groups. Bootstrap values were calculated for 1,000 replicates. Only values higher than 50 are reported

Notably, this substitution rate is at least one magnitude order lower than the one calculated for angiosperms: 1.3×10^{-8} per site per year (Ma and Bennetzen 2004).

Data from the literature showed that the amplification of the LTR-RTs in conifers occurred over long evolutionary times spanning tens of millions of years, with the most ancient extant copies inserted nearly 100 MYA (Nystedt et al. 2013; Zuccolo et al. 2015). In the case of the Ty1-*copia* family *PARTC*, the oldest extant LTR-RTs date back to 86 MYA and 74 MYA in *P. abies* and *P. glauca* genomes, respectively, with a retrotranspositional activity peak that occurred between 15 and 36 MYA in *P. abies* and 12 and 30 MYA in *P. glauca*, respectively, (Zuccolo et al. 2015). A comprehensive analysis targeting 1,200 representative LTR-RTs completed sequences collected from six gymnosperm species (*Picea abies*, *P. glauca*, *Pinus taeda*, *P. lambertiana*, *Pseudotsuga menziesii,* and *G. biloba*) demonstrated that in the genome of

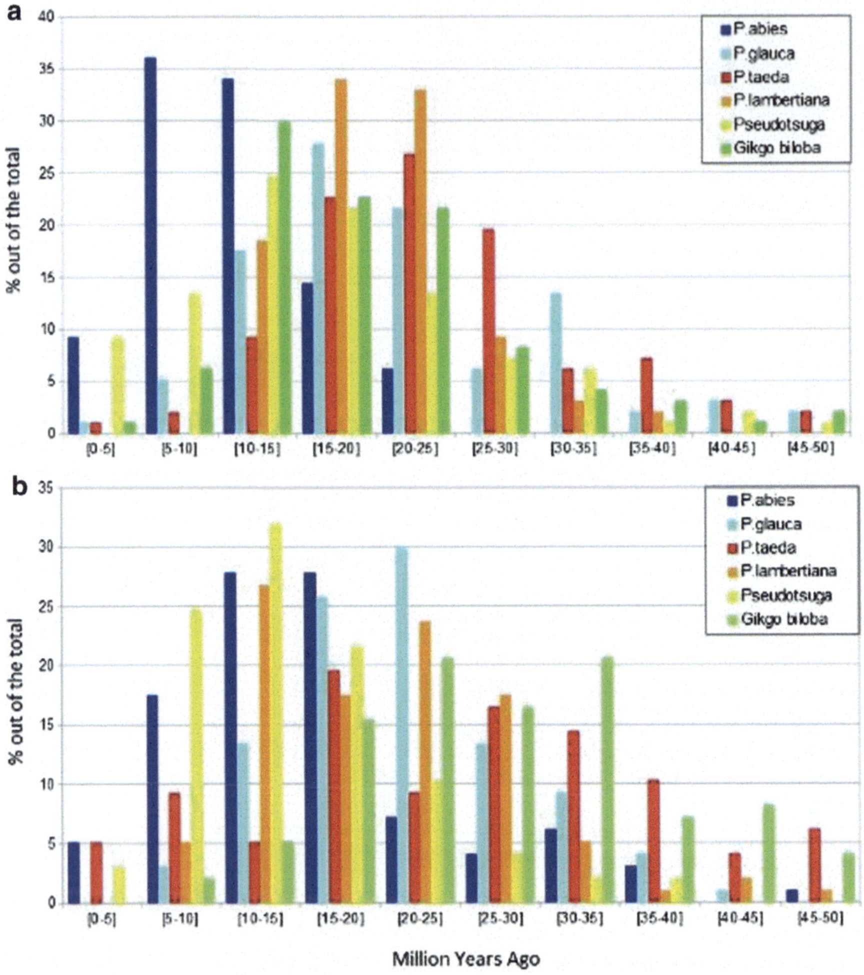

Fig. 3.4 Distributions of insertion times of Ty1-*copia* (**a**) and Ty3-*gypsy* (**b**) LTR-RTs in six gymnosperm genomes calculated using a mutation rate of 2.2×10^{-9} (per base per year)

P. abies, there is very limited evidence of recent (i.e., less than 5 MYA) retrotranspositional activity and that, more generally, LTR-RTs inserted in conifer genomes less than 4 MYA, represent about 1% of the total (Fig. 3.4).

Altogether, the phylogenetic analysis and the insertion time estimates suggest that both Ty3-*gypsy* and Ty1-*copia* LTR-RT superfamilies are ancient, mostly inactive and largely shared between conifer genomes. The striking level of conservation shown by LTR-RTs in *P. abies* and in conifers at large, could be explained by taking into account both, the limited removal mechanisms targeting LTR-RTs (see below), and the low substitution rates characterizing these elements (see above).

In sum, these features and dynamics depict a scenario for LTR-RTs in gymnosperms quite different from that characterizing the vast majority of these elements in angiosperms, where, for many species, it is possible to identify a large fraction of recently (i.e., less than 5 MYA) inserted LTR-RT elements, some of which produced LTR-RT families that sustained activity bursts in recent evolutionary times (Piegu et al. 2006; Hawkins et al. 2006; Nystedt et al. 2013).

3.4.4 LTR-RTs Removal Mechanisms

As an alternative to containing LTR-RTs activity by means of epigenetic control (see below for further details), host genomes can, at least partially, remove LTR-RTs. The two recombination mechanisms contributing to the effective removal of LTR-RTs are unequal recombination (UR) and illegitimate recombination (IR) (Devos et al. 2002; Ma et al. 2004). UR occurs between the LTRs of the same or different elements and removes one LTR, the intervening region between the two recombining LTRs and leaves behind a so-called solo-LTR (Vicient et al. 1999). UR is the major mechanism of LTR-RTs removal in angiosperm genomes where large amounts of solo-LTR were found (e.g., SanMiguel et al. 1996; Vicient et al. 1999; Dubcovsky et al. 2001; Fu and Dooner 2002; Vitte and Panaud 2003). Indeed, in Arabidopsis the ratio between solo-LTR to complete LTR-RTs is about 1:1 (Devos et al. 2002), in rice solo-LTRs outnumber complete LTR-RTs by 1.5 to 1 (Ma et al. 2004) and in *Glycine max,* the ratio is 1.18–1 (El Baidouri and Panaud 2013). The angiosperm genomes are characterized by an extremely fast turnover of LTR-RTs: the half-life of an LTR-RT element was estimated to be less than 6 MY in rice. Indeed, in the last 8 MY at least 190 Mb of LTR-RT related sequences had been removed from the rice genome (Ma et al. 2004).

Intriguingly, the preliminary analysis of UR affecting LTR-RTs in the Norway spruce genome showed a striking paucity of solo-LTRs. The analysis of three LTR-RT families for signatures of intra-element UR provided a ratio of 1:9 solo-LTRs to complete elements. This evidence led Nystedt et al. (2013) to propose that LTR-RTs removal mechanisms are less active in conifers than in most other organisms and that the *P. abies* large genome size might be explained by the slow and constant accumulation of a heterogeneous set of LTR-RTs not efficiently counterbalanced by a removal mechanism. In a recent study, Cossu et al. (2017) assessed the extent of UR on a much wider scale taking into account several LTR-RT families in four conifers (*Picea abies, P. glauca, Pinus taeda,* and *P. lambertiana*) and seven angiosperms (*Arabidopsis thaliana, Brachypodium distachyon, Oryza sativa, Vitis vinifera, Zea mays, Amborella trichopoda,* and *Populus trichocarpa*). The authors found that the lowest rates of UR were associated with the largest genomes studied, i.e., those of the four conifers and the angiosperm maize, confirming that indeed, UR is severely reduced in conifers. Ratios of complete LTR-RTs to solo-LTR were 5.6, 4, 7.5, and ~3 to 1 for *Picea abies, P. glauca, Pinus taeda,* and *P. lambertiana*, respectively.

3.4.5 Transposable Elements Methylation and Small RNA

Since heterochromatin formation in LTR-RT-rich regions via methylation reduces recombination processes (Wallberg et al. 2015; Colomé-Tatché et al. 2012; Mirouze et al. 2012; Yelina et al. 2015), Cossu et al. (2017) concluded that the epigenetic mechanisms aimed to suppress the replication of TEs might also provide the conditions for causing the reduction in recombination events that could lead to LTR-RTs removal. This concept was first suggested by Fedoroff (2012), who advanced the hypothesis that TEs can accumulate in very large quantities in the host genomes because of, not in spite of, the epigenetic mechanisms used to control their proliferation.

The Norway spruce genome showed high DNA methylation levels because of the elevated TEs content (Ausin et al. 2016). In particular, at the genome-wide level, cytosines are highly methylated in both, CG and CHG contexts, with an average methylation level of 74.7% and 69.1%, respectively. Conversely, CHH sites have an average methylation level of only 1.5%. The authors of the study pointed out how the Norway spruce methylation patterns contrast with those seen in Arabidopsis. Indeed, in *P. abies*, a robust methylation maintenance system for CHG sites seems to be in place as the bimodal distribution of cytosine methylation for the majority of CG

Table 3.1 Cytosine methylation level genomic fraction in Norway spruce for the TE and the not-TE fractions of a set of fosmid sequences (see text)

Sequence context	TEs	Not TEs
% C methylated in CpG context	74.3	41.5
% C methylated in CHG context	71.9	22.5
% C methylated in CHH context	2.2	0.9

and CHG sites being either not methylated or highly methylated suggests (Ausin et al. 2016). Similarly, as in other studied plant genomes, TEs represent a higher DNA methylation level than other genome sequences which possibly prevents their activation, transcription, and transposition. A subset of spruce sequences cloned in fosmids (Cossu et al. 2017) was masked for TEs by running RepeatMasker (Smit et al. 2015) fed with the TE library used in Nystedt et al. (2013). DNA methylation quantified in TE-related and TE-unrelated Norway spruce sequences led to report considerable differences between the two categories in all sequence contexts (Table 3.1). The methylation levels for the TE-related fraction are in very good agreement with those obtained by Ausin et al. (2016) on whole-genome scale, once more indicating how a large fraction of the spruce genome is composed by TE-related sequences.

The second line of evidence supporting the high level of methylation in TEs was collected by exploiting the Norway spruce 21- and 24-nt sRNA dataset available for needle and somatic embryogenesis tissues (Ausin et al. 2016). The mapping of 24-nt sRNA onto complete LTR-RTs was used as a "proxy" indicator of their methylation state (Slotkin and Martienssen 2007). The analysis was carried out on 84 representative Ty1-*copia* elements belonging to seven different families. All LTR-RTs appeared to be heavily targeted by both classes of small RNAs over their entire length. However, the most common mapping pattern shows that out of the different LTR-RT regions, the LTRs are usually the most heavily targeted (Fig. 3.5) suggesting that LTRs seem to be a hotspot for sRNA targeting.

3.5 Conclusions

In 2012, Dr. Nina Fedoroff proposed a suggestive and apparently counterintuitive model regarding the dynamics governing the proliferation of TEs in eukaryotic genomes. She posited that: "…it was precisely the evolution of prokaryotic epigenetic mechanisms, originally limiting recombination among horizontally exchanged sequences, to regulate homologous recombination within the eukaryotic genome that made it possible for genomes to grow" and: "What I am suggesting, then, is that TEs accumulate in eukaryotic genomes because of, not despite, epigenetic silencing mechanisms" (Fedoroff 2012). The scenario depicted by the most abundant class of TEs, i.e., LTR-RTs, in *Picea abies* and, more generally, in conifers, fits quite well into this theory. In fact, in these enormous plant genomes (up to 31 Gbp in the case of *Pinus lambertiana*), the activity of different LTR-RT families took place at a slow and steady rate and led to the expansion of the twelve ancestral chromosomes. The LTR-RTs that underwent expansion are largely shared among extant conifers. The expansion started in ancient times and it was not counteracted by any efficient removal mechanism, so many TEs remained in the genome well after their ancient burst of activity. We explained this evidence by proposing that common and effective removal mechanisms (as, e.g., those based on UR) were less active or less efficient in conifers than in most other organisms (Nystedt et al. 2013; Zuccolo et al. 2015).

The extensive use of comparative genomic approaches to contrast the evidence gained in conifers and, more generally, in gymnosperms with those gathered in angiosperms will elucidate the dramatic differences in LTR-RT behavior (proliferation control and removal strategies) in the two groups. This data, in turn, should provide valuable information for understanding the functioning of this important class of TEs in both taxonomic groups.

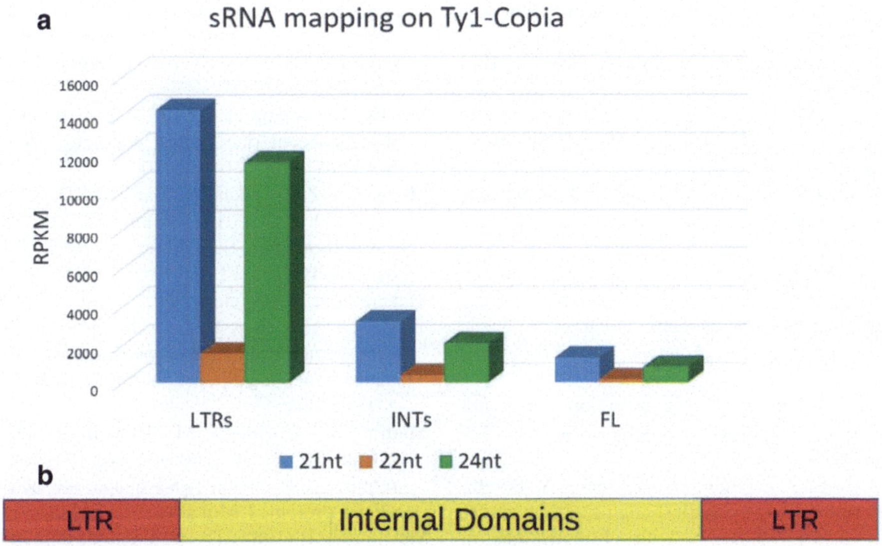

Fig. 3.5 **a** Mapping frequency of 21–22- and 24-nt small RNA onto full length (FL) LTR-RT elements and their specific regions: Long Terminal Repeats (LTR) and internal domains (INT). Values are expressed in RPKM (Reads Per Kilobase Million). **b** Simplified structure of an LTR-RT depicting the two LTRs and the internal domains

References

Arabidopsis Genome Initiative (2000) Analysis of the genome sequence of the flowering plant *Arabidopsis thaliana*. Nature 408(6814):796–815

Ausin I, Feng S, Yu C, Liu W, Kuo HY et al (2016) DNA methylome of the 20-gigabase Norway spruce genome. Proc Natl Acad Sci USA 113(50):E8106–E8113

Baucom RS, Estill JC, Chaparro C, Upshaw N, Jogi A et al (2009) Exceptional diversity, non-random distribution, and rapid evolution of retroelements in the B73 maize genome. PLoS Genet 5(11):e1000732

Birol I, Raymond A, Jackman SD, Pleasance S, Coope R et al (2013) Assembling the 20 Gb white spruce (Picea glauca) genome from whole-genome shotgun sequencing data. Bioinformatics 29(12):1492–1497

Brookfield JF (2005) The ecology of the genome-mobile DNA elements and their hosts. Nature Rev. Genet 6 (2):128–136

Bouillé M, Bousquet J (2005) Trans-species shared polymorphisms at orthologous gene loci among distant species in the conifer Picea (Pinaceae): implications for the long-term maintenance of genetic diversity in trees. Am J Bot 92:63–73

Bundock P, Hooykaas P (2005) An Arabidopsis hAT-like transposase is essential for plant development. Nature 436(7048):282–284

Charlesworth B, Borthwick H, Bartolome C, Pignatelli P (2004) Estimates of the genomic mutation rate for detrimental alleles in *Drosophila melanogaster*. Genetics 167(2):815–826

Chénais B, Caruso A, Hiard S, Casse N (2012) The impact of transposable elements on eukaryotic genomes: From genome size increase to genetic adaptation to stressful environments. Gene 509(1):7–15

Colomé-Tatché M, Cortijo S, Wardenaar R, Morgado L, Lahouze B et al (2012) Features of the Arabidopsis recombination landscape resulting from the combined loss of sequence variation and DNA methylation. Proc Natl Acad Sci USA 109(40):16240–16245

Cossu RM, Casola C, Giacomello S, Vidalis A, Scofield DG, Zuccolo A (2017) LTR retrotransposons show low levels of unequal recombination and high rates of intraelement gene conversion in large plant genomes. Genome Biol Evol 9(12):3449–3462

Cowan RK, Hoen DR, Schoen DJ, Bureau TE (2005) MUSTANG is a novel family of domesticated transposase genes found in diverse angiosperms. Mol Biol Evol 22(10):2084–2089

Dawkins R (1976) The selfish gene. Oxford University Press, New York, p 2

Devos KM, Brown JK, Bennetzen JL (2002) Genome size reduction through illegitimate recombination counteracts genome expansion in *Arabidopsis*. Genome Res 12(7):1075–1079

De Souza FS, Franchini LF, Rubinstein M (2013) Exaptation of transposable elements into novel cis-regulatory elements: is the evidence always strong? Mol Biol Evol 30(6):1239–1251

Dubcovsky J, Ramakrishna W, SanMiguel PJ, Busso CS, Yan L et al (2001) Comparative sequence analysis of colinear barley and rice bacterial artificial chromosomes. Plant Physiol 125(3):1342–1353

El Baidouri M, Panaud O (2013) Comparative genomic paleontology across plant kingdom reveals the dynamics of TE-driven genome evolution. Genome Biol Evol 5(5):954–965

Fedoroff NV (1998) Discoveries in plant biology (Kung S-D, Yang S-F (eds), vol. 1. World Scientific, Singapore, pp 89–104

Fedoroff NV (2012). Presidential address. Transposable elements, epigenetics, and genome evolution. Science 338(6108):758–767

Finnegan DJ (1989) Eukaryotic transposable elements and genome evolution. Trends Genet 5:103–107

Fu H, Dooner HK (2002) Intraspecific violation of genetic colinearity and its implications in maize. Proc Natl Acad Sci USA 99(14):9573–9578

Gaut BS, Ross-Ibarra J (2008) Selection on major components of angiosperm genomes. Science 320 (5875):484–486

Greilhuber J, Borsch T, Müller K, Worberg A, Porembski S, Barthlott (2006) Smallest angiosperm genomes found in *Lentibulariaceae*, with chromosomes of bacterial size. Plant Biol 8(6):770–777

Grzebelus D (2018) The functional impact of transposable elements on the diversity of plant genomes. Diversity 10(2):18

Guan R, Zhao Y, Zhang H, Fan G, Liu X, Zhou W, Shi C, Wang J, Liu W, Liang X, Fu Y, Ma K, Zhao L, Zhang F, Lu Z, Lee SM, Xu X, Wang J, Yang H, Fu C, Ge S, Chen W (2016) Draft genome of the living fossil Ginkgo biloba. Gigascience. 5(1):49

Hawkins JS, Kim H, Nason JD, Wing RA, Wendel JF (2006) Differential lineage-specific amplification of transposable elements is responsible for genome size variation in *Gossypium*. Genome Res 16(10):1252–1261

Hudson ME, Lisch DR, Quail PH (2003) The FHY3 and FAR1 genes encode transposase-related proteins involved in regulation of gene expression by the phytochrome A-signaling pathway. Plant J 34(4):453–471

Joly-Lopez Z, Bureau TE (2018) Exaptation of transposable element coding sequences. Curr Opin Genet Dev 49:34–42

Kolosha VO, Martin SL (2003) High-affinity, non-sequence-specific RNA binding by the open reading frame 1 (ORF1) protein from long interspersed nuclear element 1 (LINE-1). J Biol Chem 278(10):8112–8117

Leslie AB, Beaulieu JM, Rai HS, Crane PR, Donoghue MJ et al (2012) Hemisphere-scale differences in conifer evolutionary dynamics. Proc Natl Acad Sci USA 109(40):16217–16221

Lisch D (2009) Epigenetic regulation of transposable elements in plants. Annu Rev Plant Biol 60:43–66

Lisch D (2013) How important are transposons for plant evolution? Nat Rev Genet 14(1):49–61

Liu R, Bennetzen JL (2008) Enchilada redux: how complete is your genome sequence? New Phytol 179 (2):249–250

Ma J, Devos KM, Bennetzen JL (2004) Analyses of LTR-retrotransposon structures reveal recent and rapid genomic DNA loss in rice. Genome Res 14(5):860–869

Ma JX, Bennetzen JL (2004) Rapid recent growth and divergence of rice nuclear genomes. Proc Natl Acad Sci USA 101:12404–12410

Magbanua ZV, Ozkan S, Bartlett BD, Chouvarine P, Saski CA et al (2011) Adventures in the enormous: A 1.8 million clone BAC library for the 21.7 Gb genome of loblolly pine. PLoS One 6(1):e16214

Makarevitch I, Waters AJ, West PT, Stitzer M, Hirsch CN, RossElbarra J, Springer NM (2015) Transposable elements contribute to activation of maize genes in response to abiotic stress. PLoS Genet 11(1):e1004915

Martin SL, Cruceanu M, Branciforte D, Li PWL, Kwok SC et al (2005) LINE-1 retrotransposition requires the nucleic acid chaperone activity of the ORF1 protein. J Mol Biol 348(3):549–561

McClintock B (1984) The significance of responses of the genome to challenge. Science 226:792–801

Mirouze M, Lieberman-Lazarovich M, Aversano R, Bucher E, Nicolet J et al (2012) Loss of DNA methylation affects the recombination landscape in Arabidopsis. Proc Natl Acad Sci USA 109(15):5880–5885

Morse AM, Peterson DG, Islam-Faridi MN, Smith KE, Magbanua Z et al (2009) Evolution of genome size and complexity inPinus. PLoS ONE 4(2):e4332

Murray BG, Leitch IJ, Bennett MD (2004) Gymnosperm DNA C-values Database. www.kew.org/cvalues

Neale DB, Wegrzyn JL, Stevens KA, Zimin AV, Puiu D et al (2014) Decoding the massive genome of loblolly pine using haploid DNA and novel assembly strategies. Genome Biol 15(3):R59

Nystedt B, Street NR, Wetterbom A, Zuccolo A, Lin YC et al (2013) The Norway spruce genome sequence and conifer genome evolution. Nature 497(7451):579

Ohno S (1972) So much "junk" DNA in our genome. Brookhaven Symp Biol 23:366–370

Orgel LE, Crick FH (1980) Selfish DNA: the ultimate parasite. Nature 284(5757):604–607

Pellicer J, Fay MF, Leitch IJ (2010) The largest eukaryotic genome of them all? Bot J Linnean Soc 164 (1):10–15

Piegu B, Guyot R, Picault N, Roulin A, Saniyal A et al (2006) Doubling genome size without polyploidization: dynamics of retrotransposition-driven genomic expansions in *Oryza australiensis*, a wild relative of rice. Genome Res 16(10):1262–1269

Richard GF, Kerrest A, Dujon B (2008) Comparative genomics and molecular dynamics of DNA repeats in eukaryotes. Microbiol Mol Biol Rev 72(4):686–727

Rosbash M, Ford PJ, Bishop JO (1974) Analysis of the C-value paradox by molecular hybridization. Proc Natl Acad Sci USA 71(9):3746–3750

SanMiguel P, Bennetzen JL (1998) Evidence that a recent increase in maize genome size was caused by the massive amplification of intergene retrotransposons. Ann Bot 82(suppl_1):37–44

SanMiguel P, Tikhonov A, Jin YK, Motchoulskaia N, Zakharov D et al (1996) Nested retrotransposons in the intergenic regions of the maize genome. Science 274(5288):765–768

Schnable PS, Ware D, Fulton RS, Stein JC, Wei F-S, Pasternak S et al (2009) The B73 maize genome: complexity, diversity, and dynamics. Science 326 (5956):1112–1115

Slotkin RK, Martienssen R (2007) Transposable elements and the epigenetic regulation of the genome. Nat Rev Genet 8(4):272–285

Smit AFA, Hubley R, Green P (2015) RepeatMasker Open-4.0. http://www.repeatmasker.org. Accessed 10 Sept 2016

Stevens KA, Wegrzyn J, Zimin A, Puiu D, Crepeau M et al (2016) Sequence of the sugar pine megagenome. Genetics 204(4):1613–1626

Thomas CA Jr (1971) The genetic organization of chromosomes. Annu Rev Genet 5:237–256

Vicient CM, Suoniemi A, Anamthawat-Jónsson K, Tanskanen J, Beharav A et al (1999) Retrotransposon BARE-1 and its role in genome evolution in the genus Hordeum. Plant Cell 11(9):1769–1784

Vitte C, Panaud O (2003) Formation of solo-LTRs through unequal homologous recombination counterbalances amplifications of LTR retro-transposons in rice *Oryza sativa* L. Mol Biol Evol 20(4):528–540

Wallberg A, Glémin S, Webster MT (2015) Extreme recombination frequencies shape genome variation and evolution in the honeybee, Apis mellifera. PLoS Genet 11(4):e1005189

Wegrzyn JL, Liechty JD, Stevens KA, Wu LS, Loopstra C et al (2014) Unique features of the loblolly pine (*Pinus taeda L.*) megagenome revealed through sequence annotation. Genetics 196(3):891–909

Wicker T, Sabot F, Hua-Van A, Bennetzen JL, Capy P et al (2007) A unified classification system for eukaryotic transposable elements. Nat Rev Genet 8 (12):973–982

Yao JL, Dong YH, Morris BA (2001) Parthenocarpic apple fruit production conferred by transposon insertion mutations in a MADS-box transcription factor. Proc Nati Acad Sci USA 98(3):1306–1311

Yelina NE, Lambing C, Hardcastle TJ, Zhao X, Santos B et al (2015) DNA methylation epigenetically silences crossover hot spots and controls chromosomal domains of meiotic recombination in *Arabidopsis*. Gene Dev 29(20):2183–2202

Zuccolo A, Scofield DG, De Paoli E, Morgante M (2015) The *Ty1-copia* LTR retroelement family PARTC is highly conserved in conifers over 200MY of evolution. Gene 568(1):89–99

An Intact, But Dormant LTR Retrotransposon Defines a Moderately Sized Family in White Spruce (*Picea glauca*)

4

Britta Hamberger, Macaire Man Saint Yuen, Emmanuel Buschiazzo, Claire Cullis, Agnes Yuen, Carol Ritland, Jörg Bohlmann, and Björn Hamberger

Abstract

Within seed plants, the genomes of the conifer lineage are extraordinarily large and complex. While the evolutionary mechanisms driving this expansion are poorly understood, increasing evidence implicates retrotransposon activity as the driving force. We have isolated in targeted fashion and sequenced two independent white spruce genomic BAC clones for CYP701A24, involved in the biosynthesis of the phytohormone gibberellic acid. Sequence comparison showed little similarity between the two clones, one carrying the *bona fide* target *CYP701A24* and the other an intronless fragment of a *CYP701A24* pseudogene. In proximity of both *CYP701A24* loci, we detected several signatures of the long terminal repeat (LTR) retrotransposon class. Sequence characterization identified one outstanding Ty3-gypsy class element, which was termed *Picnicker1* for its size and degree of sequence conservation in the LTR. Representation of its homologous sequence in genomic amplicons and within the white spruce draft genome revealed that *Picnicker1* is the founding member of a moderately sized family. Dating of the insertion event with the synonymous substitution rate applied to the nucleotide polymorphisms of the LTR suggested an age postdating major speciation in spruce. Independent support for an evolutionary recent incident was provided by an investigation of the genomic locus in a range of spruce species with increasing relatedness to white spruce, and in white spruce for a range of geographical origins. Transcript evidence revealed that related members of the family, but not *Picnicker1* still flourish as part of the dynamic content in modern spruce genomes.

Electronic supplementary material The online version of this chapter (https://doi.org/10.1007/978-3-030-21001-4_4) contains supplementary material, which is available to authorized users.

B. Hamberger · M. M. S. Yuen · J. Bohlmann · B. Hamberger (✉)
Michael Smith Laboratories, The University of British Columbia, 2185 East Mall, Vancouver, BC V6T 1Z4, Canada
e-mail: hamberge@msu.edu

E. Buschiazzo · C. Cullis · A. Yuen · C. Ritland · J. Bohlmann
Department of Forest Sciences, The University of British Columbia, Vancouver, BC V6T 1Z4, Canada

B. Hamberger · B. Hamberger
Department of Biochemistry & Molecular Biology, Michigan State University, East Lansing, MI, USA

B. Hamberger · B. Hamberger
Molecular Plant Sciences Building, 1066 Bogue Street, East Lansing, MI 48824, USA

E. Buschiazzo
School of Natural Sciences, University of California, 5200 North Lake Road, Merced, CA 95343, USA

I. M. Porth and A. R. De la Torre (eds.), *The Spruce Genome*, Compendium of Plant Genomes,
https://doi.org/10.1007/978-3-030-21001-4_4

4.1 Introduction

Pinaceae are the largest family of conifers with the diverse genera pines (*Pinus*) and spruces (*Picea*) contributing to the major fraction. Pine and spruce have a wide geographical distribution and represent a prominent component of coastal, montane and sub-alpine forests in the Northern Hemisphere. With an average of 1C = 23.4 pg (22.89 Gb), the typical genome of *Pinaceae* is almost four times the size of angiosperm genomes. For example, with 1C = 20.2 pg (19.76 Gb), the genome of the economically and ecologically important conifer white spruce (*Picea glauca*) genome exceeds 90% of the angiosperm genomes sized to date (Bennett and Leitch 2012).

Targeting genomic sequences of retrotransposon origin identified large numbers and high diversity in conifers (LePage 2001; Stuart-Rogers and Flavell 2001; Kovach et al. 2010). Elements of the long terminal repeat classes Ty5-*copia* and Ty3-*gypsy* were found by in situ hybridization widely distributed and abundant across conifer chromosomes. While over 70% of the white spruce genome was found to consist of repetitive DNA (Rake et al. 1980), specifically contribution of retroelements to the genome content was projected at 10% (L'Homme et al. 2001; Friesen et al. 2001). Targeted and non-targeted sequencing of genomic bacterial artificial chromosomes (BAC) clones in conifers including white spruce revealed next to a ten-fold lower gene content, compared with angiosperms, the structure and composition of large segments of non-genic DNA, where retroelements were found to dominate the gene space (Hamberger et al. 2009; Kovach et al. 2010; Magbanua et al. 2011; Stival Sena et al. 2014). Importantly, these first reports of sequence assembly and annotation of large segments of conifer genomic DNA provided relevant information for the upcoming efforts to sequence the genomes of conifers (De La Torre et al. 2014a). Shotgun sequencing assembly and curation of the genomes of Norway spruce, white spruce and loblolly pine (Nystedt et al. 2013; Birol et al. 2013; Neale et al. 2014), yielded unprecedented insights into the distribution of LTR retrotransposons and the timing of their retrotranspositionally active periods. In angiosperms such as maize, wheat and *Oryza autraliensis*, a wild relative of domesticated rice, up to several million copies of LTR retrotransposons constituting 80% of the contemporary genomes were suggested to have emerged through bursts of activity (SanMiguel et al. 1998; Piegu et al. 2006; Ragupathy et al. 2010). In contrast, conifer LTR retroelements were shown to accrue over extended periods of time (Nystedt et al. 2013). Families such as the pine-specific *IFG* or *Gymny* are evolutionary ancient and are represented with up to hundreds of thousands of copies in the genome (Kamm et al. 1996; Kossack and Kinlaw 1999; L'Homme et al. 2001; Morse et al. 2009). PARTC (representative Ty1-copia element isolate in *Picea abies*), a recently discovered Ty1-*copia* element of a highly conserved family was estimated to have proliferated in white spruce from 12 to 30 million years (MY), but with extant copies dated back to 74 MY. Detected within a total of seven conifer species, PARTC was reported to cover an impressive 200 MY of progression (Zuccolo et al. 2015) and illustrates plant genome evolution at a time scale inaccessible through angiosperms.

Disrupted open reading frames of the structural elements, (LTR-)deletions, high-sequence divergence and nested insertions indicate loss-of-function of the retroelements and are part of the processes leading to the removal of retroelements, well established in conifers (Kamm et al. 1996; L'Homme et al. 2001; Rocheta et al. 2007; Morse et al. 2009).

We previously reported an approach for targeted identification and sequence characterization of BAC clones from a white spruce (genotype PG29 from Prince George, B.C., with likely introgression of Engelmann spruce, see below) genomic BAC library of approximately 3x coverage (Hamberger et al. 2009). The targets were orthologs of functionally characterized members of the large terpene synthase and cytochrome P450 families involved in conifer specialized metabolism. The charted gene space showed no evidence of (tandem-)gene duplications or higher organization in gene clusters, yet, revealed significant contribution of repetitive DNA to the composition. To compare the gene space, the library was here screened to identify BAC clones carrying the orthologue of *CYP701A24*, a cytochrome P450 of general metabolism and essential for the biosynthesis of the plant diterpene phytohormone gibberellic acid. Earlier amplification from Sitka spruce cDNA and genomic DNA suggested that *CYP701A24* is present within the genome as single copy (Hamberger and Bohlmann 2006). Intriguingly, here, the identification of a pair of independent BAC clones indicated two loci. Annotation of the shotgun sequenced gene space revealed the presence of a truncated and intronless pseudogene of *CYP710A24* in the vicinity of retroelement signatures, which prompted mapping and classification of the high-complexity repeats on both BAC sequences.

The activity of retrotransposons in plants is controlled and intact elements are typically found transpositionally dormant, i.e. transcriptionally suppressed through diverse epigenetic mechanisms (for reviews see Lisch and Bennetzen 2011; Lisch and Slotkin 2011). The discovery here, of a full-length and intact element encouraged further characterization, including estimation of the family size, investigation of the transcriptional status and age of the insertion. The findings and a distinct geographical distribution of the founding member of the new family of spruce retrotransposons are discussed in the context of recent spruce speciation.

4.2 Sequencing and Assembly of Two BAC Clones for CYP701A24

Two independent BAC clones, PGB10 and PGB17, were identified based on the presence of a target region of *CYP701A24*. Shotgun genomic fragments of the BAC clones were sequenced and assembled. Curation of PGB10 resulted in 99% of the reads falling into 11 contigs. The combined total size of the contigs at 174 kb reached the size of the genomic insert estimated by pulsed-field gel electrophoresis (PFGE, 170 kb, **supplemental table S4.1**, https://doi.org/10.5281/zenodo.3244491). The sequence of the PCR probe targeting *CYP701A24* was found on the largest contig spanning 56.5 kb within a truncated sequence of a putative pseudogene devoid of intronic sequence (Fig. 4.1A: Structure of the genomic region of PGB10). No further efforts were undertaken to assemble the remaining 10 contigs of PGB10 to completeness. In contrast, an assembly of 98.2% of the reads for PGB17 yielded a single sequence scaffold of 102 kb, close to the estimated size of 100 kb, and flanked by the left and right border of the BAC vector pINDIGO (**supplemental table S4.1**, https://doi.org/10.5281/zenodo.3244491). The gene structure of the target *CYP701A24* on PBG17 matched the previously described Sitka spruce gene model encoding the cytochrome P450 (Hamberger et al. 2011) (Fig. 4.1B: Structure of the genomic region of PGB17).

4.3 Analysis of Contents and Organization on PGB10 and PGB17 Genomic Regions

Mapping high complexity repeats (Smit et al. 2017) in PGB10 and PGB17 suggested the presence of several regions of LTR-retrotransposon origin showing varying features of degradation. These include deletions, insertions and short remaining sequences of the open reading frames (ORFs) of the structural components. We used the organization of the structural genes and sequence homology to classify the elements in the two major families of LTR-retrotransposons of the copy and paste class Ty1-*copia* and Ty3-*gypsy* (Kumar and Bennetzen 1999) (see a list of all detected elements in **supplemental table S4.2**, https://doi.org/10.5281/

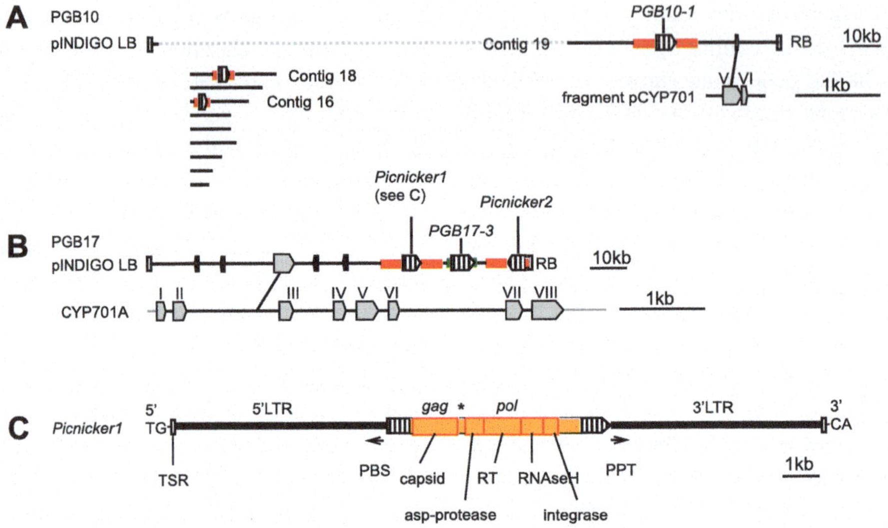

Fig. 4.1 Structure of PGB10 (A), PGB17 (B) and *Picnicker1* (C). The dashed line in A indicates the estimated size of the BAC clone based on PFGE. The position of both white spruce target regions is indicated. Transposons of the LTR class were identified with the LTR finder software (Xu and Wang 2007). Older traces of transposable elements were identified with RepeatMasker using the *viridiplantae* section of the RepBase Update as database. LB/RB left and right borders of pINDIGO; LTR flanking retroelements are shown in red (Ty3-type) or green (Ty1-type); exon numbers are given in roman letters; (p) putative pseudogene; TSR, target site repeat; PBS, primer binding site; PPT, polypurine tract

zenodo.3244491). On PGB10, a full length Ty3-*gypsy* retroelement (PGB10-1) is located 9.5 kb upstream of the putative pseudogene of *CYP701A24*. While all structural features of an autonomous LTR-transposon were identified, numerous nonsense codons disrupting the ORF of the coding regions indicate loss-of-function (Fig. 4.1). Two additional Ty3-*gypsy* retroelements were detected on contigs of PGB10, yet both were found truncated with substantial nonsense mutations and frameshifts in the structural features. In contrast, a Ty3-*gypsy* retroelement with intact structure and a remarkable size of 16.7 kb was detected on PGB17 19.5 kb downstream of the coding region of *CYP701A24*. While internal features resemble in size and composition previously described plant elements, the LTRs dominate with 5.7 kb the genomic locus harbouring the element, which was designated *Picnicker1* (Fig. 4.1C). No nonsense mutations or frameshift mutations were detected, suggesting that the structural features encoded by *Picnicker1* may have remained functional. A second homologous Ty3-*gypsy* (*Picnicker2*) retroelement with 94% identity to *Picnicker1* was located adjacent to the right border of the BAC cloning vector pINDIGO. A major fraction of the predicted sequence of the 3'LTR and the target site repeat (TSR) were truncated due to proximity to the right border of the BAC clone. LTR-retrotransposons were mapped on previously published white spruce BAC sequences PGB02, PGB04, PGB09 and PGB09 (Hamberger et al. 2009; Keeling et al. 2010). Internal duplications, deletions and nested elements indicate that with the exception of *Picnicker1* the other elements are non-functional (**supplemental table S4.2**, https://doi.org/10.5281/zenodo.3244491).

4.4 *Picnicker1* Represents a Full Length, Intact Ty3-*Gypsy* Retroelement

Picnicker1 carries all typical structural features of a full-length LTR retrotransposon, 5' and 3' LTR, TG-CA box with the nucleotides TG and CA at the 5' and 3' junction of the insertion, a target site repeat (TSR) of 6 bp flanking the element, primer binding site (PBS) near the 3' end of the 5'LTR, polypurine tract (PPT) and the domains encoding the polyproteins GAG (group-specific antigen) and POL (polyprotein). A single-stop codon separates the two ORFs of a total of 4,753 bp and constitute a combined polypeptide coding region of 1,506 amino acids, close to the typical size of 1,500 amino acids in angiosperms. Such separation has been described as a feature of the Ty3 *RIRE2* element from rice and the *BEL* element *Kamikaze* from the domestic silkmoth *Bombyx mori* (Gao et al. 2003). While ORF1 encodes the core structural protein featuring 456 amino acids, ORF2 encodes in the order characteristic for T3-*gypsy* elements: protease (PR), reverse transcriptase (RT), RNase H and integrase (INT) (Fig. 4.1C: Structural elements of *Picnicker1*). The detection of an intact retrotransposon element without degenerated structural features is considered a rare event (Ragupathy et al. 2010), which prompted further characterization.

4.5 Analysis of the *Picnicker* Family Size

To estimate representation of the family in the white spruce genome, we used PCR to target 1,106 bp of the *Picnicker1* 5' LTR in PG29 genomic DNA. Cloning of three independently generated amplicons and bidirectional sequencing a total of 60 inserts resulted in 32 distinguishable products with a range of sequence identity from 80% to 99%. Quantitatively, three clones were 100% identical to *Picnicker1*, the remaining clones resulted in 26 unique singletons and 24 products found two to six times. Ten products had no similarity to the target region and were disregarded for further analysis. The phylogenetic tree and early saturation of the *Picnicker1* LTR PCR products suggested that *Picnicker1* is a member of a family with moderate size (**supplemental figure 4.1**, https://doi.org/10.5281/zenodo.3244491). A homology-based BLAST search (-max_target_seqs 10000-max_hsps 10000-dust no-evalue 1e-20) of the *Picnicker1* 5' LTR against the recently updated draft genome assemblies (Warren et al. 2015) of white spruce PG29 and genotype WS77111, originating from a eastern geographic region, detected 22 and 167 sequences, respectively, at 80% or greater identity over at least 1,000 bp. Only 16 of the genomic contigs carried more than one region of homology to the LTR. No intact members of the family were detected (**supplemental data**, Genome-wide scan for *Picnicker1*, https://doi.org/10.5281/zenodo.3244491), consistent with a proposed mechanism reducing LTR through illegitimate recombination (Devos et al. 2002). A genomic sequence representing the full-length *Picnicker1* was not identified.

4.6 Members of the *Picnicker*-Family Are Actively Transcribed

To test if members of the picnicker-family are replication-competent and active, transcript resources were analysed. A BLAST search using the characteristic LTR of *Picnicker1* against EST databases (plant genome database, http://plantgdb.org/, option 'All EST'; NCBI EST blast.ncbi.nlm.nih.gov; accessed 08/15/2018) yielded two white spruce clones (tissue_type=" GQ040 - Shoot tip - Active growth (Normalized)") with 100% identity over short (59–87 bp) segments, and 36 EST sequences with 80–95% identity from white spruce, Sitka spruce, interior spruce and Norway spruce (**supplemental data**, Transcriptome scan for the *Picnicker* family, https://doi.org/10.5281/zenodo.3244491). These results suggest that related members of the family are actively undergoing transposition. To confirm specifically expression of *Picnicker1*, a white spruce PG29 cDNA library generated from young shoots was probed by RT-PCR. Transcript evidence for *Picnicker1* remained undetectable, indicating that this specific element is dormant under the conditions tested.

4.7 Dating of the *Picnicker1* Insertion Time

The intact ORFs and the highly homologous LTR imply that *Picnicker1* is the result of a recent transposition event. The LTR sequences flanking the ORFs are identical at the time of insertion and have diverged over time with the accumulation of nucleotide polymorphisms. This phenomenon permits estimating the timing of the insertion event using the neutral, or synonymous, substitution rate (SanMiguel et al. 1998), and the accuracy of the estimation critically depends on the length. The LTR sequences in *Picnicker1* were determined as 0.0091 ± 0.0012 (standard error) divergent (or over 99% identical) over their unusually high length of 5,663 bp. The divergence is approximately ten-fold lower than the average nucleotide divergence (0.0885) determined between LTR pairs in a Norway spruce fosmid library (Nystedt et al. 2013) and indicates a relatively young event. With a conservative average synonymous substitution rate (μ_S) of 0.68×10^{-9} substitutions per site and year (Buschiazzo et al. 2012), the age of the insertion was estimated at 6.69 ± 1.75 MY. Where determined, the molecular clock of angiosperms ticks with substitution rates (per site and year) from 0.65×10^{-8} for maize to 1.5×10^{-8} in Arabidopsis (SanMiguel et al. 1998; Koch et al. 2000) approximately ten times faster than the conifer clock. Retrotransposon activity has been associated with an increase of the angiosperm genomes over the last 3–6 MY in maize, 2.6–5 MY in wheat and 2–11 MY in grapevine (SanMiguel et al. 1998; Moisy et al. 2008; Ragupathy et al. 2010). Hence, the transposition of *Picnicker1* occurred within a similar time frame and marks an exceptionally recent evolutionary event in conifers (Fig. 4.2).

4.8 Molecular Palaeontology: Tracking of *Picnicker1* in Spruce Species

Estimates of speciation within the spruce family and rates of divergence of spruce species vary depending on the methodology employed (Rungis et al. 2004; Ran et al. 2006; Chen et al. 2009). More specifically, approximation of the speciation of white spruce (*Picea glauca* (Moench) Voss), black spruce (*P. mariana* (Mill.) B.S.P.) and the North Eurasian Norway spruce (*P. abies* (L.) Karst.) ranges from 13–16 MY to 8.5–15 MY (Bouillé and Bousquet 2005; Chen et al. 2009). Applying plastidial, mitochondrial and nuclear sequences, the most common ancestor of living spruce species was dated to 21–37 MY and the speciation of white spruce and Engelmann spruce to 5 MY (Lockwood et al. 2013). As the location of a retrotransposition event is the result of an indiscriminate insertion mechanism, the shared presence of a retrotransposon locus in two individuals is an indicator of relatedness with minimal phylogenetic error (Shedlock et al. 2004). To test for presence of *Picnicker1* in relatives of white spruce, the genomic locus was amplified and sequenced from *P. abies*, *P. wilsonii*, *P. bicolor*, *P. jeozensis* and *P. sitchensis*. In species closely related to the white spruce clonal line PG29 (origin Prince George, BC), the locus was amplified in four Engelmann spruce individuals (*P. engelmannii*) (East Kootenays), a clonal line of white spruce (Canada, QC, Québec) and eighteen individuals of white spruce originating from three different geographical locations in British Columbia (Fort Nelson, Prince George and East Kootenays). *Picnicker1* was not detected in the locus in *P. abies*, *P. wilsonii*, *P. bicolor*, *P. jeozensis* and

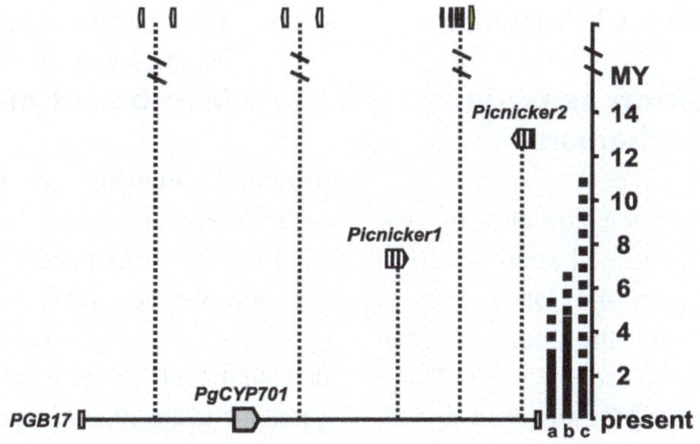

Fig. 4.2 Timing of the Picnicker1 insertion event. The physical structure of PGB17 is given on the bottom, with dashed lines indicating the positions of the major transposons. The time frame of the insertion event is indicated in million years (MY) on the right. A high degree of divergence prevented dating on the elements indicated on the top. As reference, recent LTR retrotransposon activity in maize (**a**), wheat (**b**) and grapevine (**c**) is given next to the time scale (SanMiguel et al. 1998; Moisy et al. 2008; Ragupathy et al. 2010). The solid bar indicates ongoing activity, while the dashed bar shows the time frame where early activity was detected in those species, contributing to increases of the genome sizes

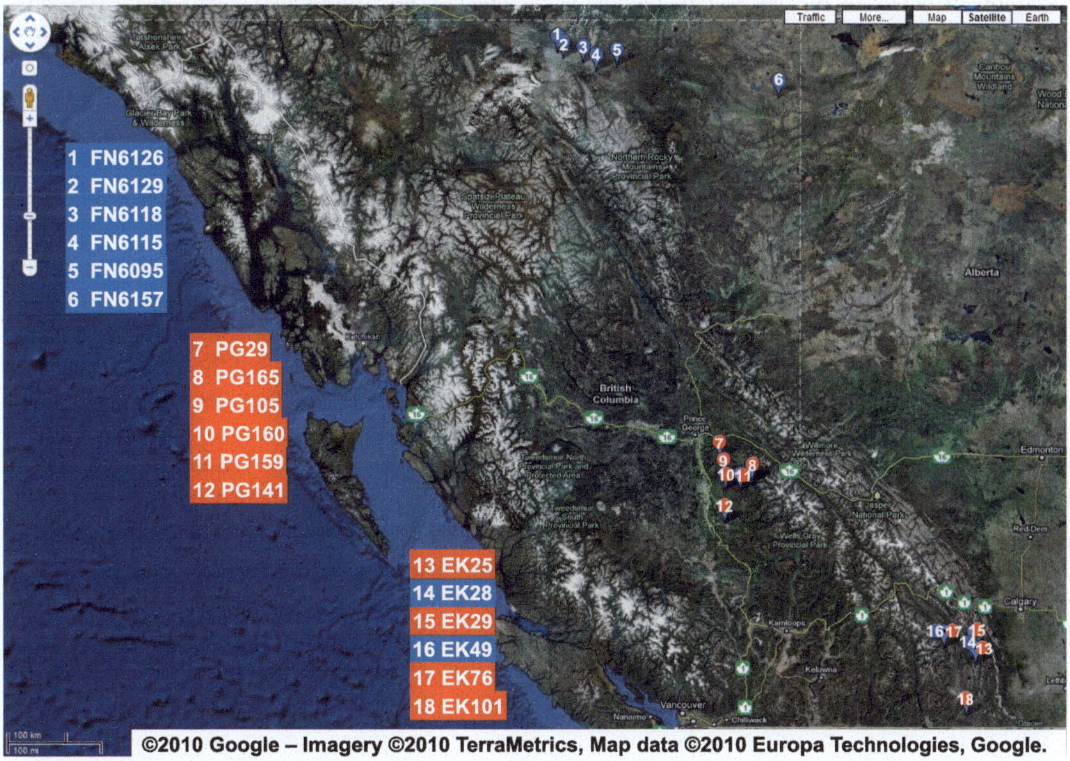

Fig. 4.3 Geographical distribution of white spruce tested for *Picnicker1*. Highlighted in blue are trees lacking *Picnicker1*, while trees carrying *Picnicker1* are indicated in red

P. sitchensis, the white spruce line of East Canadian origin and the white spruce trees originating from the Fort Nelson area. In contrast, three out of the four Engelmann spruce individuals carried *Picnicker1* in the genomic locus. All white spruce trees from Prince George and four of the six tested from the East Kootenays also carried *Picnicker1* (Fig. 4.3).

4.9 Two BAC Library *CYP701A24* Loci Support Retrotransposition as Gene Duplication Mechanism

Here, we report the sequence contents of two BAC clones, isolated in a targeted approach from a white spruce PG29 genomic DNA library. A truncated region of the target gene *CYP701A24* lacking intronic sequence was found in close distance to an LTR retrotransposon, indicating that retrotransposition may have facilitated partial translocation concurrent with intron loss. While gene duplication through retrotransposition is well established in mammalian genomes, evidence for this mechanism is only recently emerging in plants (Thibaud-Nissen et al. 2009; Koduri et al. 2010). The expression of genes of the general metabolism, and intrinsically linked, their copy number is tightly controlled. Indeed, the functionally related genes of the general metabolism were suggested to be represented by very few copies or as single genes (Hamberger and Bohlmann 2006; Keeling et al. 2010). The sequence fragment of *CYP701A24* on PGB10 shows a high degree of degradation supporting an earlier suggested accelerated rate of divergence of retrotransposed DNA (Jun et al. 2009). The draft sequence of the white spruce genome provided further insights into the expansion and contraction of the gene families involved in general and specialized metabolism, i.e. terpene synthases and cytochrome P450s of diterpenoid biosynthesis. Specifically, Warren and co-workers identified nine *CYP701* gene models (including five putative pseudogenes) in contrast to the far more abundant models for *CYP720Bs* of specialized diterpene resin acid biosynthesis (50, including 22 putative pseudogenes) (Warren et al. 2015).

4.10 The Large LTR Retroelement *Picnicker1* as Founding Member of an Active Family

Structural analysis of the 102-kb spanning genomic region of the BAC clone PGB17 carrying the functional copy of *CYP701A24* yielded two homologous LTR retrotransposons of the Ty3-*gypsy* class. Compared to the previously identified LTR retrotransposons in conifers both appeared to be recent integration events with an extraordinary degree of sequence conservation in their LTR regions. The striking length of the LTRs dominates the footprint of *Picnicker1* in the white spruce genomic sequence. At a total size of over 16 kb, the element rivals previously described gigantic elements such as *Grande-1* in maize (13.5 kb), *Cyclops-2* in pea (12.5 kb) or *Athila1-1* in Arabidopsis (10.5 kb) (Martínez-Izquierdo et al. 1997; Chavanne et al. 1998; Wright and Voytas 1998). Distinct divergence within the LTR sequences of both *Picnicker* elements implies independent insertion events. Even though all structural elements of a functional retrotransposon are encoded by two intact open reading frames on *Picnicker1*, it is unclear whether a capacity for autonomous retrotransposition remains. Complete elements with non-degenerated internal domains are exceptional (Ragupathy et al. 2010), but the lack of transcript support suggests that the element is quiescent under the conditions tested.

A threshold of 80% sequence identity has been suggested for unified classification of elements into families, putatively mobilized by the same set of proteins (Wicker et al. 2007). Hence, 94% identity between *Picnicker1* and its homolog *Picnicker2* show that they share a common ancestor and fall into the same family. Probing the size of the family, we found early saturation of PCR amplified sequences and limited numbers of homologous sequences in the recent draft genome of white spruce, supporting that the *Picnicker* family is moderate in size in contrast to the pine *IFG* and *Gymny* families present at an estimated 10^4–10^6 copies in the genome (Rocheta et al. 2007; Morse et al. 2009; Magbanua et al. 2011). We analysed a total of 0.89 Mb of BAC-derived white spruce genomic DNA sequences and detected, including *Picnicker*, a total of 23 transposons of the LTR type covering 26% of the sequence space. While the distribution into *Ty1* and *Ty3* classes was found with nine and eleven elements roughly equal, no additional members of the *Picnicker* family were detected. Despite two white spruce EST clones with a short identical match to *Picnicker1*, independent transcript support from cDNA could not be revealed. However, based on EST sequences with 80–95% identity to *Picnicker1* from white spruce, Sitka spruce and interior spruce, it appears that members of the family are actively transcribed and continue to proliferate as dynamic part of the modern spruce genome.

4.11 Dating the Insertion Event with a Faster Substitution Rate

The exceptional conservation and the length of the LTR region of *Picnicker1* allowed us to date the transposon integration into a period of plant evolution throughout which bursts of retrotransposition resulted in increased genome sizes and phenotypical complexities in the much faster evolving angiosperm lineages. Independent calculations of the substitution rate based on orthologous LTR retroelements in Norway spruce and white spruce and an estimation of their divergence based on fossil records at 13 MY resulted in a substantially faster rate of $\mu = 2.20 \times 10^{-9}$ substitutions per site and year (Nystedt et al. 2013). Applying this substitution rate to white spruce LTR elements in the genome assembly was used to suggest major sustained transposition activity between 16 and 30 MY (Nystedt et al. 2013). This context would move the insertion time for *Picnicker1* from 6.69 MY to a recent 2.1 MY and with that on evolutionary scale past the speciation of white spruce and Engelmann spruce at 5 MY (Lockwood et al. 2013), implying the origin of the marker in either of these species.

4.12 *Picnicker1* Is Limited to Closely Related Individuals of White Spruce and Engelmann Spruce

The low degree of penetration of *Picnicker1* in the spruce species tested is in agreement with an age of the insertion postdating the evolution of the major spruce lineages. The detection in individuals of populations of white spruce and its closest known relative Engelmann spruce in central and southern British Columbia complicates an unequivocal assignment of the species where the event occurred. White spruce and Engelmann spruce are known to form the interior spruce hybrid complex (*P. glauca x engelmannii*). Even though the genotype PG29 of white spruce was the basis for previously establishing the genomic BAC library, pure white spruce was described in the northern, Fort Nelson area, while the population from Prince George was reassigned in 2014 as hybrid population with a high level of admixture and introgression (De La Torre et al. 2014b). Hence, the presence of *Picnicker1* in both species could be the result of gene flow during hybridization, with an origin in either white spruce or Engelmann spruce plausible. In contrast, an origin in a southern population of white spruce during the last ice age could lend support to the suggestion, based on chloroplast DNA, that British Columbia was populated by bidirectional postglacial migration of white spruce originating from a north-west (Beringia) refugium and independent southern populations (Anderson et al. 2006; de Lafontaine et al. 2010).

4.13 Perspectives

Identification of *Picnicker1* on a BAC clone, yet not in the recently updated draft of the white spruce genome indicates that the targeted isolation and sequencing of BAC clones gave an adequate resolution and allowed targeting of local loci with substantial repetitive sequences such as LTR retroelements. Indeed, 50 kb of the region of PGB17 harbouring CYP701A24 upstream of *Picnicker1* (to position PGB17 53,243) were detected on the whole genome shotgun sequence (ALWZ04S2151859.1; starting position *Picnicker1*, PGB17 62,542), while matching sequence for *Picnicker1*, element *PGB17-3* and *Picnicker2* could not be identified.

Picnicker1 represents a genomic DNA phylogenetic marker suitable for the study of population dynamics on a very short evolutionary scale. The moderate size and divergence of the family makes other members such as *Picnicker2* with about twice the age of *Picnicker1* promising markers for elucidation of a broader range of evolutionary events. Furthermore, the features of the family are suited for the development of retrotransposon-based markers for genotyping studies such as sequence-specific amplification polymorphisms and retrotransposon microsatellite amplified polymorphism (Kalendar and Schulman 2007; Syed and Flavell 2007).

Specifically, for biotechnological applications, the unusually large size of the *Picnicker* LTR regions could allow installation of foreign genes. Well established in yeast, replication-competent elements may provide opportunities for future engineering of gene delivery and genome engineering.

4.14 Materials and Methods

4.14.1 Targeted Sequencing of White Spruce BAC Clones

The previously described non-arrayed genomic BAC library of white spruce PG29 (Hamberger et al. 2009) was screened to identify BAC clones for the white spruce orthologue of *CYP701A24* with the following oligonucleotides (given in 5' to 3' direction)

BAC_CYP701 F1 GATATTGGTAGTGGATCCCATGAAAGG

BAC_CYP701 R1 CAGCGCTCAGATATGGTAGTTTGG;

BAC_CYP701 F2 GAAGAAACGCTGTGCTTCGAACTC

BAC_CYP701 R2 CATTTGTAGGACTTGAGTGCCCATGT.

Cloning, sequencing and assembly of large continuous segments of white spruce genomic DNA was performed as previously described (Hamberger et al. 2009).

4.14.2 *Amplification, Cloning and Sequencing of Orthologous Genomic Loci Corresponding to* Picnicker1*, the LTR Homologous Region in Spruce and Detection of* Picnicker1 *Insertion*

Isolation of genomic DNA, RNA and synthesis of cDNA was done as previously described (Hamberger et al. 2011). The geographical position of the white spruce trees is given in **supplemental table S4.3** (https://doi.org/10.5281/zenodo.3244491). The following oligonucleotides (given in 5' to 3' direction) were used for amplification of the genomic locus 01 GAGGTGAAGACAGATCTTCACAAAGTGG and 02 CATGACCTAGGTGCCTGTTTTAGTCC and in combination with 03 CTAACATCAAAAAGGGTCTAGATCACTTGC for the region flanking the 5' LTR and 04 GGTTCAAGTGGTCATGTAGGATCGAG and 05 CCTGTCAACTCAACATCAGTTGCATC for amplification of the region flanking the 3' LTR, respectively, for detection of *Picnicker1*. 06 GAAGACTTTGGTAGAACGGCAAGGA and 07 CGCACATGATCAACTCAATCAGCAAC were used for the amplification of the specific region on the 5' LTR. For testing of transcript accumulation 08 CCTAGTGTCTATCTCTGCCTACAGCC and 09 TTCTGTGACAGTGTTGTGCCTAACC were used in combination with 10 CGCA-CATGATCAACTCAATCAGCAAC. PCR reactions on genomic DNA were performed using Phusion polymerase (Finnzymes, now Thermo Fisher Scientific) following the manufacturer's protocol and 45 cycles. All products were cloned (pJET, Fermentas, now CloneJET Thermo Fisher Scientific) and sequence verified.

4.14.3 Sequence Analysis

The white spruce genomic BAC sequences were screened for putative full-length LTR retrotransposons using the LTR finder (Xu and Wang 2007) with the following non-default parameters: max LTR length 6,000 bp; detection of at least two of TSR, PBS and PPT; tRNA database *Arabidopsis thaliana* (2004). Estimation of evolutionary divergence *k* between 5'-LTR and 3'-LTR sequences was performed on manually curated sequence alignments (Bioedit version 7.0.5 http://www.mbio.ncsu.edu/BioEdit/bioedit.html) and divergence analyses were conducted using the maximum composite likelihood and Kimura 2-parameter methods implemented in MEGA4 (Tamura et al. 2007). Results are based on the pairwise analysis of 2 sequences and standard error estimate(s) were obtained by a bootstrap procedure (10,000 replicates). Distance calculation for *Picnicker1* used a total of 5,663 positions in the dataset. Phylogenetic analyses of aligned sequences of homologous LTR regions (dialign2; http://bioweb.pasteur.fr; manually curated; Bioedit) were tested for consistency and the reconstructed maximum likelihood tree was bootstrapped [PhyML; http://atgc.lirmm.fr/phyml; four rate substitution categories, γ shape parameter optimized, JTT (Jones–Taylor–Thornton) substitution model, BioNJ starting tree and 100 bootstrap repetitions]. The tree topology was supported by TREE-PUZZLE (http://www.tree-puzzle.de).

The sequences of PGB10 and PGB17 have been deposited at GenBank under the accession numbers MH813438 and MH813439.

Acknowledgements We are grateful to Armand Séguin of the Canadian Forest Service, Laurentian Forestry Centre, Quebec City, Quebec, Canada) for shared material of the white spruce clonal line PG653 and Barry Jaquish of the B.C. Ministry of Forests and Range and for tissue of Engelmann spruce. Bj.H. acknowledges current support, which permitted finishing this manuscript, by the U. S. Department of Energy-Great Lakes Bioenergy Research Center Cooperative Agreement DE-FC02-07ER64494 and DE-SC0018409, the Michigan State University Strategic Partnership Grant program 'Plant-inspired Chemical Diversity', recruitment funding from the Department of Molecular Biology and Biochemistry, Michigan State University and support from Michigan State University AgBioResearch (MICL02454). Bj.H. is in part supported by the National Science Foundation under Grant Number 1737898. Any opinions, findings, and conclusions or recommendations expressed in this material are those of the author(s) and do not necessarily reflect the views of the National Science Foundation.

References

Anderson LL, Hu FS, Nelson DM et al (2006) Ice-age endurance: DNA evidence of a white spruce refugium in Alaska. Proc Natl Acad Sci 103:12447 LP–12450

Bennett M, Leitch I (2012) Angiosperm DNA C-values database (release 8.0, Dec. 2012). http://www.kew.org/cvalues/

Birol I, Raymond A, Jackman SD et al (2013) Assembling the 20 Gb white spruce (*Picea glauca*) genome from whole-genome shotgun sequencing data. Bioinformatics 29:1492–1497

Bouillé M, Bousquet J (2005) Trans-species shared polymorphisms at orthologous nuclear gene loci among distant species in the conifer Picea (Pinaceae): implications for the long-term maintenance of genetic diversity in trees. Am J Bot 92:63–73

Buschiazzo E, Ritland C, Bohlmann J, Ritland K (2012) Slow but not low: genomic comparisons reveal slower evolutionary rate and higher dN/dS in conifers compared to angiosperms. BMC Evol Biol 12:8

Chavanne F, Zhang D-X, Liaud M-F, Cerff R (1998) Structure and evolution of Cyclops: a novel giant retrotransposon of the Ty3/Gypsy family highly amplified in pea and other legume species. Plant Mol Biol 37:363–375

Chen J, Källman T, Gyllenstrand N, Lascoux M (2009) New insights on the speciation history and nucleotide diversity of three boreal spruce species and a Tertiary relict. Heredity (Edinb) 104:3

De La Torre A, Birol I, Bousquet J et al (2014a) Insights into conifer giga-genomes. Plant Physiol 166:1724–1732

De La Torre AR, Roberts DR, Aitken SN (2014b) Genome-wide admixture and ecological niche modelling reveal the maintenance of species boundaries despite long history of interspecific gene flow. Mol Ecol 23:2046–2059

de Lafontaine G, Turgeon J, Payette S (2010) Phylogeography of white spruce (*Picea glauca*) in eastern North America reveals contrasting ecological trajectories. J Biogeogr 37:741–751

Devos KM, Brown JKM, Bennetzen JL (2002) Genome size reduction through illegitimate recombination counteracts genome expansion in Arabidopsis. Genome Res 12:1075–1079

Friesen N, Brandes A, Heslop-Harrison JS (2001) Diversity, origin, and distribution of retrotransposons (gypsy and copia) in conifers. Mol Biol Evol 18:1176–1188

Gao X, Havecker ER, Baranov PV et al (2003) Translational recoding signals between gag and pol in diverse LTR retrotransposons. RNA 9:1422–1430

Hamberger B, Bohlmann J (2006) Cytochrome P450 mono-oxygenases in conifer genomes: discovery of members of the terpenoid oxygenase superfamily in spruce and pine. Biochem Soc Trans 34:1209 LP–1214

Hamberger B, Hall D, Yuen M et al (2009) Targeted isolation, sequence assembly and characterization of two white spruce (*Picea glauca*) BAC clones for terpenoid synthase and cytochrome P450 genes involved in conifer defence reveal insights into a conifer genome. BMC Plant Biol 9

Hamberger B, Ohnishi T, Hamberger B et al (2011) Evolution of diterpene metabolism: sitka spruce CYP720B4 catalyzes multiple oxidations in resin acid biosynthesis of conifer defense against insects. Plant Physiol 157:1677–1695

Jun J, Ryvkin P, Hemphill E, Nelson C (2009) Duplication mechanism and disruptions in flanking regions determine the fate of mammalian gene duplicates. J Comput Biol 16:1253–1266

Kalendar R, Schulman AH (2007) IRAP and REMAP for retrotransposon-based genotyping and fingerprinting. Nat Protoc 1:2478

Kamm A, Doudrick RL, Heslop-Harrison JS, Schmidt T (1996) The genomic and physical organization of Ty1-copia-like sequences as a component of large genomes in *Pinus elliottii var. elliottii* and other gymnosperms. Proc Natl Acad Sci 93:2708 LP–2713

Keeling CI, Dullat HK, Yuen M et al (2010) Identification and functional characterization of monofunctional *ent*-copalyl diphosphate and *ent*-kaurene synthases in white spruce reveal different patterns for diterpene synthase evolution for primary and seconda. Plant Physiol 152:1197 LP–1208

Koch MA, Haubold B, Mitchell-Olds T (2000) Comparative evolutionary analysis of chalcone synthase and alcohol dehydrogenase loci in Arabidopsis, Arabis, and related genera (Brassicaceae). Mol Biol Evol 17:1483–1498

Koduri PKH, Gordon GS, Barker EI et al (2010) Genome-wide analysis of the chalcone synthase superfamily genes of *Physcomitrella patens*. Plant Mol Biol 72:247–263

Kossack DS, Kinlaw CS (1999) IFG, a gypsy-like retrotransposon in Pinus (Pinaceae), has an extensive history in pines. Plant Mol Biol 39:417–426

Kovach A, Wegrzyn JL, Parra G et al (2010) The Pinus taeda genome is characterized by diverse and highly diverged repetitive sequences. BMC Genomics 11:420

Kumar A, Bennetzen JL (1999) Plant retrotransposons. Annu Rev Genet 33:479–532

L'Homme Y, Seguin A, Tremblay F (2001) Different classes of retrotransposons in coniferous spruce species

LePage B (2001) New species of Picea A. Dietrich (Pinaceae) from the middle Eocene of Axel Heiberg Island, Arctic Canada. Bot J Linn Soc 135:137–167

Lisch D, Bennetzen JL (2011) Transposable element origins of epigenetic gene regulation. Curr Opin Plant Biol 14:156–161

Lisch D, Slotkin RK (2011) Strategies for silencing and escape: the ancient struggle between transposable elements and their hosts (Jeon KWBT-IR of C and MB (ed)). Academic Press, pp 119–152

Lockwood JD, Aleksić JM, Zou J et al (2013) A new phylogeny for the genus Picea from plastid, mitochondrial, and nuclear sequences. Mol Phylogenet Evol 69:717–727

Magbanua Z V, Ozkan S, Bartlett BD, et al (2011) Adventures in the Enormous: A 1.8 Million Clone BAC Library for the 21.7 Gb Genome of Loblolly Pine. PLoS One 6:e16214

Martínez-Izquierdo JA, García-Martínez J, Vicient CM (1997) What makes Grande1 retrotransposon different? Genetica 100:15–28

Moisy C, Garrison KE, Meredith CP, Pelsy F (2008) Characterization of ten novel Ty1/copia-like retrotransposon families of the grapevine genome. BMC Genomics 9:469

Morse AM, Peterson DG, Islam-Faridi MN et al (2009) Evolution of genome size and complexity in Pinus. PLoS ONE 4:e4332

Neale DB, Wegrzyn JL, Stevens KA et al (2014) Decoding the massive genome of loblolly pine using haploid DNA and novel assembly strategies. Genome Biol 15:R59

Nystedt B, Street NR, Wetterbom A et al (2013) The Norway spruce genome sequence and conifer genome evolution. Nature 497:579

Piegu B, Guyot R, Picault N et al (2006) Doubling genome size without polyploidization: dynamics of retrotransposition-driven genomic expansions in Oryza australiensis, a wild relative of rice. Genome Res 16:1262–1269

Ragupathy R, Banks T, Cloutier S (2010) Molecular characterization of the Sasanda LTR copia retrotransposon family uncovers their recent amplification in

Triticum aestivum (L.) genome. Mol Genet Genomics 283:255–271
Rake AV, Miksche JP, Hall RB, Hansen KM (1980) DNA reassociation kinetics of four conifers. Can J Genet Cytol 22:69–79
Ran J-H, Wei X-X, Wang X-Q (2006) Molecular phylogeny and biogeography of Picea (Pinaceae): implications for phylogeographical studies using cytoplasmic haplotypes. Mol Phylogenet Evol 41:405–419
Rocheta M, Cordeiro J, Oliveira M, Miguel C (2007) PpRT1: the first complete gypsy-like retrotransposon isolated in *Pinus pinaster*. Planta 225:551–562
Rungis D, Bérubé Y, Zhang J et al (2004) Robust simple sequence repeat markers for spruce (Picea spp.) from expressed sequence tags. Theor Appl Genet 109:1283–1294
SanMiguel P, Gaut BS, Tikhonov A et al (1998) The paleontology of intergene retrotransposons of maize. Nat Genet 20:43
Shedlock AM, Takahashi K, Okada N (2004) SINEs of speciation: tracking lineages with retroposons. Trends Ecol Evol 19:545–553
Smit AFA, Hubley R, Green P (2017) 1996–2010. RepeatMasker Open-3.0
Stival Sena J, Giguère I, Boyle B et al (2014) Evolution of gene structure in the conifer Picea glauca: a comparative analysis of the impact of intron size. BMC Plant Biol 14:95
Stuart-Rogers C, Flavell AJ (2001) The evolution of Ty1-copia group retrotransposons in gymnosperms. Mol Biol Evol 18:155–163
Syed NH, Flavell AJ (2007) Sequence-specific amplification polymorphisms (SSAPs): a multi-locus approach for analyzing transposon insertions. Nat Protoc 1:2746
Tamura K, Dudley J, Nei M, Kumar S (2007) MEGA4: molecular evolutionary genetics analysis (MEGA) software version 4.0. Mol Biol Evol 24:1596–1599
Thibaud-Nissen F, Ouyang S, Buell CR (2009) Identification and characterization of pseudogenes in the rice gene complement. BMC Genomics 10:317
Warren RL, Keeling CI, Saint Yuen MM et al (2015) Improved white spruce (*Picea glauca*) genome assemblies and annotation of large gene families of conifer terpenoid and phenolic defense metabolism. Plant J 83:189–212
Wicker T, Sabot F, Hua-Van A et al (2007) A unified classification system for eukaryotic transposable elements. Nat Rev Genet 8:973
Wright DA, Voytas DF (1998) Potential retroviruses in plants: Tat1 is related to a group of *Arabidopsis thaliana* Ty3-gypsy retrotransposons that encode envelope-like proteins. Genetics 149:703 LP–715
Xu Z, Wang H (2007) LTR_FINDER: an efficient tool for the prediction of full-length LTR retrotransposons. Nucl Acids Res 35:W265–W268
Zuccolo A, Scofield DG, De Paoli E, Morgante M (2015) The Ty1-copia LTR retroelement family PARTC is highly conserved in conifers over 200MY of evolution. Gene 568:89–99

The Pliable Genome: Epigenomics of Norway Spruce

5

Igor Yakovlev, Marcos Viejo, and Carl Gunnar Fossdal

Abstract

Recent discoveries have highlighted multiple mitotically and meiotically inherited alterations in gene expression that could not be explained solely by changes in the DNA sequence but were acknowledged as epigenetic. The modern view on epigenetics considers it as an integral part of genetics. Epigenetic mechanisms are encoded by genes in the genome and contribute to an essential part of genomic diversity, significantly extending its regulatory abilities. Epigenetic mechanisms involve molecular chromatin alterations through DNA methylation and histone modifications, as well as, complex non-coding RNAs and related enzyme machinery leading to changes in gene expression and resulting in changing phenotypes. In plants, epigenetic mechanisms may occur over their lifetime and across multiple generations, and can contribute substantially to phenotypic plasticity, stress responses, disease resistance, acclimation and adaptation to habitat conditions. In this review, we summarize recent advances with regards to Norway spruce epigenomics. We first consider the large size of the spruce genome that is linked to epigenetic mechanisms and why epigenomics is vitally important for spruce. Then, we discuss the molecular machinery supporting epigenetic mechanisms in Norway spruce and putative gene models involved. We presume substantial extension of gene families of epigenetic regulators and non-coding RNAs, especially in reproductive tissues. Norway spruce was the first species among forest trees in which epigenetic memory and epigenetic mechanisms were studied. The induction of an epigenetic memory during sexual reproduction and somatic embryogenesis has been described in Norway spruce. We discuss the latest results of epigenomic variation and epigenetic memory studies in Norway spruce and define the future perspectives for epigenetic studies. However, there is still a long way to decipher how the epigenetic mechanisms are involved in maintaining the stability of the spruce epigenome, how the epigenome is set to produce the epigenetic memory phenomenon and how these may result in an increased rate of adaptation to a changing environment.

Electronic supplementary material **The online version of this chapter (https://doi.org/10.1007/978-3-030-21001-4_5) contains supplementary material, which is available to authorized users.**

I. Yakovlev (✉) · M. Viejo · C. G. Fossdal
Division of Forestry and Forest Resources, Norwegian Institute for Bioeconomy Research, Pb 115, 1431 Ås, Norway
e-mail: Igor.Yakovlev@nibio.no

M. Viejo
Department of Plant Sciences, Norwegian University of Life Sciences, Pb 5003, 1432 Ås, Norway

I. M. Porth and A. R. De la Torre (eds.), *The Spruce Genome*, Compendium of Plant Genomes,
https://doi.org/10.1007/978-3-030-21001-4_5

5.1 Introduction

Prior to the recent surge in epigenetic studies, the language of heredity has largely been viewed as a relatively simple one, written on and translated through the DNA sequences, i.e., the genome. This simplistic view championed by classical genetics is also reflected in contemporary genome sequencing studies. However, multiple mitotically and meiotically inherited changes in gene function occur outside of the realm of classical genetics, through the modification of DNA or chromatin structures. These changes are epigenetic. Despite being of a reversible nature, these epigenetic marks can persist through the whole lifespan of an organism, from birth to death, impacting the first crucial steps that govern embryonic development, genome stability, interaction with the environment and thus, altering gene expression which is crucial in determining the phenotype.

There are numerous definitions of epigenetics varying in focus from the molecular to the transgenerational aspects of this phenomenon (Nicoglou and Merlin 2017). The consensus definition considers epigenetics as any stable inheritable changes in the phenotype of an organism that are resulting from chromosome changes without alteration in the primary DNA sequence (Berger et al. 2009; Franks and Hoffmann 2012; Johnson and Tricker 2010). Other definitions focus on the heritable changes in gene expression and gene function that are not directly due to variation in the underlying DNA sequence (Donohue 2014; Bräutigam et al. 2013). The potential of switching genes on and off falls under the reach of epigenetics. Epigenetic mechanisms have the ability to modulate plant growth and development, and have been implicated in plant adaptation to environmental changes and stress. The modern view of epigenetics should consider it as a component of the field of genetics and not as a completely separate layer above the genes. Epigenetic mechanisms are encoded in the genome by genes and non-protein coding sequences and form an integral part of genomic diversity, significantly extending the regulatory abilities of genomes.

Currently, there is a burst of epigenetic literature beyond model organisms, focusing on the importance of epigenetics in agricultural crops and economically important forest tree species (Bräutigam et al. 2013; Pascual et al. 2014; Lämke and Bäurle 2017; Kalavacharla et al. 2017; Yadav et al. 2018; Sow et al. 2018). The authors state very cautiously that despite no direct evidences that epigenetic mechanisms are involved in plant evolution in response to biotic and abiotic stresses and adaptation to changing environment, epigenetics still is the best explanation for resolving the multiple gaps and controversies which for many years had not been solved based on solely classical genetics approaches. Recently, all the hopes and credits, which have been given to genomics and transcriptomics to solve most basic and applied tasks in plants sciences are transferred to an increased interest in epigenomics as the importance of the environmental component is increasing. This shift is fueled by the quest for a more in-depth understanding of gene expression mechanisms, the role of chromatin modifications and non-coding RNAs in the regulation of gene expression, and the realization that to solve the conundrum of phenotypic plasticity, integrative studies combining genomics and epigenomics are needed to understand the mechanisms that determine the phenotype. Nowadays, there is real hope that the more classical genetics-inclined plant researchers will also consider the importance of epigenetic mechanisms in their research projects related to adaptation to the environment, and in applied studies such as breeding, conservation biology, and forest management.

In this chapter, we aim to provide an introduction and an evaluation of the current state of epigenetics and epigenomics in the economically and ecologically important forest tree species Norway spruce (*Picea abies*). We discuss the potential contribution of epigenetics in promoting the evolution and fast adaptation to changing environments that helped spruce recolonize and cover enormous areas following glaciation periods, and prosper for millions of years. Spruce epigenomics is still in its very early stage and our purpose is to draw attention to the importance of epigenetics in genomic studies not only for long lived forest trees but for plants in general.

5.2 Linked with the Genome: Why Epigenomic Is Vitally Important for Spruce

The most spectacular genomic feature of Nor spruce, and other conifers, is the enormous size of its genome. The spruce genome size is around 20 Gb and among the largest (De La Torre et al. 2014). Other main features of the spruce genome is the presence of very long introns and large amounts of apparent pseudogenes and non-coding transcripts, an abundance of repeat-rich content, mostly in the form of transposable elements as well as a smaller contribution of tandem repeats. Transposable elements are mainly in the form of long terminal repeat-retrotransposons (LTR-RTs), along with fewer non-LTR-RTs, such as long interspersed nuclear elements (LINEs) and short interspersed nuclear elements (SINEs), and DNA transposons (class II TEs) (De La Torre et al. 2014). Additionally, the spruce genome is characterized by a high level of gene turnover, which conspicuously contrasts with the low rates of nucleotide substitution (De La Torre et al. 2017). These opposite trends are puzzling but may be explained by elevated frequencies of gene copy number variant (CNV) formation, rapid gene family expansion and/or contraction, or a combination of the two. High levels of CNV formation may be created by illegitimate recombination between the abundant repetitive sequences, especially retroelements (Casola and Koralewski 2018). There is no clear evidence of recent whole-genome duplication in gymnosperms and the chromosome number is almost invariant in conifers. Only one ancient whole-genome duplication event that predated the angiosperm–gymnosperm split (approximately 350 million years ago) is reported to have occurred (De La Torre et al. 2014). So, it means that conifer genomes, including Norway spruce, appear to have been remarkably stable for several hundred millions of years as compared to angiosperms.

There are two main points of view on the reason for the existence of such large genomes. The first one is the "mechanistic," and considers the lack or limited (ineffective) mechanisms of removal of repetitive elements (mostly retrotransposons) from the genome through unequal recombination (Nystedt et al. 2013; Michael 2014). Another explanation for the large size of plant genomes is the "epigenetic" and has been proposed in angiosperm species. Here, it is considered that adaptation and immune memory in plants are somehow boosted through the retention of silenced TEs. TEs are thus maintained in the genome as the important generator of "immune memory" helping to generate siRNAs that can trigger *trans*-silencing of active relatives, as a defense against the invasion of new TEs that have sequence homology to existing genomic TEs and also against TEs that have escaped silencing and need to be resilenced (Bousios and Gaut 2016; Roessler et al. 2018). Effective DNA methylation (over many generations) in plants may require the presence of transposons that act to recruit and maintain the methylation repertoire (Turck and Coupland 2014). It could be further speculated that the whole repetitive parts of plant genomes could even play a role as a type of "memory" accumulated spanning centuries, with temporarily deactivated information from (redundant) genes and pathways involved in the tree's responses to biotic and abiotic stresses, resembling the CRISPR-based immune system in prokaryotes.

Fedoroff (2012) developed a hypothesis to explain the accumulation of large amounts of repetitive elements in eukaryote genomes. She suggests that TEs can accumulate in large quantities because of, not in spite of, the epigenetic mechanisms used to control their expression. These epigenetic mechanisms maintain heterochromatin where repeats are rich; suppresses the expression and transposition of TEs, while simultaneously reduces recombinational events that could lead to TE removal by unequal recombination. Recent results indicate that heterochromatin affects the resolution of recombination, by favoring gene conversion over crossing-over. Control of LTR-RT proliferation via formation of heterochromatin would likely facilitate large genomes in eukaryotes carrying high LTR-RT content (Cossu et al. 2017).

In any case, it is clear that supporting the stability of such large genomes seems impossible without epigenetic mechanisms (Kovalchuk 2018) and there should be evolutionary benefits of large amount of repetitive elements and pseudogenes. The genome may be viewed as an ecosystem inhabited by diverse communities of TEs, which seek to propagate and multiply through sophisticated interactions with each other and with other cell components. TEs are rarely, if ever, randomly distributed in the genome and TEs can have both positive and negative effects on host fitness (Bourque et al. 2018). Despite the obvious evolutionary success of TEs, the plant controls their proliferation by suppressing TE activity both before and after transcription. Posttranscriptional modifications relies mainly on *RNAi* that recognizes and degrades TE mRNA and may also trigger DNA methylation. Transcriptional silencing is achieved through epigenetic modifications like DNA methylation (Ikeda and Nishimura 2015) and reinforced by the RNA-dependent DNA methylation (RdDM) pathway. RdDM relies on 24-nt small interfering RNAs and ultimately establishes TEs in a quiescent state (Bousios and Gaut 2016; Roessler et al. 2018). Numerous molecular studies have characterized the *RNAi* and RdDM pathways, even with implications to spruce (reviewed in Matzke et al. 2015). In *Arabidopsis*, with very low levels of TE, repetitive elements silencing is still maintained using two silencing systems that likely act synergistically. It is tempting to propose that in species with large genomes and many TEs there could be more than only two silencing systems or that the two systems are expended in the numbers of paralogs to make these systems more effective. Despite the fact that some researchers consider the complexity of silencing systems growing from *Physcomitrella* through gymnosperms towards *Arabidopsis* (Matzke et al. 2015), it does not rule out that some conifers like Norway spruce have their own "gimmicks" to deal with such elements. The lack of 24-nt sRNAs in vegetative tissues of spruce, which are normally produced by the RdDM pathways and known to preferentially participate in TE silencing, also supports the existence of unknown mechanisms of TE silencing in spruce.

The level of silencing of TEs could also impact the expression of neighboring genes. Hollister and Gaut (2009) showed that epigenetic silencing of TE could cause a conflict between reducing transposition and interfering with the expression of adjacent genes. Since epigenetically silenced TEs have a specific chromatin state that differs from active functional DNA sequences, the spreading of chromatin modifications, initially intended to silence a TE insertion, could affect neighboring genes and their *cis*-regulatory regions and therefore silence or activate these genes (Hirsch and Springer 2017).

Another evolutionary implication of this complex TE silencing is that it is likely to be very costly energetically. The cost for the production, maintenance and activation of multiple polymerases, methylases, protein complexes, and siRNAs with each cell division must have definite benefits for the plant beyond the silencing of promiscuous DNA elements. The larger the genome, the higher this cost is expected to be. From an evolutionary point of view, a plant would not maintain such an energetically expensive process unless it contributes significantly to fitness and is beneficial throughout evolution. It is "quite probable that eukaryotic genome evolution is driven from within, not just by the gentle breeze of the genetic mechanisms that replicate and repair DNA, but by the stronger winds (with perhaps occasional gale-force gusts) of transposon activity and epigenetic mechanisms that are of great importance here" (Fedoroff 2012). Such much larger research questions, however, should still be explored in the future.

In conclusion, epigenetic mechanisms are vitally important for spruce. Not only do they underpin the management of transcription and chromosome structure, but seem to be the keys to understand the evolutionary causes for genome size and organization in spruce and other eukaryotes. Therefore, spruce may well prove to be a suitable model species for conifers in this context.

5.3 Molecular Mechanisms of Epigenetics

5.3.1 Epigenetic Performers in Plants

Epigenetic mechanisms are based on the fact that DNA does not freely and independently exist in the nucleus, but is tightly packaged into chromatin. Chromatin is a highly dynamic nucleoprotein complex and consists first and foremost of DNA and also has a protein component mostly made of histones. Chromatin regulation plays a central role in how and when DNA is copied and transcribed into RNA. Owing to the variety of histone variants and of possible amino acid residue modifications, histones might be viewed as the primary source of epigenomic complexity.

The major epigenetic regulatory mechanisms (Fig. 5.1) include: DNA modifications, multiple histone modifications (methylation, acetylation, ubiquitination, phosphorylation, etc.), chromatin remodeling, incorporation of histone variants, and also non-coding RNA (both small and long)-based mechanisms (Saze 2008; Boyko and Kovalchuk 2010; Angel et al. 2011; Heo and Sung 2011; Suganuma and Workman 2011; Satake and Iwasa 2012; Heo et al. 2013; Arnaudo and Garcia 2013; Van Oosten et al. 2014; Avramova 2015; Saroha 2017). All of these modifications, referred to as epigenetic marks, are reversible, have different stability and are achieved directly or indirectly with the aid of specific enzymes. Numerous plant research labs have already identified such enzymes and homologs in a variety of land plant species that participate in the regulation of the plant epigenomes (Chen et al. 2010; Pikaard and Mittelsten Scheid 2014). Other epigenetic patterns such as imprinting and paramutation are also essential for plant inheritance and development. The roles played by smRNAs in plants have been reviewed in several comprehensive articles (Voinnet 2009; Carthew and Sontheimer 2009).

Much of the current information about epigenetic modification is coming from biomedical and animal research and substantial knowledge of the varieties of epigenetic marks and enzymes derive from such animal systems (Tarakhovsky 2010; Lu 2013; Arrowsmith et al. 2012). This knowledge is

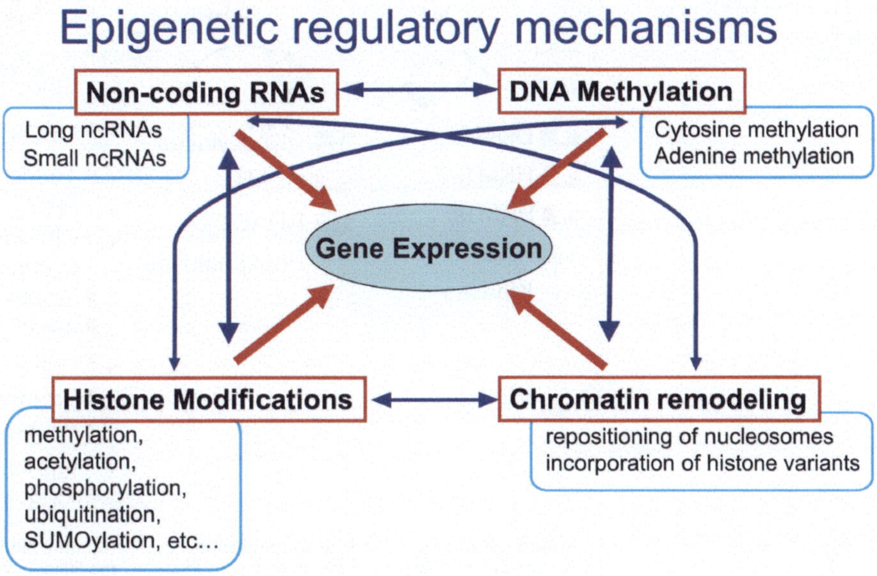

Fig. 5.1 The major epigenetic regulatory mechanisms

partially applicable also to homologou—and analogous—mechanisms found in plants. The epigenetic tools, enzymes, and proteins involved in epigenetic mechanisms, can conveniently be categorized as *writers*, *erasers*, and *readers* of epigenetic marks (Nicholson et al. 2015; Biswas and Rao 2018). The enzymes responsible for the establishment and the maintenance of epigenetic marks, i.e., catalyzing the addition of chemical residual groups, onto either histone tails or DNA itself, are considered epigenetic "writers." Specific proteins containing domains able to recognize and then being recruited to specific marks on histones or nucleotides and initiate subsequent signaling pathways, are named epigenetic "readers" (Fig. 5.2). Epigenetic modifications, however how long they last, are per definition not permanent; they can be removed by a group of enzymes known as epigenetic "erasers" in order to restore or alter gene expression. In plants, the current understanding of the epigenetic machinery repertoire has primarily come from genetic screenings in *Arabidopsis* and maize (Chen et al. 2010; Pikaard and Mittelsten Scheid 2014) and from studies in non-model species such as Norway spruce (Yakovlev et al. 2016).

Despite the fact that more than 100 distinct epigenomic features may exist, experimental studies usually concentrate on a handful of epigenetic marks. The typically profiled marks with respect to gene silencing include increased methylation of DNA, lysine 9 and 27 of histone H3, as well as the absence of histone acetylation. More in-depth studies of enzymes, involved in epigenetic mechanisms, revealed that their activity is often needed to ensure normal development, and such enzymes tend rarely to be specific but rather display a broad effect and can act in many parts of the genome. For example, the chromatin-modifying enzymes may alter not only the histones but a number of other proteins (Bultmann and Stricker 2018). Epigenetic marks are diverse, interdependent, and often overlapping in terms of the effect on genes′ expression, therefore only a small fraction of single modifications possess high regulative power while

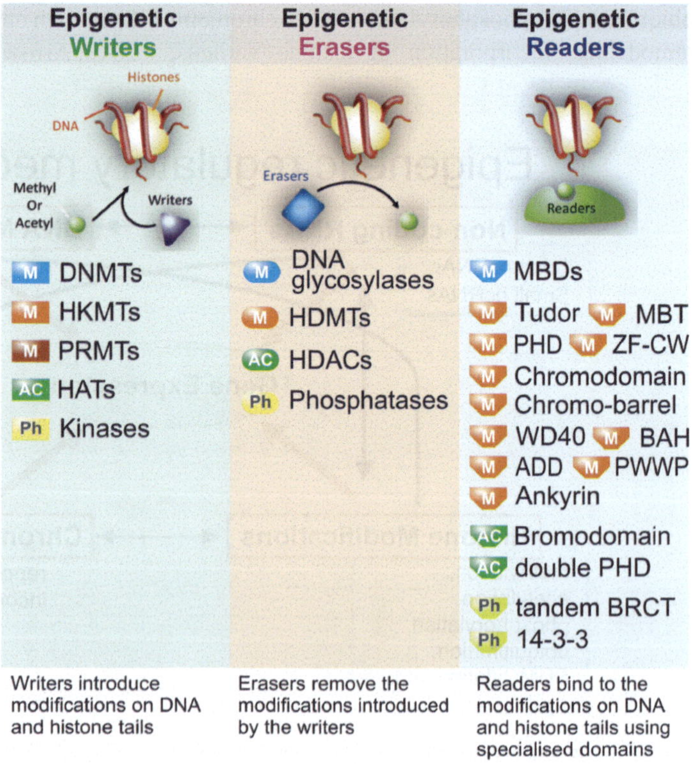

Fig. 5.2 Epigenetic tools—a representation of epigenetic writers, readers, and erasers. These enzymes and protein domains carry out most of the epigenetic modifications on DNA and histone tails. For domains and protein descriptions, see text

others are counteracted by the effect of other marks. Taking into account the highly dynamic nature of epigenetic marks, it constitutes a real challenge to in silico "convert" epigenomic profiles into easily predictable functional responses (Stricker et al. 2016).

Among the in silico defined Norway spruce genes, there are currently more than 800 gene models, putatively involved in this species' epigenetic machinery. Mostly, this classification is based on the possession of one or more functional domains and/or sequence similarity to genes known or predicted to be involved in epigenetic regulation in other species (Table 5.1).

It should be highlighted that there are no specific functional studies for most of the gene models defined in spruce. All epigenetic tools in Norway spruce and their related gene products presented, is based on the best knowledge of matching gene models and their putative functional annotations in other species, and their absolute number and classification may significantly change when additional genomic and functional data are obtained in the near future.

5.3.2 Epigenetic "Writers" and "Erasers"

DNA methylation is one of the most important epigenetic processes in eukaryotic species. It results in a direct modification of DNA nucleotides and can be retained through mitosis and meiosis. DNA methylation occurs at the 5′-position of cytosine residues within three DNA sequence contexts: symmetrical, at CpG sites or CHG, or in nonsymmetrical CHH context, where H is any nucleotide except G. DNA methylation plays important roles in transposon silencing, heterochromatin organization, genomic imprinting, and gene expression (Zhang et al. 2006; Suzuki and Bird 2008). DNA methylation in repeat sequences, such as transposable elements, is believed to prevent the activation of promiscuous and repetitive elements that could transpose or otherwise disrupt normal genome function (Chan et al. 2005).

In plants, DNA methylation patterns can be inherited over many generations (Schmitz et al. 2011; Seymour and Becker 2017), and such patterns may persist largely unchanged in reproductively isolated populations (Hagmann et al. 2015). In the case of symmetric methylation patterns, methyltransferase MET1 copies CG methylation faithfully to the daughter strand after DNA replication (i.e., semi-conservative); methylation patterns of CHG and CHH sites are established and maintained by enzymes CMT3 (chromodomain methyltransferase 3) and CMT2, respectively (Stroud et al. 2013; Saze 2008). In addition, CMT3 could also bind and perform histone H3 lysine 9 methylation (H3K9me) and lysine 27 methylation (H3K27me) directly, suggesting CHG methylation does not necessarily occur in a DNA replication-coupled manner but might be achieved (indirectly) through preexisting histone modification. In addition, for all presented sequence contexts (CG, CHG, and CHH), methylation can be achieved de novo by cytosine DNA methyltransferase DRM2 (Domain Rearranged Methyltransferase 2), which is recruited to the DNA by the RNA-directed DNA methylation pathway (RdDM) (Law and Jacobsen 2010). Since theory tells that maintenance of asymmetric cytosine methylation in a semi-conservative manner is impossible, cytosine methylation at asymmetric sites must be maintained by continuous de novo methylation activities; their methylation therefore depends on other surrounding epigenetic marks such as CG/CHG methylation, histone modifications, small RNAs, or perhaps even on the primary DNA structure itself (Saze 2008).

Active **DNA demethylation** in all sequence contexts requires the enzymatic removal of methylated cytosine. This process in plants is initiated by a family of DNA glycosylases including Demeter (DME), Demeter-like 2 (DML2), and repressor of silencing (ROS1) through a base excision repair (BER) pathway (Li et al. 2018; Law and Jacobsen 2010). Moreover, an ROS3 protein containing an RNA recognition motif has been found to bind to ROS1. This may be evidence of the existence of RNA-directed demethylation, possibly as part of the RdDM pathway (Zheng et al. 2008).

Table 5.1 General characterization of putative epigenetic regulators in Norway spruce (involving a total of 805 gene models)

Epigenetic mechanism	"Writer" enzymes	Number of genes	"Reader" enzymes	Number of genes	"Eraser" enzymes	Number of genes
DNA methylation	DNA Methyltransferases	30	Methyl-CpG–Binding domain Hydroxymethyl-CpG–binding domain	18 1	DNA glycosylases (demethylases)	8
Histone acetylation	Histone acetyltransferases (HAT)	43	Bromodomain; Bromodomain associated; PH (Pleckstrin Homology) domain	16 30 9	Histone deacetylases	30
Histone lysine and arginine methylation	SET domain (KMT); SPRY domain Polycomb Arginine Methyltransferases (PRMTs)	38 2 13 5	Chromodomain; Tudor domain; PHD Fingers; ZF-CW domain; PWWP; WD40 domain; Bromo-adjacent homology (BAH)	2 16 64 10 7 22 12	Histone demethylases (JmjC)	15
Histone phosphorylation	Ser/Thr Kinases (AURORA, CDK, PRK, MSK, MST)	66	BRCT domain; 14-3-3 domain (phosphor. Serine)	18 11	Protein (Serine/Threonine) Phosphatases	32
Histone ubiquiti-nation	Ubiquitin E1 activases; Ubiquitin E2 conjugases—Ubc2; E3 Ligases	15 61	Unknown/interact with histone deacetylation and histone methylation	–	Ubiquitin proteases (UBPs)	17
Histone sumoyolation	SUMO E1, E2 (Ubc9) and E3 enzymes SUMO conjugating enzyme 1	5 3	Unknown	–	ULP-related proteases	8
Chromatin remodeling	SNF2 family; SWIB/MDM2 domain; BAF1/ABF1 chromatin reorganizing factor; Wings apart-like	54 20 2 2				
sRNA biogenesis	Dicer-like Argonaut-like (PAZ, PIWI-domain) miRNA biogenesis pathway ta-siRNAs and RNA-directed DNA methylation (RdDM)	12 19 39 30				
Total		459		236		110

Controlled methylation and demethylation activities is required to ensure long-term stability in the inheritance of this type of epigenetic information (Williams and Gehring 2017). Around 30 DNA methyltransferases and 8 demethylases have been identified in the spruce genome and could provide high flexibility in DNA methylation under different conditions or even have tissue- or cell type-specific roles (Table 5.1).

Chromatin is clearly not an inert or rigid structure, but rather an instructive DNA scaffold for proteins and that behaves like a coil in response to external cues in order to regulate the many tasks of DNA in relation to gene expression. The principal protein component of chromatin that plays a lead role in the accessibility of the DNA to *cis*- and *trans*-acting factors is the modification of histones. Amino acids on N-peptide tails protruding from the nucleosome of H3 and H4 are easier to modify than other histone amino acids. There is an ever-growing list of the types of histone modifications, including acetylation, methylation, phosphorylation, ubiquitination and sumoylation, deimination, ADP ribosylation, biotinylation, to give the most well-documented examples (Bannister and Kouzarides 2011; Van Oosten et al. 2014).

Histone methylation is associated with either transcriptional repression or activation. Methylation of histone lysines plays a role in the regulation of stem cell maintenance and differentiation found in both plants and animals (Liu et al. 2010). Methylation of histones at both lysine and arginine residues is catalyzed by histone methyltransferases (HMT), lysine methyltransferases (KMT), and arginine methyltransferases (PRMTs). Methylation of lysine can be at three different levels, depending on how many methyl groups are deposited—mono-, di-, and trimethylation, thus proving an elaborate indexing system in which specific methylations can point to particular gene activity states, thereby partitioning the genome (Barth and Imhof 2010). Lysine methyltransferases (KMTs) mostly belong to the SET (Su(var) 3–9, Enhancer of zeste, trithorax) domain (PF00856) family proteins (Dillon et al. 2005). On the other hand, Polycomb group (PcG) proteins are members of the POLYCOMB REPRESSIVE COMPLEX 2 (PRC2) catalyzing histone H3 lysine 27 trimethylation (H3K27me3), a repressive chromatin mark.

Histone demethyltransferases (HDMT) directly remove methyl marks from histones, and include Jumonji-C (JmjC) domain-containing proteins (PF02373; PF02375) and LSD lysine demethylases (LSD). The JmjC family of histone demethylases have the ability to remove all types of methylation with the help of cofactors $Fe2+$ and α-ketoglutarate and normally contain the JmjC domain, hydroxylase (PF02373), F-box-like (PF12937), and Cupin-like (PF13621) domains (Klose et al. 2006).

Among the different types of histone modifications, **acetylation** is the most extensively characterized (Waterborg 2011; Fehér 2014). Histone acetylation involves the activity of histone acetyltransferases (HATs) and deacetylases (HDACs). histone (HATs), and, more specifically, Lysine acetyltransferases (KATs) attach an acetyl group to the ε-amino group of lysine residues taking acetyl-coenzyme A as a cofactor; the modification can be reverted by histone deacetylases (HDACs). While HDACs create a condensed chromatin conformation that represses transcription, KAT-mediated acetylation reduces the strength of the interaction between histones and DNA, promoting transcription. HATs and HDACs have quite a complicated classification. There are three groups of HATs: GNAT family acetyltransferase (PF00583; PF13302; PF13673; PF13508), MYST-like histone acetyltransferases (PLN00104) and histone acetyltransferases subunit 3 (PF10198). In the spruce genome, genes coding for proteins belong to two large groups of HATs—GNAT-like and MYST-like histone acetyltransferases and just two types of HDACs —containing histone deacetylase domain (PF00850) or the Sin3 binding region of histone deacetylase complex subunit SAP30 (PF13867). Most of HATs in spruce epitype embryos were not differentially expressed (Yakovlev et al. 2016), considering this modification most important to general embryo development.

Histone phosphorylation is regulated by the combined action of specific protein kinases

(PKs) and phosphatases (PPs). Histone phosphorylation was attributed initially to mitotic events that start with the hyperphosphorylation of linker histone H1 and histone H3 at the onset of mitosis. However, some phosphorylations occur outside of mitosis, at specific promoters, where they regulate the occurrence of additional modifications on the histone tail and are linked to transcription activation, apoptosis, and DNA damage repair (Houben et al. 2007; Rossetto et al. 2012). As an example, in mammalian cells, the histones in promoters of some very early genes such as *Jun* and *Fos* are phosphorylated on H3S10 by members of the mitogen- and stress-activated protein kinase (MSK1/2) family in response to serum stimulation (Soloaga et al. 2003). This phosphorylation was tightly linked to the acetylation of K9 and K14 in the same histone tail, showing that histone phosphorylation can modulate, and even promote, subsequent modifications (Rossetto et al. 2012).

Although identified 28 years ago, **histone ubiquitination** remains one of the least understood histone modifications. It has been showed that histone ubiquitination occurs in the largely monoubiquitinated form and is not linked to degradation. Unlike other reversible histone modifications, in which addition or removal of a group from a histone molecule results in opposing transcriptional effects, sequential ubiquitination and deubiquitination could be both involved in transcriptional activation or repression, depending on its genomic and gene location. The effect of ubiquitination and deubiquitination signals is likely mediated not directly, but through histone methylation (Zhang 2003). PRC1-mediated monoubiquitination of H2AK119 in plants by several *Arabidopsis* RING finger proteins was shown to be required for H3K27me3 marking of target genes to maintain the repression (Fehér 2014). Histone ubiquitination is a reversible modification, similar to that of acetylation and phosphorylation. Addition of an ubiquitin moiety to a protein involves the sequential action of the *E1-E2-E3 enzymatic* cascade. Some specific ubiquitin conjugating enzymes, like Ubc2 (homolog of the yeast RAD6 gene) or RING finger E3 ubiquitin ligase were reported to be involved in histone modifications (Wang and Dasso 2009; Nathan et al. 2003). Removing the ubiquitin moiety, on the other hand, is achieved through the action of the deubiquitinating enzymes (DUB) consisting of ubiquitin C-terminal hydrolases (UCH) and ubiquitin-specific processing proteases (UBP) or isopeptidases. The Ubp8 enzyme (subunit of the SAGA acetyltransferase complex) participates in deubiquitination and is required for transcription activation (Kouzarides 2007).

Small ubiquitin-related modifiers (SUMOs) are ubiquitin-like polypeptides that become covalently conjugated to cellular proteins in a manner similar to ubiquitynation (Wang and Dasso 2009). Sumoylation represents a very large modification and has been shown to take place on all four core histones, and specific sites have been identified on H4, H2A, and H2B. Sumoylation antagonizes both acetylation and ubiquitynation, which occur on the same lysine residue, and consequently, this modification represses transcription as shown in yeast (Kouzarides 2007). Both SUMOylation and deSUMOylation (SUMO deconjugation) are highly dynamic processes.

SUMOylation involves processing by specific activating (E1), conjugating (E2), and ligating (E3) enzymes. SUMO is first activated in an ATP-dependent reaction by formation of a thioester bond with an E1 (SUMO-activating—SAE1) enzyme and then transferred to the SUMO-conjugating (E2) enzyme Ubc9 (plant homolog SCE1—EMBRYO DEFECTIVE 1637). SUMO ligases (like SIZ1) facilitate the majority of SUMOylation under physiological conditions. Ubiquitin-like protein-specific proteases (Ulps) and sentrin-specific proteases (SENPs) mediate deSUMOylation.

Chromatin-remodeling enzymes use ATP hydrolysis' energy to mobilize, disrupt, or change the histone composition of nucleosomes, thus facilitating nearly each nuclear event (Manning and Peterson 2013). Several types of chromatin remodeling complexes have been described to perform a pivotal function in the regulation of eukaryotic gene expression. The different catalytic subunits of well-characterized remodeling complexes comprise several distinct ATPase subfamilies that contain, in addition to their core ATPase-homology domain, characteristic N- or C-terminal sequence motifs: Snf2 (SWI2/SNF2), ISWI, CHD1, SWR1, and INO80 (Clapier and Cairns 2009; Manning and Peterson 2013; Narlikar Geeta et al. 2013). Normally, SWI/SNF complexes are able to induce ATP-dependent disruption of nucleosome structure, facilitating the binding of transcription factors to their target sites on nucleosome templates. In contrast, ISWI members relocate nucleosomes by sliding the histone octamers along the DNA template. While the SWI/SNF subfamily mainly functions in transcriptional activation or repression, the ISWI subfamily has additional roles in maintaining stable higher-order chromatin structure and chromatin assembly (Rosa and Shaw 2013). Furthermore, mutations in the SWI/SNF2- type chromatin-remodeling ATPases (SPLAYAED (SYD) and BRAHMA (BRM)) of *Arabidopsis* were shown to cause defects in the maintenance of both the shoot and the root apical meristem cells (Sang et al. 2012). Another type of chromatin remodeling factors, an *Arabidopsis* homologue of the animal CHD3 ATPase, PICKLE (PKL), was identified, where its mutation (pkl) prompted seed-specific traits to be expressed in seedlings (Fehér 2014).

We found around 330 Norway spruce gene models that could be related to epigenetic writers and erasers (see Table 5.1). What is true for all identified spruce genes, in general, also these putative genes require more precise annotations and further functional studies. Because acetylation and phosphorylation belong to highly mobile histone modifications (Barth and Imhof 2010), they probably provide less likely stable and long-lived epigenetic memory-related marks. At the same time, we found a large number of genes encoding for proteins involved in DNA and histone methylation, a number which largely exceeds the amount of gene homologs found in poplar and *Arabidopsis*, including a substantial number of DEGs (differentially expressed genes) among them. These findings suggest that DNA and chromatin methylation during embryo formation could be strongly involved in the formation of epigenetic memory during spruce embryogenesis (Yakovlev et al. 2016). All methylation modifications, however, have a much slower turnover rate compared to acetylation or phosphorylation (Barth and Imhof 2010) and, therefore, they are more suited as long-term memory marks.

5.3.3 Epigenetic "Readers"

The epigenetic "readers" are proteins that do not alter the histone or the DNA but rather bind to or, in some other fashion, detect epigenetic modifications. Several protein domains have been identified from cells nuclei and which recognize and bind to specific epigenetic marks on histones or nucleotides and then recruit or stabilize various components of the nuclear transcription machinery at the specific genomic sites (Musselman et al. 2012; Yun et al. 2011). At the same time, numerous chromatin modifiers act as epigenetic readers due to the presence of specialized domains that can identify and bind different covalent modifications present on DNA and histones (Biswas and Rao 2018).

Methyl-CpG-binding domain (PF01429) proteins (MBPs) bind and interpret methylated DNA that contains one or more symmetrically methylated CpGs. MBPs decipher the epigenetic information encoded by DNA methylation and provide the link between DNA methylation, chromatin structure modification and gene silencing, and serve as important linchpins in an envisaged self-propagating system that mediates the epigenetic states memory (Woo et al. 2008).

"Readers" of methylated lysine residues at histone tails have been thoroughly categorized, consisting of various proteins that harbor specialized domains facilitating recognition of these modifications (Biswas and Rao 2018). So far, the identified readers of methyl lysine residues include tudor domain, tandem tudor domain (TTD), chromodomain, chromobarrel, double chromodomain (DCD), malignant brain tumor (MBT), PWWP (Pro-Trp-Trp-Pro), plant homeodomain (PHD), WD40, ADD (ATRX-DNMT3-DNMT3L), Ankyrin, bromo-adjacent homology (BAH) and zinc finger CW (zf-CW) domain (Musselman et al. 2012). Compared to lysine methylation, information on writing and especially erasing and recognition of methylated arginine is rather limited; however, there are some indications that this mark is read by histone effectors, and a number of readers show sensitivity to the methylation state of arginine. Recognition of methylated arginines in histone proteins can also influence transcriptional processes (Musselman et al. 2012).

Acetylation of histones uncoils the DNA wrapped around the nucleosome and leads to transcriptional activation. In this context, various protein domains have so far been identified that bind specifically to acetylated histones; these include bromodomain (BRD), double PHD finger and Yeats domains (Biswas and Rao 2018).

Similar to lysine acetylation, phosphorylation of serine and threonine alters the electrostatic and topographic properties of histones by adding a bulky, negatively charged group in residue modification. These modifications are essential in DNA damage response pathways, mitosis and transcriptional regulation. It was described that some members of the 14-3-3 family of proteins and tandem BRCT domain proteins could recognize phosphorylated serine, threonine, and tyrosine residues (Musselman et al. 2012). Until now, no specific histone ubiquitination "readers" are known. Similarly, histone sumoylation may also lack specific "readers" given that sumoylation triggers a direct association of histone deacetylation and histone methylation activities that lead to transcription repression (Nathan et al. 2003; Shiio and Eisenman 2003).

Although the list of newly identified histone readers is growing, reports mostly come from human and animal research. We have detected and defined more than 230 gene models of epigenetic readers in Norway spruce (see again Table 5.1). However, for Norway spruce and other plants, there is still little known about the activities of these domains in the context of the nucleosome or chromatin fiber. Moreover, we are only beginning to understand the multifaceted contacts between epigenetic marks and readers, the synergistic or antagonistic nature of their interactions, and the engagement of other components of the epigenetic machinery.

5.3.4 The MiRNA and Other SRNA Biogenesis Pathway

The largest part of transcriptomes of all plants and animal species consists of non-coding RNA sequences (Costa 2010). Non-coding RNA sequences can be divided into two major groups: short non-coding RNAs or small RNAs shorter than 100 bp (sRNA), and long non-coding RNAs, altogether longer than 100 bp (lncRNA) (Lee 2012). Despite the fact that we are able to detect their presence, the function and activity of non-coding RNAs remains poorly understood. It is clear that both lncRNAs and sRNAs are core components of gene regulation (Costa 2010) and sRNAs are by far the most studied non-coding RNAs. Multiple pathways of sRNA formation lead to a number of distinguishable types of sRNAs, which are divided into four large classes in plants: microRNA (miRNA), trans-acting small interfering RNA (ta-siRNA), natural antisense transcripts RNA (nat-siRNA), and repeat-associated sequences RNA (ra-siRNA) (Vazquez and Hohn 2013). Axtell (2013), on the other hand, has divided all small RNAs in plants into two main groups: those derived from single-stranded precursors with a hairpin structure, referred to as hairpin RNAs (hpRNAs) as opposed to those derived from double-stranded RNA (dsRNA) precursors, referred to as small interfering RNAs (siRNAs). The hpRNAs in plants can further be divided into two secondary groups: microRNAs and those that are not microRNAs. The currently known siRNAs fall mostly into one of three secondary groups: heterochromatic siRNAs, secondary siRNAs, and natural antisense transcript siRNAs. Tertiary subdivisions can be identified within many of the secondary classifications as well, e.g., siRNAs, which can silence other genes *in trans* were identified to be ta-siRNAs. In any case, new ways of grouping sRNAs are likely to happen and even new types of sRNAs will probably be discovered in the near future.

A large set of genes coding for proteins involved in the processing, maturation, and targeting of miRNAs in plants have been described. In total, we detected close to 100 gene models involved in sRNA processing for Norway spruce (see Table 5.1). MicroRNAs are the most studied clade of small RNAs in plants: canonical miRNAs are produced by the DNA-dependent RNA polymerase II (Pol II) interacting with AT-NEGATIVE ON TATALESS 2 (NOT2) proteins and VIRE2-INTERACTING PROTEIN 2 (VIP2). The pre-miRNAs are believed to be stabilized subsequently by an RNA-binding protein named DAWDLE (DDL) and duly processed into stem loop pre-miRNAs by a combinatorial action of the RNase III enzyme Dicer-like 1 (DCL1), the double-stranded RNA (dsRNA)-binding protein HYPONASTIC LEAVES (HYL1), the C2H2-zinc finger protein SERRATE (SE), and the nuclear cap-binding protein complex (CBP20 and CBP80). The miRNA/miRNA* RNA duplexes are then methylated by the Hua enhancer 1 (HEN1) and transported from the nucleus to the cytoplasm by the plant exportin-5 orthologue HASTY (HST) and other unknown factors. The guide strand of the miRNA/miRNA* RNA duplexes is then incorporated into an RNA-induced silencing complex (RISC). The Argonaute family (PAZ, PIWI-domain) proteins play a central role in the RNA silencing processes as essential components of the RISC. Some proteins like the microtubule-severing enzyme KATANIN, SQUINT (SQN), the decapping component VARICOSE (VCS)/Ge-1, the 3-hydroxy-3-methylglutaryl CoA reductase HMG1, the sterol C-8 isomerase HYDRA1, and SUO are shown to be also required for miRNA-mediated translational repression. The single-stranded miRNA, however, can be additionally degraded by a family of exoribonucleases called SMALL RNA DEGRADING NUCLEASE (SDN) and HEN1 SUPPRESSOR1 (HESO1) (Fuchs Wightman et al. 2018). Recent studies also identified some new components involved in miRNA biogenesis. These are ERECTA mRNA UNDER-EXPRESSED (EMU), the flowering time control protein FCA, TOUGH (TGH), STABILIZED1 (STA1), SICKLE (SIC), and MODIFIER OF SNC1 (MOS2). However, their exact function in miRNA biogenesis is still not well understood (Axtell et al. 2011; Wu 2013).

In addition, plants are rich in another type of well-studied small RNAs that are different from miRNAs: the trans-acting siRNAs (ta-siRNAs). ta-siRNAs originate from the loci that give rise to non-coding transcripts which are themselves targets of miRNAs. The miRNA-mediated cleavage of the transcripts recruits an RNA-dependent RNA polymerase 6 (RdR6) together with SUPPRESSOR OF GENE SILENCING 3 (SGS3) in order to use the cleaved transcripts as templates in generating long double-stranded RNAs (dsRNAs), which, in turn, serve as the source for multiple ta-siRNAs produced by DCL4 (Allen et al. 2005).

An important epigenetic process that involves both short and long noncoding RNAs, represents the RNA-directed DNA methylation (RdDM). RdDM directs in all sequence contexts the de novo DNA methylation based on RNA–DNA sequence homology (Matzke and Mosher 2014; Matzke et al. 2015). The sources and targets of RdDM appear to mainly reside, with high density, in heterochromatic regions of transposons and repetitive sequence, and, with less density, in euchromatic regions. The heterochromatic targets may be kept silent permanently by methylation but the euchromatic targets may exhibit reversible silencing and hence are subjected to the actions of methyltransferases and demethylases. Mainly, RdDM may silence transposons, but the potential reversibility suggests its much broader epigenetic roles including genome defense, imprinting, paramutations, and gene expression control in response to growth, developmental, or stress signals (Mahfouz 2010).

The induction of RdDM and the generation of these RNAs rely on complex transcriptional machineries comprising two plant-specific RNA polymerase II (Pol II)-related RNA polymerases known as Pol IV and Pol V, as well as a host of auxiliary factors. Pol IV is tightly coupled to RDR2 (RNA-DEPENDENT RNA POLYMERASE 2), which transcribes complex transcripts newly synthesized by Pol IV–DTF1/SHH1 and generates double-stranded RNAs (dsRNAs). This process is probably aided by the putative ATP-dependent chromatin-remodeling factor CLSY1 (CLASSY1). The dsRNAs are diced by DICER-LIKE 3 (DCL3) into 24-nt siRNAs with 3′ overhangs, which are then methylated by HUA ENHANCER 1 (HEN1).

Single-stranded 24-nt siRNAs are finally loaded into ARGONAUTE 4 (AGO4) or ARGONAUTE 6 (AGO6) to form the RdDM effector complex (Zhang et al. 2013a). DTF1/SHH1 (DNA TRANSCRIPTION FACTOR 1/SAWADEE HOMEODOMAIN HOMOLOG 1) is thought to contribute to the target specificity of Pol IVA protein. Among the closest spruce gene models homologs of SHH1 and annotated as SHH2, two of them (out of three) were DEGs between epitypes. Some of the specialized Pol V pathway components were also found in spruce. Comparative evolution studies in basal, gymnosperm, and angiosperm species showed increased complexity in the Pol V subunit structure from *Physcomitrella* towards *Arabidopsis*. Mosses and gymnosperms seem to lack the DeCL motif in the Pol V subunit, and spruce also lacks the Ago hook (Matzke et al. 2015). Nevertheless, the apparent absence of these domains could be a result of the incomplete annotation of the current spruce reference genome, since we identified three separate homologs of DEFECTIVE CHLOROPLASTS AND LEAVES (DeCLs) in other spruce gene models.

5.3.5 Temperature Sensing Pathways Genes

The epigenetic mechanisms behind the induction of epigenetic memory must, in some way, be impacted by factors sensing the external environment, such as the temperature sum, and transferring this information into epigenetic modifications. Despite some few examples of thermosensors, e.g., in *Listeria* (Kamp and Higgins 2011), the pathways by which temperature is sensed and responded to are simply not known. The main efforts made to study temperature stress response are those leading to heat/cold acclimation (Browse and Xin 2001; Penfield 2008; Knight and Knight 2012). It is believed that the primary temperature sensing mechanism of plants relies on changes in membrane fluidity, which then activates calcium channels. The inward flux of calcium that follows, is thought to activate signal transduction events, based on calcium signaling kinases, phosphatases, and ROS (reactive oxygen scavengers) and alters plant's metabolism in a physiological process called acclimation (Mittler et al. 2012).

Among eukaryotes, multiple classes of multi-pass transmembrane ion channel proteins have emerged as key mediators of thermal sensing. These include members of the Transient Receptor Potential (TRP) family of cation channels, the Anoctamin family of Ca^{2+}-activated chloride channels and the TREK-1 potassium channel. In addition, *Drosophila* RHODOPSIN has been suggested as a possible thermosensor (Sengupta and Garrity 2013). Also, it had previously been established that chromatin has a key role to play in the detection of changes in ambient temperature (Kumar and Wigge 2010). Experiments revealed that the eviction of the histone variant H2A.Z from nucleosomes performs a central role in plant thermosensory perception (Franklin 2010). It is very likely that accurate plant temperature sensing involves complex integration of chromatin-remodeling events together with altered membrane fluidity, in addition to the more direct effects on protein stability and activity. Some additional factors, like ACTIN RELATED PROTEIN6 (ARP6) and PHOTOPERIOD-INDEPENDENT EARLY FLOWERING 1 PROTEIN (PIE1)—subunits of the SWR1 chromatin-remodeling complex, required for the deposition of the histone variant H2A.Z in nucleosomes, could also be involved in chromatin-based temperature sensing (Kumar and Wigge 2010). It has been shown that the incorporation of the H2A.Z histone variant that protects genes from DNA methylation activates gene expression; therefore, temperature-dependent histone variants could play an important general role in regulating gene expression (Zilberman et al. 2008). More recent work has reported that the basic helix-loop-helix transcription factor PHYTOCHROME INTERACTING FACTOR 4 (PIF4) is highly temperature-dependent and required for plant architectural adaptations to warmer temperatures (Wigge 2013).

In total, for Norway spruce, we have identified over 30 gene models putatively involved in temperature sensing and signal transduction based on sequence similarity. There are four closest homologs of PHYTOCHROME INTERACTING FACTOR4 (PIF4), five homologs of H2A.Z, and more than 10 of potassium channel-related proteins. The study of thermosensing pathways and the related epigenetic changes is still in its initial phase. However, a high number of DEGs among selected gene models could indicate their involvement in epigenetic memory formation. Advanced analyses of interrelations between thermosensing genes and epigenetic "writers" are required in the future.

5.4 Epigenetic Variation in Spruce

Norway spruce and its closely related Siberian spruce cover the vast Eurasian area, extending over 30° in latitude (from 70° to 40°N) and 140° in longitude (from 4° to 140°E), growing at altitudes up to 1700 m and shaping many boreal ecosystems. Spruce is characterized by a high genetic diversity and possesses multiple edaphic, climatic and geographical ecotypes, showing clear latitudinal, longitudinal, or drought clines related to many adaptive traits. Growing across regions with substantial changes within and between seasons, this species complex also displays phenotypic plasticity. As epigenetic mechanisms represent a possible tool for achieving a rapid response to changing environmental conditions in long-lived tree species, the level of epigenetic diversity (variation) in the form of epigenetic memory could be of great value for the species' adaptive capacity. Recently, an increasing number of studies have shown that phenotypic plasticity and adaptation to environmental stress is mediated through epigenetic mechanisms creating epigenetic variation (Richards 2006; Richards et al. 2010; Zhang et al. 2013b; Duncan et al. 2014; Van Oosten et al. 2014; Herman and Sultan 2016; Fortes and Gallusci 2017; Kumar 2018; Ecker et al. 2018). Epigenetic variation may enable plants to remember their former interactions with the environment, learn from those interactions, and apply the acquired knowledge to improve their performance in future environments (Latzel et al. 2016).

Epigenetic variation (polymorphisms at the level of DNA methylation, histone modifications, and sRNAs repertoire) modulates chromatin accessibility, which has an impact on transcriptional activity and thus spur phenotypic variation (Seymour and Becker 2017). Similarly to using alleles in the evaluation of genetic variation (i.e., allelic diversity), epigenetic variation could be characterized by comparing epialleles. Epialleles represent alternative chromatin states (i.e., affect the transcriptional level) at any given locus (Johannes et al. 2008).

In contrast to the genetic alleles, the epialleles can be formed and thus, these might react more quickly to environmental change, even though they can be reversible, and persist for only a few generations (Kalisz and Purugganan 2004). If a new epiallele were to cause only a mild phenotype variation through alterations of gene expression, it will probably undergo less strong purifying selection pressure than a loss-of-function sequence mutation (Kalisz and Purugganan 2004), and thus, this epiallele can persist and enable rapid, yet fine-tuned, trait modifications. Epialleles could arise in a directional way, after environmental stress (Rapp and Wendel 2005; Verhoeven et al. 2016) or simply stochastically, due to imperfect genome maintenance (Johannes and Schmitz 2018; Kovalchuk 2018). Ultimately, the significance of adaptive epialleles within wild populations will depend on their frequency and stability (Rohde and Junttila 2008).

The initial discovery of the importance of epialleles was through the unusual patterns of non-Mendelian inheritance related to clearly visible abnormalities in morphology of specific traits, e.g., in pea (Santo et al. 2017), and in the famous KARMA epimutant that drastically reduces yield and has largely halted efforts to clone elite hybrids for oil production in palm (Ong-Abdullah et al. 2015). Later, epialleles were discovered in many crop and model species, e.g., *Arabidopsis* (Silveira et al. 2013), maize (Hollick et al. 1995), tomato (Quadrana et al. 2014), and rice (Zhang et al. 2017).

DNA methylation is the most widely studied epigenetic modification (Seymour and Becker

2017) for a review), and the only epigenetic mark studied so far in spruce. The first genome-wide high coverage single-base resolution methylation study of Norway spruce (*Picea abies*) was performed using bisulfite sequencing (BS) of needles and somatic embryogenesis cultures (Ausin et al. 2016). It was based on the uncompleted spruce genome and allowed to characterize preliminary global DNA methylation patterns of both genic and intergenic regions in needles and somatic embryogenic cultures. The global average methylation level of CG, CHG, and CHH contexts in needles was 74%, 69%, and 1.5%, respectively. Compared to needles, somatic embryogenesis (SE) culture cells showed similar patterns but with slightly reduced methylation levels of CG and CHG sites and increased CHH methylation (68%, 62%, and 2.3%, respectively). On average, DNA methylation levels of CG and CHG were higher in Norway spruce than in other previously studied plants. Transposons were highly methylated in both CG and CHG contexts, and showed a higher level of CHH methylation than within genes. There were also high methylation levels upstream and downstream of annotated TEs, which is likely due to the high TE content and the frequency of nested or adjacent TEs.

To avoid uncertainties due to the uncompleted spruce genome sequence, a targeted bisulfite sequencing (TBS) approach was used to study methylation patterns in the Norway spruce (Heer et al. 2018). Here, more than 26,000 genes were targeted using TBS and gene body methylations were characterized. Methylation levels were reported to be relatively constant across gene bodies, with 21, 11, and 1.3% in the CG, CHG, and CHH contexts, respectively, implying that methylation detected in gene bodies was considerably lower than previously found using the whole genome approach (compared to ≈60, 25, and 0.8% in Ausin et al. 2016) and was also lower than gene body methylation data reported from other gymnosperm species (Takuno et al. 2016; Bewick et al. 2017). Such low level of methylation in gene bodies would therefore indicate an extremely high level of methylation in the repetitive parts of the spruce genome.

Reversed-phase HPLC coupled with tandem mass spectrometry method and stable isotope-labeled standards were used for assessing the levels of 5-methylcytosine (5mC) along with other oxidatively induced DNA modifications within the genomic DNA of Norway spruce during bud set (A. Ruzov, personal communication, January 18, 2018). These data show that, on average, the global level of cytosine methylation is around 36 methylated cytosines per 1,000 nucleotides. Very roughly, this could provide an estimate that 18% of all cytosines within spruce's genomic DNA were methylated. There is yet no information about sequence contexts, and these new data showed so far the lowest levels of DNA methylation in comparison with BS approaches. Additionally, the products of further consecutive 5mC oxidations to 5-hydroxymethylcytosine (5-HmC), 5-formyl cytosine (5-fC) and 5-carboxylcytosine (5-caC) were analyzed: levels of 5-HmC and 5-fC modifications were detectable in non-negligible quantities, approximately in 0.8 and 0.13 per 10^6 nucleosides, respectively. Unexpectedly, no 5-caC modification was detected in our samples. These results reveal the presence of a specific range of noncanonical DNA bases in this conifer genome, implying potential roles for these modifications in plant development and homeostasis.

The observed differences in DNA methylation levels in Norway spruce between different research groups could be attributed to the analysis of distinctly different developmental stages, different genetic plant material, and/or the use of different techniques. Ultimately, future studies must resolve to what extent the DNA methylation levels differ between individuals and the different developmental stages and cyclic times during the seasons within this species to better understand the epigenetic variation in Norway spruce.

There is little published information on chromatin modifications variation beyond DNA methylation in Norway spruce. ChIP analyses confirmed the enrichment in novel chromatin modification, acetylation at Lys-36 in histone H3 (H3K36ac) in the 5′-region of active genes as in many plant species, including Norway spruce (Mahrez et al. 2016). Some ChIP-seq studies are now ongoing at the Norwegian Institute for Bioeconomy research (Magerøy M., pers. comm.) within the research projects EPIDEFENCE (FRIMEDBIO #249958, https://www.nibio.no/en/projects/beyond-the-genome-epigenetics-of-defense-priming-and-climatic-adaptation-in-plants?locationfilter=true) and EPICLIMA—Molecular basis of epigenetic memory formation in somatic embryos of Norway spruce (https://www.researchgate.net/project/Molecular-basis-of-epigenetic-memory-formation-in-somatic-embryos-of-Norway-spruce). Hence, new data may be available soon.

A general overview of small non-coding RNAs found in Norway spruce showed that all types of sRNAs are present within the spruce genome (see Sect. 5.3.4 above). Some earlier reports stated the absence of 24-nt sRNAs in coniferous species (Dolgosheina et al. 2008), but after subsequent narrow (Yakovlev et al. 2010) and also wider scale (Nystedt et al. 2013) sRNA analyses, it was confirmed that 24-nt sRNAs exist in gymnosperms but that these are mainly expressed in embryos.

Embryos have a more abundant pool of sRNA types in comparison to vegetative tissues. The distribution of read length in sRNA libraries for the combined Norway spruce embryonic tissues sample is shown in Fig. 5.3. Three clear read length peaks were found for the embryonic sRNA pool—24–23 nt, 21–20 nt, and 31–32 nt. This last group of longer sRNAs, ranging from 29 to 37-nt lengths with prevalent size of 32 nt, was discovered initially in seeds and embryos of Norway spruce and is also present in many other gymnosperm species tested but is almost absent in all vegetative tissues analyzed so far. The discovery of this 29–37-nt (32 sRNA) group is based on more than 14,000 sRNAs tags, and is supported by more than 3 million reads.

In embryogenic tissues, we detected close to 1,600 different 32 sRNA families containing large amounts of isoforms differing in their terminal nucleotides. Most of these 32 sRNAs families were conserved among the nine gymnosperm species being tested. More than half of these families may have originated from the non-coding part of the transcriptome, with more than 30% of 32 sRNA families putatively originating from tRNAs, while others may have arisen from enzyme-coding transcripts. Transcriptomics found around 1,100 of these 32 sRNA families differentially expressed in spruce embryos at different epitype-inducing temperatures. Additionally, most of the 32 sRNA families were differentially expressed between different types of tissues, suggesting high tissue specificity and high level of transcriptional control for the 32 sRNAs expressed in Norway spruce. It is tempting to speculate that these 32 sRNAs are not only involved in epigenetic memory processes but also central to inhibiting the expression of repetitive elements during embryogenesis. Therefore, repetitive elements

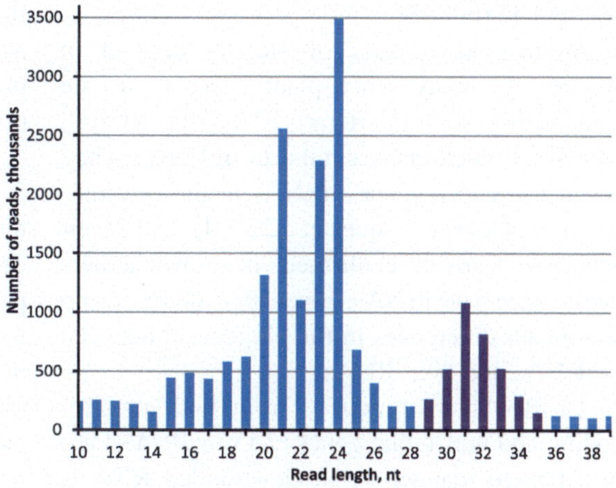

Fig. 5.3 Distribution of read length of sRNAs in Norway spruce embryonic tissues (based on nine libraries). The sRNA reads with lengths between 28 and 36 nt are shown in dark blue color

are being analyzed as putative targets of 32 sRNAs but these studies are yet to be concluded.

MicroRNAs are a specific class of predominantly 20–22-nt sRNAs, characteristically generated from hairpin-folded single-stranded RNA precursor structures (Meyers et al. 2008). They play critical roles in gene regulation during plant development and stress responses (Chen 2009; Sunkar et al. 2012; Li and Zhang 2016; Singh et al. 2018). Based on the spruce reference genome project small RNA datasets, Nystedt et al. (2013) reported 2,719 de novo miRNA gene annotations in spruce using the University of East Anglia (UEA) sRNA workbench tools. Later, Xia et al. (2015) reanalyzed the small RNA sequencing data from the 22 different tissue samples (~ 352 million reads) published in the spruce genome paper (Nystedt et al. 2013). Their comparatively more stringent pipeline for the de novo annotation generated a set of 585 miRNA loci. Nearly half of these miRNA loci were novel and without any known homologs in the miRBase. Subsequent comparison of spruce mature sequences with ginkgo sRNAs indicated that approximately one-third of the newly identified loci may be spruce specific.

In silico analysis of miRNAs in Norway spruce embryos allowed the identification of 1,053 highly expressed miRNAs, including 523 conserved and 530 novel (Yakovlev and Fossdal 2017). For nearly all conserved miRNA families, we found large numbers of miRNAs (isomiRs). On average, there were 15 isomiRs per family, but the number of variants ranged widely from one to 102 (miR166).

Another large class of sRNAs defined in spruce and many other plant species are the *trans*-acting small interfering RNAs (tasiRNAs), which are functionally similar to miRNAs. They act *in trans* to regulate mRNAs at the posttranscriptional level (Vazquez et al. 2004). Until their acting *in trans* is confirmed, most researchers name them phasiRNAs, given that there are no significant differences in the biogenesis between tasiRNAs and phasiRNAs.

TasiRNAs are produced from both protein-coding and noncoding genes (*TAS* or *PHAS* loci) in a phased manner. A single-stranded RNA is initially transcribed from a *TAS* locus; the resulting transcript is then cleaved by a phase initiator, usually being a miRNA, but in some cases, also a tasiRNA suffices for this procedure. The resulting fragments are recognized by the RNA-dependent RNA polymerase 6 (RDR6). This converts the remaining single-stranded precursor into a dsTAS precursor by initiating dsRNA polymerization via the 3' miRNA/tasiRNA cleavage fragment duplex. The DICER-Like 4 enzyme then acts on the dsRNA to produce 21-nt phased dsRNA tasiRNAs that exhibit a symmetrical 2-nt overhang. In *Arabidopsis*, dsRNA TAS3 is often precisely phased. The generated tasiRNAs are subsequently loaded into the ARGONAUTE proteins to form RISC complexes which then guide the cleavage of targeted mRNAs (Allen and Howell 2010; Yoshikawa 2013).

TasiRNAs are a major class of sRNAs in plants. Källman et al. (2013) reported that phasiRNAs derived from the *NBS-LRR* (*nucleotide-binding site leucine-rich repeat*) family are the main source of sRNAs in Norway spruce, but their study did not assess the overall picture of phasiRNAs for this gymnosperm. Later, Xia et al. (2015) characterized an impressively high number of *PHAS* loci (more than 2 thousands) in spruce, triggered by as many as 41 miRNA families, for example, with only 19 miRNA families capable of targeting over 750 *NBS-LRR* genes to generate phasiRNAs. Similarly, the major protein-coding *PHAS* genes in spruce belong to tandem repeats protein families, aside from the aforementioned NBS-LRRs, also PPRs, Myb-like families, etc. This spruce miRNA–phasiRNA network seems to be greater than others formerly characterized in plants.

There are several thousand miRNAs encoded in the spruce genome, which is a relatively large number compared to the typical range (~ 100–750) of miRNA annotations reported for diverse plant species listed within miRBase 21. Therefore, spruce´s "small RNAome" appears to be unique and differs from what had been described (at least so far) for other plant species, mostly for angiosperms. Since there exists an extensive network of miRNAs and phasiRNAs and other sRNAs in the gymnosperm Norway spruce, there must also be still unknown mechanisms involved in the fine-tuning of sRNAs production and in the targeting of different tissues and developmental stages, responsible for developmental regulation and epigenetic memory formation in Norway spruce. Further studies should reveal and characterize novel types of sRNAs, focusing on their precise mode of function.

With regards to studies on long non-coding RNA (lncRNA) for Norway spruce, they are largely absent from the current scientific literature, with the exception of the putative lncRNA catalogue, which was included in the first Norway spruce whole-genome assembly (Nystedt et al. 2013).

5.5 Epigenetic Memory in Spruce

The genomic machinery provides plants with the necessary phenotypic plasticity allowing them to adjust to changes in the environment. Given their sessile lifestyle and often long generation periods, it would be beneficial for plants to create reversible response mechanisms to process the information acquired during their lifetimes about the local environment, and furthermore, to pass this information on to their progeny. The likelihood that their progeny will experience a similar environment is rather high, since seeds are often dispersed locally. Thus, passing such information on to the progeny in the form of epigenetic marks, would have an adaptive advantage, and epigenetic inheritance is the prime candidate for its transmission (Grossniklaus et al. 2013). Many recent studies have focused on the epigenetic inheritance of responses to biotic and abiotic stresses. Experiments undertaken in several species indicate that effects of stress or other environmental conditions cause induced responses such as acquired resistance and other forms of epigenetic-based molecular memory that might be inherited and persists for several generations even without any subsequent stress exposure (Molinier et al. 2006; Whittle et al. 2009; Verhoeven et al. 2010). It is not surprising then that the transgenerational epigenetic inheritance of environmentally induced changes has attracted much attention.

The epigenetic memory is defined as a heritable change in gene expression or behavior that is induced by a previous stimulus (D'Urso and Brickner 2014). The stimulus can be developmental, biotic or environmental. The basic molecular apparatus needed to form epigenetic marks such as histone modifications and DNA methylation is in place. However, the exact molecular mechanisms behind epigenetic memory are currently under active investigation, mostly based on model organisms with short generation times (Thellier and Lüttge 2013; D'Urso and Brickner 2014; Iwasaki and Paszkowski 2014). A molecular model for epigenetic memory generally includes external and/or internal cues transferred into sequence-specific signals, which transforms into specific epigenetic modifications, which, in turn, should be maintained through mitosis- and/or meiosis-related cell divisions. These modifications are further implemented in transcriptional changes and pathway shifts altering the phenotype. It has also been put forward that when necessary, the mentioned modifications should be removable (Thellier and Lüttge 2013; Saze 2008).

The inheritance of epigenetic memory is one of the most pivotal research questions for many scientists, and the positive effect of the adaptive response it generates and its stability is ultimately of interest to practitioners (foresters). In general, there are two main modes to maintain epigenetic memory in the progeny by "rough propagation," i.e., the mitotically and the meiotically heritable epigenetic memory (Iwasaki and Paszkowski 2014).

The *mitotically heritable epigenetic memory* could be of more immediate importance in forest tree propagation, which is normally characterized by trees' long lifespan and long generational intervals. During their lifetime, forest trees should withstand seasonal changes and stressful environmental impacts, pest and diseases attacks, such that any remembrance of successful response to biotic and abiotic stresses could help these organisms to counteract those stressors in the future. Such memory includes, for example, the priming or memory of pathogen attacks (also coined acquired resistance) and the epigenetic memory passing through somatic embryogenesis (SE) or epialleles maintained after genotypes' cloning by ways of cuttings and the like for propagation.

One of the most well-described systems in plants, which controls the responses to pathogen or herbivore attacks, is called defense priming (Martinez-Medina et al. 2016; Conrath et al. 2006) or systemic acquired resistance (SAR) (Vlot et al. 2008). Priming was further applied to induce plant responses to abiotic stresses as well (Friedrich et al. 2018). The first exposure of a plant to a pathogen, to beneficial microbes, or arthropods, as well as to chemicals and abiotic cues can induce long-lasting, systemic immunity against subsequent pathogen attacks (Mauch-Mani et al. 2017; Pastor et al. 2013; Espinas et al. 2016). Defense priming has been reported for a range of plant taxa, including wild species and cultivated varieties, and from herbaceous to long-lived woody plants (Hilker et al. 2016). Priming is the phenomenon that enables cells to already respond to very low levels of a stimulus in a more rapid and robust manner than non-primed cells. Thus, primed plants subsequently show faster and/or stronger activation of defense responses after challenged by microbes, insects, or abiotic stress, and this is frequently linked to the development of local and systemic immunity and stress tolerance (Conrath 2011; He and Li 2018; Lämke and Bäurle 2017).

The molecular mechanisms of priming are still much under investigation. During priming, the transcription of salicylic acid (SA)-responsive genes or other analogous signaling pathways (depending on type of stress and species) is activated, including genes encoding antimicrobial pathogenesis-related proteins (PR). Elevated levels of SA may induce changes in chromatin modification at these target genes. It has been suggested that DNA methylation, histone modification, and/or histone replacement by histone variants may prime pathogen-responsive genes for rapid activation during subsequent pathogen attacks (Avramova 2015; Espinas et al. 2016). For example, the levels of H3 acetylation, H4 acetylation, and H3K4 methylation are increased at the *PR-1* promoter (Jaskiewicz et al. 2011). In addition, the transcription start sites of genes involved in defense are enriched with one of the most conserved eukaryotic histone variants, H2A.Z. Since deposition of H2A.Z is associated with transcriptional memory and rapid reactivation of genes, H2A.Z may be important for priming. It is still not clear to which extent such modifications may contribute to the stability of SAR in terms of enhanced memory of the initial pathogen attack (Iwasaki and Paszkowski 2014).

As of 2019, we have an ongoing project at the Norwegian Institute of Bioeconomy Research and in which we study the epigenetic modifications involved in defense priming against pests and pathogens and the climatic adaptation in plants with an emphasis on Norway spruce and the wild forest strawberry (Krokene 2016). The project aims at deciphering the epigenetic mechanisms involved in priming in response to plant stress and the epigenetic memory induced during embryogenesis; the goal is to elucidate those mechanisms that are in common between both and which aspects differ since the longevity of these phenomena differs as well. We expect to generate novel insights into defense priming related to angiosperms and gymnosperms through the studies of woodland strawberry and Norway spruce.

In contrast to the previously discussed mitotically inherited epigenetic memory, the *meiotically heritable epigenetic memory* goes through the formation of epialleles, i.e., when certain loci are converted to alternative and relatively stable epigenetic states that are transmitted between generations in the form of heritable epialleles. The epigenetic memory acts as a dominant trait, and can be transmitted maternally and paternally to successive generations (Iwasaki and Paszkowski 2014). There are no data on epialleles for forest trees. Additionally, the underlying mechanism of the epigenetic memory transmission, and the mediator of the epigenetic memory are still to be studied further (Kinoshita and Seki 2014; He and Li 2018; Kumar 2018). There are several studies showing transgenerational maintenance of methylation patterns and the construction of epigenetic recombinant inbred lines (epiRILs) (Law and Jacobsen 2010; Johannes et al. 2009). In addition, the heritable chromatin-based modifications (Friedrich et al. 2018) and the small RNAs (Bond and Baulcombe 2014) are also considered as the main mechanisms for transgenerational epigenetic memory.

In spruce, we have a well-defined experimental system to study the phenomenon of *environmentally induced epigenetic memory* in a long-lived species. The crux of such a system is the stability of the epitypes. In Norway spruce, the temperature sum during zygotic and SE induces epitypes that are stable regarding their impact on bud phenology (Rohde and Junttila 2008). There are distinct life periods when the epigenome is more sensitive or amendable to modifications, as during early development and transmission to reproductive development. The embryogenesis is likely the most sensitive period related to the influence of environmental factors impacting the creation and deletion of lasting epigenetic changes, due to the high rates of DNA synthesis and (re)establishments of the epigenome (Patkin and Sofronov 2013; Eckardt 2006).

Norway spruce is one of the few forest tree species, for which epigenetic memory phenomena have already been firmly established. Earlier studies on quantitative phenotypic traits showed an extremely rapid adaptation to the local climatic conditions, which may take place even after only one generation, and such phenomenon can only be accounted for by phenotypic plasticity processes such as epigenetics (for a review see Skrøppa et al. 2010).

It has been proven that environmental conditions during embryo formation affect the performance of newly formed trees for several related adaptive traits, such as timing of bud set, cold acclimation, spring dehardening and flushing. In common garden experiments, the timing for these traits was shown to be delayed if the seeds were produced at warmer than normal temperatures and vice versa, advanced if seeds were produced at colder than normal temperature. On the other hand, the conditions during pre-zygotic stage and fertilization did not affect phenology (Johnsen et al. 2005b). The effect of temperature also interacts with the day-length effects; plants subjected to high temperature and long days and those subjected to low temperature and short days expressed characteristics similar to cold-subjected individuals from the previously mentioned study: autumn bud set and spring flushing were earlier and they were more frost-hardy in autumn (Johnsen et al. 2005a). The observed phenomenon was confirmed by SE experiments, when genetically uniform somatic tissues were grown at different temperatures during subculturing and maturation, and later the regenerated clones were tested in common garden (Kvaalen and Johnsen 2008). It could be concluded that the environmental conditions during embryogenesis resulted in the formation of epitypes, and that the epigenetic memory was mitotically propagated resulting in significant and long-lasting phenotypic changes in seasonal timing of bud phenology and other adaptive traits. This epitype-testing experiment undertaken in Hoxmark (Norway) has shown how somatic embryogenesis can mimic the process ongoing during zygotic embryo development and has become a very effective experimental system to study the molecular basis of epigenetic memory formation, including transcriptional and microRNA changes between epitypes (Yakovlev et al. 2010, 2014, 2016; Yakovlev and Fossdal 2017).

Using RNAseq approaches in Norway spruce, our studies revealed a wide range of differences in gene expression in clonal embryos at different epitype-inducing (EpI) temperatures, and in the resulting epitype trees. In embryos, the EpI conditions induced differential expression of genes encoding different DNA- and histone-modifying proteins, proteins recognizing specific epigenetic changes in histones and DNA, small regulatory RNA molecules, transcription factors, and proteins involved in temperature and stress responses. For spruce, there were indications of a significant expansion of the epigenetic regulator gene families compared to their orthologous gene families in Populus and *Arabidopsis* (Yakovlev et al. 2016). More than 10,000 differentially expressed genes (DEGs) were identified across EpI temperature conditions in clonal embryos developing under 18, 23, and 28 °C. We found that 329 DEGs, responding to these embryogenesis temperatures, out of the overall 735 putative orthologs of epigenetic regulators, were indeed genes putatively involved in epigenetic regulation based on their annotations. Among these were genes encoding DNA methyltransferases, histone acetyltransferases (HAT), polycomb arginine methyltransferases, SUMO E1, E2, and E3 enzymes, SUMO-conjugated enzyme 1, and SNF2 family ABF1 of chromatin reorganizing factors. The highest number of epigenetic regulators DEG included genes encoding enzymes creating DNA methylation marks and histone modifications as well as small RNA pathway-related genes. These results demonstrate that a wide range of genes associated with epigenetic regulation are DEGs during epitypes formation in response to the temperature sum during embryogenesis. Our results support that methylation of DNA and histones, as well as sRNAs, are pivotal for the establishment of the epigenetic memory (Yakovlev et al. 2016).

A variety of the predicted miRNAs showed distinct EpI temperature-dependent expression patterns. These putative EpI miRNAs target spruce genes with a wide range of functions, including genes known to be involved in epigenetic regulation, which in turn could provide a feedback process leading to the formation of epigenetic marks. We found 654 miRNAs to be differentially expressed (DE miRNAs, DEMs) depending on the EpI temperature. Putative gene targets were identified for 470 DEMs which could target around 1,139 gene models, including 930 annotated gene models from ca. 212 gene families and 209 unknown gene models without matches in the NCBI databases. The targets were represented mostly by transcripts of multiple-repeats proteins, such as TIR, NBS-LRR, PPR and TPR repeats, Clathrin/VPS proteins, Myb-like, AP2, etc. Notably, 124 DE miRNAs targeted 203 differentially expressed epigenetic regulators. We found high redundancies in miRNA–target mRNAs pairs and in the genes for most miRNAs origins: in multiple cases, one gene could be targeted by several miRNAs, and one miRNA could target several gene models. Since these many tandem-repeats protein transcripts were among DEG identified under sterile SE conditions, we suggest that TIR-, NBS- and LRR-domain-containing proteins are involved not only in plant defense, but could fulfill more general functions in the transduction of signals from external environmental stimuli into a molecular response (Yakovlev and Fossdal 2017).

In field-grown, clonal epitype trees, originating from embryos developed under 18 or 28 °C, a wide range of DEGs between epitypes occurred in buds harvested in late winter/early spring and long before any signs of budburst were evident (Carneros et al. and Viejo et al., manuscripts in prep.). Such DEGs included a variety of epigenetically related genes but also a wide range of putative phenology-related target genes of the epigenetic machinery, including cell cycle-control-, hormone- and stress-related genes, transcription factors, and other genes involved in budburst, bud set and frost tolerance, such as *EARLY BUD BURST* (*EBB*), *FTL2,* and *DEHYDRIN* (*DHN*) genes. These observations support the notion that the epigenetic machinery is indeed affected by temperature during embryogenesis which in turn may modulate the activity of genes involved in the control of phenology/climatic adaptation in the resulting trees. With the specific isolation of mother cells (stem cells), leaf initials, and other cell types of shoot tips/winter buds, using the laser dissection cryomicroscopy technology, substantial cell and tissue type-specific gene expression differences could be revealed: e.g., 8,681 and 1,861 DEGs were detected between the epitypes in leaf initials and mother cells, respectively (Carneros et al. & Viejo et al., in prep). In the mother cells, unique DEGs (not found in other cell types) between epitypes included a variety of transcriptional regulators (Viejo et al., in prep). Spring budburst and autumnal bud set require the activation and inactivation of cell division in the SAM (shoot apical meristem), and the mother cells are the ultimate stem cells giving rise to all other cells of the growing shoot tip. Thus, the epigenetic status in mother cells must be of particular importance for the mitotic as well as the hypothesized meiotic transfer of the epigenetic memory. Thus, it is not surprising that coordinated differences in expression of phenology-related genes in the epitypes were observed in the buds but not the previous year`s needles during the time course leading to and during budburst (Carneros et al. 2017).

The surrounding environmental conditions during germination and early growth can affect phenology in a similar way. In a nursery experiment with Norway spruce and European larch, Gömöry et al. (2015) found that provenances sown in a warm nursery delayed budburst consistently compared to those grown in a cold nursery, independent to wherever they were later transplanted (Gömöry et al. 2017). Similar data were obtained in European black poplar (*Populus nigra*). Maternal temperature during seed maturation in full-sib families can considerably impact seed germination and the phenological responses of even 2-year-old seedlings in the common garden experiment, suggesting the existence of a memory of maternal temperature during seed maturation. Warmer (ΔT: +10 °C) maternal temperatures decreased the seed germination success. Seedlings from warmer maternal environment also displayed later budburst and earlier bud set. Seedlings resulting from colder maternal environment (ΔT: −10 °C) grew taller than those from warmer environment during the first, but not the second, growing season (Dewan et al. 2018). In common oak and beech, warming of the seedlings advanced budburst in both species. For oak seedlings, however, for which higher temperatures were experienced by their mother trees during their reproductive phase, budburst was delayed in control conditions (Dewan et al. 2019).

Epigenetic memory capability seems to be species-dependent and differs by various factors, yet, there does not exist much comparative research. In general, oak displays more plastic responses to temperatures than beech (Dewan et al. 2019). Similarly, Norway spruce seedlings were more responsive to temperature treatments than European larch (Gömöry et al. 2015). Therefore, various plant species could differ in their ability to remember specific environmental impacts and change their performance accordingly.

Another finding described the response of flowering phenology to soil temperature in the perennial mouse-ear chickweed (*Cerastium fontanum*), plants appear to memorize the soil temperature of their origin. In a common garden, these plants responded to spring warming in a counter-gradient fashion; plants originating from warmer microclimates flowered later, relative to those from colder sites (Valdés et al. 2018). In this study, however, the authors explained the observed results exclusively through classical genetic-based adaptation models and did not consider that epigenetic influences could play a major role. Future genetic end epigenetic studies of their materials are needed before the true nature of this intriguing *Cerastium fontanum* model system can be verified.

Therefore, the concept of epigenetic memory is moving from being a working hypothesis to becoming a developed theory of molecular memory, as the molecular mechanisms underlying environment-dependent phenotypic changes are slowly being unraveled (Iwasaki and Paszkowski 2014; D'Urso and Brickner 2014). Nonetheless, much work still remains, both phenotypic and molecular in nature, in order to better understand the mechanisms and the exact players of the epigenetic machinery behind the observed effects.

5.6 Concluding Remarks and Future Perspectives

The induction of an epigenetic memory during sexual reproduction and somatic embryogenesis has been demonstrated for Norway spruce. Various environmental cues (i.e., temperature, salinity, or drought) have been found to generate a significant impact on the stability of epigenetic settings through sexual and asexual reproduction for a number of species, mostly herbaceous, with a limited impact on the successive generations. The retention of such memory is subject to the maintenance of stress conditions. In Norway spruce, although the temperature regime during embryogenesis is known to be a driving force resulting in a fixed and non-erasable epigenetic memory, the extreme long generation times (15–30 years) have hitherto hampered the study on the transgenerational transmission of this memory. However, for the first time in this species, it will be possible to carry out these groundbreaking studies using our access to a unique set of clonal epitype trees, now sufficiently old to be forced to flowering through hormone application. Such experiments are planned for the spring of 2019 and their implementation depends on the successful induction of flowering in order to perform the needed crosses to produce seeds from these epitypes.

In addition, technological developments and methods have the potential to facilitate a breakthrough in the identification of key epigenetic players behind the epigenetic memory. Such advances include: (1) more complete and accurate genome sequences based on longer read sequencing methods and improved bioinformatics tools; (2) sequence capture methods for studies of DNA methylation and histone modifications in large sets of specific genes and promoters (Kurdyukov and Bullock 2016); (3) laser dissection microscopy of stem cells of the SAM, combined with single-cell type molecular applications; (4) methods for targeted epigenetic modification in specific loci using a CRISPR (clustered regularly interspaced short palindromic repeats) system with non-cutting dCAS9 fused to an epigenetic effector domain (e.g., DNA demethylation/methylation), combined with a recently developed functional transformation system (Xie et al. 2018).

Clearly, there is still a long way to decipher how the epigenetic mechanisms are involved in maintaining the stability of the spruce epigenome, how the epigenome is set to produce the epigenetic memory phenomenon, and how all of this is linked to providing a more rapid adaptive response to the changing environment.

References

Allen E, Howell MD (2010) miRNAs in the biogenesis of trans-acting siRNAs in higher plants. Semin Cell Dev Biol 21(8):798–804. https://doi.org/10.1016/j.semcdb.2010.03.008

Allen E, Xie Z, Gustafson AM, Carrington JC (2005) MicroRNA-directed phasing during trans-acting siRNA biogenesis in plants. Cell 121(2):207–221. https://doi.org/10.1016/j.cell.2005.04.004

Angel A, Song J, Dean C, Howard M (2011) A polycomb-based switch underlying quantitative epigenetic memory. Nature 476(7358):105–108. https://doi.org/10.1038/nature10241

Arnaudo A, Garcia B (2013) Proteomic characterization of novel histone post-translational modifications. Epigenetics & Chromatin 6(1):24

Arrowsmith CH, Bountra C, Fish PV, Lee K, Schapira M (2012) Epigenetic protein families: a new frontier for drug discovery. Nat Rev Drug Discov 11(5):384–400

Ausin I, Feng S, Yu C, Liu W, Kuo HY, Jacobsen EL, Zhai J, Gallego-Bartolome J, Wang L, Egertsdotter U, Street NR, Jacobsen SE, Wang H (2016) DNA methylome of the 20-gigabase Norway spruce genome. PNAS. https://doi.org/10.1073/pnas.1618019113

Avramova Z (2015) Transcriptional 'memory' of a stress: transient chromatin and memory (epigenetic) marks at stress-response genes. Plant J 83(1):149–159. https://doi.org/10.1111/tpj.12832

Axtell MJ (2013) Classification and comparison of small RNAs from plants. Annu Rev Plant Biol 64(1):137–159. https://doi.org/10.1146/annurev-arplant-050312-120043

Axtell M, Westholm J, Lai E (2011) Vive la difference: biogenesis and evolution of microRNAs in plants and animals. Genome Biol 12(4):221

Bannister AJ, Kouzarides T (2011) Regulation of chromatin by histone modifications. Cell Res 21(3):381–395. https://doi.org/10.1038/cr.2011.22

Barth TK, Imhof A (2010) Fast signals and slow marks: the dynamics of histone modifications. Trends Biochem Sci 35(11):618–626

Berger SL, Kouzarides T, Shiekhattar R, Shilatifard A (2009) An operational definition of epigenetics. Genes Dev 23(7):781–783. https://doi.org/10.1101/gad.1787609

Bewick AJ, Niederhuth CE, Ji L, Rohr NA, Griffin PT, Leebens-Mack J, Schmitz RJ (2017) The evolution of CHROMOMETHYLASES and gene body DNA methylation in plants. Genome Biol 18(1):65. https://doi.org/10.1186/s13059-017-1195-1

Biswas S, Rao CM (2018) Epigenetic tools (The Writers, The Readers and The Erasers) and their implications in cancer therapy. Eur J Pharmacol 837:8–24. https://doi.org/10.1016/j.ejphar.2018.08.021

Bond DM, Baulcombe DC (2014) Small RNAs and heritable epigenetic variation in plants. Trends Cell Biol 24(2):100–107. https://doi.org/10.1016/j.tcb.2013.08.001

Bourque G, Burns KH, Gehring M, Gorbunova V, Seluanov A, Hammell M, Imbeault M, Izsvák Z, Levin HL, Macfarlan TS, Mager DL, Feschotte C (2018) Ten things you should know about transposable elements. Genome Biol 19(1):199. https://doi.org/10.1186/s13059-018-1577-z

Bousios A, Gaut BS (2016) Mechanistic and evolutionary questions about epigenetic conflicts between transposable elements and their plant hosts. Curr Opin Plant Biol 30:123–133. https://doi.org/10.1016/j.pbi.2016.02.009

Boyko A, Kovalchuk I (2010) Transgenerational response to stress in *Arabidopsis thaliana*. Plant Signal Behav 5 (8):995–998

Bräutigam K, Vining KJ, Lafon-Placette C, Fossdal CG, Mirouze M, Marcos JG, Fluch S, Fraga MF, Guevara MÁ, Abarca D, Johnsen Ø, Maury S, Strauss SH, Campbell MM, Rohde A, Díaz-Sala C, Cervera M-T (2013) Epigenetic regulation of adaptive responses of forest tree species to the environment. Ecol Evol 3 (2):399–415. https://doi.org/10.1002/ece3.461

Browse J, Xin Z (2001) Temperature sensing and cold acclimation. Curr Opin Plant Biol 4(3):241–246. https://doi.org/10.1016/S1369-5266(00)00167-9

Bultmann S, Stricker SH (2018) Entering the post-epigenomic age: back to epigenetics. Open Biol 8 (3):180013. https://doi.org/10.1098/rsob.180013

Carneros E, Yakovlev I, Viejo M, Olsen JE, Fossdal CG (2017) The epigenetic memory of temperature during embryogenesis modifies the expression of bud burst-related genes in Norway spruce epitypes. Planta 246:553–566. https://doi.org/10.1007/s00425-017-2713-9

Carthew RW, Sontheimer EJ (2009) Origins and Mechanisms of miRNAs and siRNAs. Cell 136(4):642–655

Casola C, Koralewski TE (2018) Pinaceae show elevated rates of gene turnover that are robust to incomplete gene annotation. Plant J 95(5):862–876. https://doi.org/10.1111/tpj.13994

Chan SWL, Henderson IR, Jacobsen SE (2005) Gardening the genome: DNA methylation in *Arabidopsis thaliana*. Nat Rev Genet 6(5):351–360

Chen X (2009) Small RNAs and their roles in plant development. Annu Rev Cell Dev Biol 25:21–44. https://doi.org/10.1146/annurev.cellbio.042308.113417

Chen M, Lv S, Meng Y (2010) Epigenetic performers in plants. Dev Growth Diff 52(6):555–566. https://doi.org/10.1111/j.1440-169X.2010.01192.x

Clapier CR, Cairns BR (2009) The biology of chromatin remodeling complexes. Annu Rev Biochem 78 (1):273–304. https://doi.org/10.1146/annurev.biochem.77.062706.153223

Conrath U (2011) Molecular aspects of defence priming. Trends Plant Sci 16(10):524–531. https://doi.org/10.1016/j.tplants.2011.06.004

Conrath U, Beckers GJM, Flors V, García-Agustín P, Jakab G, Mauch F, Newman M-A, Pieterse CMJ, Poinssot B, Pozo MJ, Pugin A, Schaffrath U, Ton J, Wendehenne D, Zimmerli L, Mauch-Mani B (2006) Priming: getting ready for battle. Mol Plant-Microbe Interact 19(10):1062–1071. https://doi.org/10.1094/MPMI-19-1062

Cossu RM, Casola C, Giacomello S, Vidalis A, Scofield DG, Zuccolo A (2017) LTR retrotransposons show low levels of unequal recombination and high rates of intraelement gene conversion in large plant genomes. Genome Biol Evol 9(12):3449–3462. https://doi.org/10.1093/gbe/evx260

Costa FF (2010) Non-coding RNAs: meet thy masters. BioEssays 32(7):599–608. https://doi.org/10.1002/bies.200900112

D'Urso A, Brickner JH (2014) Mechanisms of epigenetic memory. Trends Genet 30(6):230–236. https://doi.org/10.1016/j.tig.2014.04.004

De La Torre AR, Birol I, Bousquet J, Ingvarsson PK, Jansson S, Jones SJM, Keeling CI, MacKay J, Nilsson O, Ritland K, Street N, Yanchuk A, Zerbe P, Bohlmann J (2014) Insights into conifer giga-genomes. Plant Physiol 166(4):1724–1732. https://doi.org/10.1104/pp.114.248708

De La Torre AR, Li Z, Van de Peer Y, Ingvarsson PK (2017) Contrasting rates of molecular evolution and patterns of selection among gymnosperms and flowering plants. Mol Biol Evol 34(6):1363–1377. https://doi.org/10.1093/molbev/msx069

Dewan S, Vander Mijnsbrugge K, De Frenne P, Steenackers M, Michiels B, Verheyen K (2018) Maternal temperature during seed maturation affects seed germination and timing of bud set in seedlings of European black poplar. For Ecol Manag 410:126–135. https://doi.org/10.1016/j.foreco.2018.01.002

Dewan S, De Frenne P, Leroux O, Nijs I, Vander Mijnsbrugge K, Verheyen K (2019) Phenology and growth of Fagus sylvatica and Quercus robur seedlings in response to temperature variation in the parental versus offspring generation. Plant Biol 0(0). https://doi.org/10.1111/plb.12975

Dillon S, Zhang X, Trievel R, Cheng X (2005) The SET-domain protein superfamily: protein lysine methyltransferases. Genome Biol 6(8):227

Dolgosheina EV, Morin RD, Aksay G, Sahinalp SC, Magrini V, Mardis ER, Mattsson J, Unrau PJ (2008) Conifers have a unique small RNA silencing signature. RNA 14(8):1508–1515. https://doi.org/10.1261/rna.1052008

Donohue K (2014) The epigenetics of adaptation: focusing on epigenetic stability as an evolving trait. Evolution 68(3):617–619. https://doi.org/10.1111/evo.12347

Duncan EJ, Gluckman PD, Dearden PK (2014) Epigenetics, plasticity, and evolution: How do we link epigenetic change to phenotype? J Exp Zool Part B: Mol Dev Evol 322(4):208–220. https://doi.org/10.1002/jez.b.22571

Eckardt NA (2006) Genetic and epigenetic regulation of embryogenesis. Plant Cell 18(4):781–784. https://doi.org/10.1105/tpc.106.042440

Ecker S, Pancaldi V, Valencia A, Beck S, Paul DS (2018) Epigenetic and transcriptional variability shape phenotypic plasticity. BioEssays 40(2):1700148. https://doi.org/10.1002/bies.201700148

Espinas NA, Saze H, Saijo Y (2016) Epigenetic control of defense signaling and priming in plants. Front Plant Sci 7(1201). https://doi.org/10.3389/fpls.2016.01201

Fedoroff NV (2012) Transposable elements, epigenetics, and genome evolution. Science 338(6108):758–767. https://doi.org/10.1126/science.338.6108.758

Fehér A (2014) Somatic embryogenesis—stress-induced remodeling of plant cell fate. Biochim Biophys Acta (0). http://dx.doi.org/10.1016/j.bbagrm.2014.07.005

Fortes A, Gallusci P (2017) Plant stress responses and phenotypic plasticity in the Epigenomics era: perspectives on the grapevine scenario, a model for perennial crop plants. Front Plant Sci 8(82). https://doi.org/10.3389/fpls.2017.00082

Franklin KA (2010) Plant chromatin feels the heat. Cell 140(1):26–28. https://doi.org/10.1016/j.cell.2009.12.035

Franks SJ, Hoffmann AA (2012) Genetics of climate change adaptation. Annu Rev Genet 46(1):185–208. https://doi.org/10.1146/annurev-genet-110711-155511

Friedrich T, Faivre L, Bäurle I, Schubert D (2018) Chromatin-based mechanisms of temperature memory in plants. Plant Cell Environ 0(0). https://doi.org/10.1111/pce.13373

Fuchs Wightman F, Giono LE, Fededa JP, de la Mata M (2018) Target RNAs strike back on MicroRNAs. Front Genet 9:435. https://doi.org/10.3389/fgene.2018.00435

Gömöry D, Foffová E, Longauer R, Krajmerová D (2015) Memory effects associated with early-growth environment in Norway spruce and European larch. Eur J For Res 134(1):89–97. https://doi.org/10.1007/s10342-014-0835-1

Gömöry D, Hrivnák M, Krajmerová D, Longauer R (2017) Epigenetic memory effects in forest trees: a victory of “Michurinian biology”? Cent Eur For J 63 (4):173–179. https://doi.org/10.1515/forj-2017-0024

Grossniklaus U, Kelly WG, Ferguson-Smith AC, Pembrey M, Lindquist S (2013) Transgenerational epigenetic inheritance: how important is it? Nat Rev Genet 14:228. https://doi.org/10.1038/nrg3435

Hagmann J, Becker C, Müller J, Stegle O, Meyer RC, Wang G, Schneeberger K, Fitz J, Altmann T, Bergelson J, Borgwardt K, Weigel D (2015) Century-scale methylome stability in a recently diverged arabidopsis thaliana lineage. PLoS Genet 11(1):e1004920. https://doi.org/10.1371/journal.pgen.1004920

He Y, Li Z (2018) Epigenetic environmental memories in plants: establishment, maintenance, and reprogramming. Trends Genet 34(11):856–866. https://doi.org/10.1016/j.tig.2018.07.006

Heer K, Ullrich KK, Hiss M, Liepelt S, Schulze Brüning R, Zhou J, Opgenoorth L, Rensing SA (2018) Detection of somatic epigenetic variation in Norway spruce via targeted bisulfite sequencing. Ecol Evol 0:1–11. https://doi.org/10.1002/ece3.4374

Heo JB, Sung S (2011) Encoding memory of winter by noncoding RNAs. Epigenetics 6(5):544–547. https://doi.org/10.4161/epi.6.5.15235

Heo J, Lee Y-S, Sung S (2013) Epigenetic regulation by long noncoding RNAs in plants. Chromosom Res 21 (6–7):685–693. https://doi.org/10.1007/s10577-013-9392-6

Herman JJ, Sultan SE (2016) DNA methylation mediates genetic variation for adaptive transgenerational plasticity. Proc R Soc B: Biol Sci 283(1838). https://doi.org/10.1098/rspb.2016.0988

Hilker M, Schwachtje J, Baier M, Balazadeh S, Bäurle I, Geiselhardt S, Hincha DK, Kunze R, Mueller-Roeber B, Rillig MC, Rolff J, Romeis T, Schmülling T, Steppuhn A, van Dongen J, Whitcomb SJ, Wurst S, Zuther E, Kopka J (2016) Priming and memory of stress responses in organisms lacking a nervous system. Biol Rev 91(4):1118–1133. https://doi.org/10.1111/brv.12215

Hirsch CD, Springer NM (2017) Transposable element influences on gene expression in plants. Biochim Biophys Acta 1860(1):157–165. https://doi.org/10.1016/j.bbagrm.2016.05.010

Hollick JB, Patterson GI, Coe EH, Cone KC, Chandler VL (1995) Allelic interactions heritably alter the activity of a metastable maize pl allele. Genetics 141(2):709–719

Hollister JD, Gaut BS (2009) Epigenetic silencing of transposable elements: A trade-off between reduced transposition and deleterious effects on neighboring gene expression. Genome Res 19(8):1419–1428. https://doi.org/10.1101/gr.091678.109

Houben A, Demidov D, Caperta AD, Karimi R, Agueci F, Vlasenko L (2007) Phosphorylation of histone H3 in plants—a dynamic affair. Biochim et Biophys Acta (BBA) - Gene Struct Expr 1769(5–6):308–315. http://dx.doi.org/10.1016/j.bbaexp.2007.01.002

Ikeda Y, Nishimura T (2015) The role of DNA methylation in transposable element silencing and genomic imprinting. In: Pontes O, Jin H (eds) Nuclear functions in plant transcription, signaling and development. Springer New York, pp 13–29. https://doi.org/10.1007/978-1-4939-2386-1_2

Iwasaki M, Paszkowski J (2014) Epigenetic memory in plants. EMBO J 33(18):1987–1998. https://doi.org/10.15252/embj.201488883

Jaskiewicz M, Conrath U, Peterhänsel C (2011) Chromatin modification acts as a memory for systemic acquired resistance in the plant stress response. EMBO Rep 12(1):50–55. https://doi.org/10.1038/embor.2010.186

Johannes F, Schmitz RJ (2018) Spontaneous epimutations in plants. New Phytol 0(0). https://doi.org/10.1111/nph.15434

Johannes F, Colot V, Jansen RC (2008) Epigenome dynamics: a quantitative genetics perspective. Nat Rev Genet 9:883. https://doi.org/10.1038/nrg2467

Johannes F, Porcher E, Teixeira FK, Saliba-Colombani V, Simon M, Agier N, Bulski A, Albuisson J, Heredia F, Audigier P, Bouchez D, Dillmann C, Guerche P, Hospital F, Colot V (2009) Assessing the impact of transgenerational epigenetic variation on complex traits. PLoS Genet 5(6):e1000530. https://doi.org/10.1371/journal.pgen.1000530

Johnsen Ø, Daehlen OG, Østreng G, Skrøppa T (2005a) Daylength and temperature during seed production interactively affect adaptive performance of *Picea abies* progenies. New Phytol 168(3):589–596. https://doi.org/10.1111/j.1469-8137.2005.01538.x

Johnsen Ø, Fossdal CG, Nagy N, Molmann J, Dælen OG, Skrøppa T (2005b) Climatic adaptation in *Picea abies* progenies is affected by the temperature during zygotic embryogenesis and seed maturation. Plant Cell Environ 28(9):1090–1102. https://doi.org/10.1111/j.1365-3040.2005.01356.x

Johnson LJ, Tricker PJ (2010) Epigenomic plasticity within populations: its evolutionary significance and potential. Heredity 105(1):113–121. https://doi.org/10.1038/hdy.2010.25

Kalavacharla V, Subramani M, Ayyappan V, Dworkin MC, Hayford RK (2017) Plant epigenomics. In: Tollefsbol TO (ed) Handbook of epigenetics, 2nd edn. Academic Press, pp 245–258. https://doi.org/10.1016/B978-0-12-805388-1.00016-X

Kalisz S, Purugganan MD (2004) Epialleles via DNA methylation: consequences for plant evolution. Trends Ecol Evol 19(6):309–314. https://doi.org/10.1016/j.tree.2004.03.034

Källman T, Chen J, Gyllenstrand N, Lagercrantz U (2013) A significant fraction of 21 nt sRNA originates from phased degradation of resistance genes in several perennial species. Plant Physiol 162:741–754. https://doi.org/10.1104/pp.113.214643

Kamp HD, Higgins DE (2011) A protein thermometer controls temperature-dependent transcription of flagellar motility genes in listeria monocytogenes. PLoS Pathog 7(8): e1002153. https://doi.org/10.1371/journal.ppat.1002153

Kinoshita T, Seki M (2014) Epigenetic memory for stress response and adaptation in plants. Plant Cell Physiol 55:1859–1863. https://doi.org/10.1093/pcp/pcu125

Klose RJ, Kallin EM, Zhang Y (2006) JmjC-domain-containing proteins and histone demethylation. Nat Rev Genet 7(9):715–727

Knight MR, Knight H (2012) Low-temperature perception leading to gene expression and cold tolerance in higher plants. New Phytol 195(4):737–751. https://doi.org/10.1111/j.1469-8137.2012.04239.x

Kouzarides T (2007) Chromatin modifications and their function. Cell 128(4):693–705. https://doi.org/10.1016/j.cell.2007.02.005

Kovalchuk I (2018) Role of epigenetics in transgenerational changes: genome stability in response to plant stress. In: Baluska F, Gagliano M, Witzany G (eds) Memory and learning in plants. Springer International Publishing, Cham, pp 79–109. https://doi.org/10.1007/978-3-319-75596-0_5

Krokene P (2016) Carbon castles and insect invaders: dissecting physical defences in conifer stems. Plant Cell Environ 39(8):1643–1645. https://doi.org/10.1111/pce.12687

Kumar S (2018) Epigenomics of plant responses to environmental stress. Epigenomes 2(1):6

Kumar SV, Wigge PA (2010) H2A.Z-containing nucleosomes mediate the thermosensory response in *Arabidopsis*. Cell 140(1):136–147. http://dx.doi.org/10.1016/j.cell.2009.11.006

Kurdyukov S, Bullock M (2016) DNA methylation analysis: choosing the right method. Biology 5(1):3. https://doi.org/10.3390/biology5010003

Kvaalen H, Johnsen O (2008) Timing of bud set in *Picea abies* is regulated by a memory of temperature during zygotic and somatic embryogenesis. New Phytol 177(1):49–59. https://doi.org/10.1111/j.1469-8137.2007.02222.x

Lämke J, Bäurle I (2017) Epigenetic and chromatin-based mechanisms in environmental stress adaptation and stress memory in plants. Genome Biol 18(1):124. https://doi.org/10.1186/s13059-017-1263-6

Latzel V, Rendina González AP, Rosenthal J (2016) Epigenetic memory as a basis for intelligent behavior in clonal plants. Front Plant Sci 7:1354. https://doi.org/10.3389/fpls.2016.01354

Law JA, Jacobsen SE (2010) Establishing, maintaining and modifying DNA methylation patterns in plants and animals. Nat Rev Genet 11(3):204–220

Lee JT (2012) Epigenetic regulation by long noncoding RNAs. Science 338(6113):1435–1439. https://doi.org/10.1126/science.1231776

Li C, Zhang B (2016) MicroRNAs in control of plant development. J Cell Physiol 231(2):303–313. https://doi.org/10.1002/jcp.25125

Li Y, Kumar S, Qian W (2018) Active DNA demethylation: mechanism and role in plant development. Plant Cell Rep 37(1):77–85. https://doi.org/10.1007/s00299-017-2215-z

Liu C, Lu F, Cui X, Cao X (2010) Histone methylation in higher plants. Annu Rev Plant Biol 61(1):395–420. https://doi.org/10.1146/annurev.arplant.043008.091939

Lu D (2013) Epigenetic modification enzymes: catalytic mechanisms and inhibitors. Acta Pharm Sin B 3(3):141–149. https://doi.org/10.1016/j.apsb.2013.04.007

Mahfouz MM (2010) RNA-directed DNA methylation. Plant Signal Behav 5(7):806–816. https://doi.org/10.4161/psb.5.7.11695

Mahrez W, Arellano MST, Moreno-Romero J, Nakamura M, Shu H, Nanni P, Köhler C, Gruissem W, Hennig L (2016) H3K36ac is an evolutionary conserved plant histone modification that marks active genes. Plant Physiol 170(3):1566–1577. https://doi.org/10.1104/pp.15.01744

Manning BJ, Peterson CL (2013) Releasing the brakes on a chromatin-remodeling enzyme. Nat Struct Mol Biol 20(1):5–7. https://doi.org/10.1038/nsmb.2482

Martinez-Medina A, Flors V, Heil M, Mauch-Mani B, Pieterse CMJ, Pozo MJ, Ton J, van Dam NM, Conrath U (2016) Recognizing plant defense priming. Trends Plant Sci 21(10):818–822. https://doi.org/10.1016/j.tplants.2016.07.009

Matzke MA, Mosher RA (2014) RNA-directed DNA methylation: an epigenetic pathway of increasing complexity. Nat Rev Genet 15(6):394–408. https://doi.org/10.1038/nrg3683

Matzke MA, Kanno T, Matzke AJM (2015) RNA-directed DNA methylation: the evolution of a complex epigenetic pathway in flowering plants. Annu Rev Plant Biol 66(1):243–267. https://doi.org/10.1146/annurev-arplant-043014-114633

Mauch-Mani B, Baccelli I, Luna E, Flors V (2017) Defense priming: an adaptive part of induced resistance. Annu Rev Plant Biol 68(1):485–512. https://doi.org/10.1146/annurev-arplant-042916-041132

Meyers BC, Axtell MJ, Bartel B, Bartel DP, Baulcombe D, Bowman JL, Cao X, Carrington JC, Chen X, Green PJ, Griffiths-Jones S, Jacobsen SE, Mallory AC, Martienssen RA, Poethig RS, Qi Y, Vaucheret H, Voinnet O, Watanabe Y, Weigel D, Zhu J-K (2008) Criteria for annotation of plant MicroRNAs. Plant Cell tpc.108.064311. https://doi.org/10.1105/tpc.108.064311

Michael TP (2014) Plant genome size variation: bloating and purging DNA. Brief Funct Genomics. https://doi.org/10.1093/bfgp/elu005

Mittler R, Finka A, Goloubinoff P (2012) How do plants feel the heat? Trends Biochem Sci 37(3):118–125. https://doi.org/10.1016/j.tibs.2011.11.007

Molinier J, Ries G, Zipfel C, Hohn B (2006) Transgeneration memory of stress in plants. Nature 442:1046. https://doi.org/10.1038/nature05022

Musselman CA, Lalonde M-E, Côté J, Kutateladze TG (2012) Perceiving the epigenetic landscape through histone readers. Nat Struct Mol Biol 19(12):1218–1227. https://doi.org/10.1038/nsmb.2436

Narlikar Geeta J, Sundaramoorthy R, Owen-Hughes T (2013) Mechanisms and functions of ATP-dependent chromatin-remodeling enzymes. Cell 154(3):490–503. https://doi.org/10.1016/j.cell.2013.07.011

Nathan D, Sterner DE, Berger SL (2003) Histone modifications: now summoning sumoylation. Proc Natl Acad Sci USA 100(23):13118–13120. https://doi.org/10.2307/3148100

Nicholson TB, Veland N, Chen T (2015) Writers, readers, and erasers of epigenetic marks. In: Gray SG (ed) Epigenetic cancer therapy. Academic Press, Boston, pp 31–66. https://doi.org/10.1016/B978-0-12-800206-3.00003-3

Nicoglou A, Merlin F (2017) Epigenetics: A way to bridge the gap between biological fields. Stud Hist Philos Sci Part C: Stud Hist Philos Biol Biomed Sci https://doi.org/10.1016/j.shpsc.2017.10.002

Nystedt B, Street NR, Wetterbom A, Zuccolo A, Lin Y-C, Scofield DG, Vezzi F, Delhomme N, Giacomello S, Alexeyenko A, Vicedomini R, Sahlin K, Sherwood E, Elfstrand M, Gramzow L, Holmberg K, Hallman J, Keech O, Klasson L, Koriabine M, Kucukoglu M, Kaller M, Luthman J, Lysholm F, Niittyla T, Olson A, Rilakovic N, Ritland C, Rossello JA, Sena J, Svensson T, Talavera-Lopez C, Theiszen G, Tuominen H, Vanneste K, Wu Z-Q, Zhang B, Zerbe P, Arvestad L, Bhalerao R, Bohlmann J, Bousquet J, Garcia Gil R, Hvidsten TR, de Jong P, MacKay J, Morgante M, Ritland K, Sundberg B, Lee Thompson S, Van de Peer Y, Andersson B, Nilsson O, Ingvarsson PK, Lundeberg J, Jansson S (2013) The Norway spruce genome sequence and conifer genome evolution. Nature 497:579–584. https://doi.org/10.1038/nature12211

Ong-Abdullah M, Ordway JM, Jiang N, Ooi S-E, Kok S-Y, Sarpan N, Azimi N, Hashim AT, Ishak Z, Rosli SK, Malike FA, Bakar NAA, Marjuni M, Abdullah N, Yaakub Z, Amiruddin MD, Nookiah R, Singh R, Low E-TL, Chan K-L, Azizi N, Smith SW, Bacher B, Budiman MA, Van Brunt A, Wischmeyer C, Beil M, Hogan M, Lakey N, Lim C-C, Arulandoo X, Wong C-K, Choo C-N, Wong W-C, Kwan Y-Y, Alwee SSRS, Sambanthamurthi R, Martienssen RA (2015) Loss of Karma transposon methylation underlies the mantled somaclonal variant of oil palm. Nature 525(7570):533–537. https://doi.org/10.1038/nature15365

Pascual J, Cañal MJ, Correia B, Escandon M, Hasbún R, Meijón M, Pinto G, Valledor L (2014) Can epigenetics help forest plants to adapt to climate change? In: Epigenetics in plants of agronomic importance: fundamentals and applications: transcriptional regulation and chromatin remodelling in plants. Springer International Publishing, Cham, pp 125–146. https://doi.org/10.1007/978-3-319-07971-4_8

Pastor V, Luna E, Mauch-Mani B, Ton J, Flors V (2013) Primed plants do not forget. Environ Exp Bot 94:46–56. https://doi.org/10.1016/j.envexpbot.2012.02.013

Patkin EL, Sofronov GA (2013) Population epigenetics, ecotoxicology, and human diseases. Russ J Genet: Appl Res 3(5):338–351. https://doi.org/10.1134/s2079059713050079

Penfield S (2008) Temperature perception and signal transduction in plants. New Phytol 179(3):615–628. https://doi.org/10.1111/j.1469-8137.2008.02478.x

Pikaard CS, Mittelsten Scheid O (2014) Epigenetic regulation in plants. Cold Spring Harb Perspect Biol 6(12):a019315. https://doi.org/10.1101/cshperspect.a019315

Quadrana L, Almeida J, Asís R, Duffy T, Dominguez PG, Bermúdez L, Conti G, Corrêa da Silva JV, Peralta IE, Colot V, Asurmendi S, Fernie AR, Rossi M, Carrari F (2014) Natural occurring epialleles determine vitamin E accumulation in tomato fruits. Nat Commun 5:4027. https://doi.org/10.1038/ncomms5027

Rapp RA, Wendel JF (2005) Epigenetics and plant evolution. New Phytol 168(1):81–91. https://doi.org/10.1111/j.1469-8137.2005.01491.x
Richards EJ (2006) Inherited epigenetic variation—revisiting soft inheritance. Nat Rev Genet 7:395. https://doi.org/10.1038/nrg1834
Richards CL, Bossdorf O, Verhoeven KJF (2010) Understanding natural epigenetic variation. New Phytol 187(3):562–564. https://doi.org/10.1111/j.1469-8137.2010.03369.x
Roessler K, Bousios A, Meca E, Gaut BS (2018) Modeling interactions between transposable elements and the plant epigenetic response: a surprising reliance on element retention. Genome Biol Evol 10(3):803–815. https://doi.org/10.1093/gbe/evy043
Rohde A, Junttila O (2008) Remembrances of an embryo: long-term effects on phenology traits in spruce. New Phytol 177(1):2–5. https://doi.org/10.1111/j.1469-8137.2007.02319.x
Rosa S, Shaw P (2013) Insights into chromatin structure and dynamics in plants. Biology 2(4):1378–1410
Rossetto D, Avvakumov N, Côté J (2012) Histone phosphorylation: a chromatin modification involved in diverse nuclear events. Epigenetics 7(10):1098–1108. https://doi.org/10.4161/epi.21975
Sang Y, Silva-Ortega CO, Wu S, Yamaguchi N, Wu M-F, Pfluger J, Gillmor CS, Gallagher KL, Wagner D (2012) Mutations in two non-canonical Arabidopsis SWI2/SNF2 chromatin remodeling ATPases cause embryogenesis and stem cell maintenance defects. Plant J: For Cell Mol Biol 72(6):1000–1014. https://doi.org/10.1111/tpj.12009
Santo T, Pereira R, Leitão J (2017) The Pea (Pisum sativum L.) Rogue paramutation is accompanied by alterations in the methylation pattern of specific genomic sequences. Epigenomes 1(1):6
Saroha M, Singroha G, Sharma M, Mehta G, Gupta O, Sharma P (2017) sRNA and epigenetic mediated abiotic stress tolerance in plants. Indian J Plant Physiol 22(4):458–469. https://doi.org/10.1007/s40502-017-0330-z
Satake A, Iwasa Y (2012) A stochastic model of chromatin modification: cell population coding of winter memory in plants. J Theor Biol 302:6–17. https://doi.org/10.1016/j.jtbi.2012.02.009
Saze H (2008) Epigenetic memory transmission through mitosis and meiosis in plants. Semin Cell Dev Biol 19 (6):527–536. https://doi.org/10.1016/j.semcdb.2008.07.017
Schmitz RJ, Schultz MD, Lewsey MG, O'Malley RC, Urich MA, Libiger O, Schork NJ, Ecker JR (2011) Transgenerational epigenetic instability is a source of novel methylation variants. Science 334(6054):369–373. https://doi.org/10.1126/science.1212959
Sengupta P, Garrity P (2013) Sensing temperature. Curr Biol 23(8):R304–R307. https://doi.org/10.1016/j.cub.2013.03.009
Seymour DK, Becker C (2017) The causes and consequences of DNA methylome variation in plants. Curr Opin Plant Biol 36:56–63. https://doi.org/10.1016/j.pbi.2017.01.005
Shiio Y, Eisenman RN (2003) Histone sumoylation is associated with transcriptional repression. Proc Natl Acad Sci USA 100(23):13225–13230. https://doi.org/10.1073/pnas.1735528100
Silveira AB, Trontin C, Cortijo S, Barau J, Del Bem LEV, Loudet O, Colot V, Vincentz M (2013) Extensive natural epigenetic variation at a de novo originated gene. PLoS Genet 9(4):e1003437. https://doi.org/10.1371/journal.pgen.1003437
Singh A, Gautam V, Singh S, Sarkar Das S, Verma S, Mishra V, Mukherjee S, Sarkar AK (2018) Plant small RNAs: advancement in the understanding of biogenesis and role in plant development. Planta 248(3):545–558. https://doi.org/10.1007/s00425-018-2927-5
Skrøppa T, Tollefsrud MM, Sperisen C, Johnsen Ø (2010) Rapid change in adaptive performance from one generation to the next in Picea abies—central European trees in a Nordic environment. Tree Genet Genom 6(1):93–99. https://doi.org/10.1007/s11295-009-0231-z
Soloaga A, Thomson S, Wiggin GR, Rampersaud N, Dyson MH, Hazzalin CA, Mahadevan LC, Arthur JSC (2003) MSK2 and MSK1 mediate the mitogen- and stress-induced phosphorylation of histone H3 and HMG-14 22(11):2788–2797. https://doi.org/10.1093/emboj/cdg273
Sow MD, Allona I, Ambroise C, Conde D, Fichot R, Gribkova S, Jorge P, Le-Provost G, Pâques L, Plomion C, Salse J, Sanchez L, Segura V, Tost J, Maury S (2018) Epigenetics in forest trees: state of the art and potential implications for breeding and management in a context of climate change. In: Gallusci P, Bucher E, Mirouze M (eds) Plant epigenetics coming of age for breeding applications, vol 88. Academic Press, pp 387–453. https://doi.org/10.1016/bs.abr.2018.09.003
Stricker SH, Köferle A, Beck S (2016) From profiles to function in epigenomics. Nat Rev Genet 18:51. https://doi.org/10.1038/nrg.2016.138
Stroud H, Do T, Du J, Zhong X, Feng S, Johnson L, Patel DJ, Jacobsen SE (2013) Non-CG methylation patterns shape the epigenetic landscape in Arabidopsis. Nat Struct Mol Biol 21:64. https://doi.org/10.1038/nsmb.2735
Suganuma T, Workman JL (2011) Signals and combinatorial functions of histone modifications. Annu Rev Biochem 80(1):473–499. https://doi.org/10.1146/annurev-biochem-061809-175347
Sunkar R, Li Y, Jagadeeswaran G (2012) Functions of microRNAs in plant stress. Trends Plant Sci 17:196–203
Suzuki MM, Bird A (2008) DNA methylation landscapes: provocative insights from epigenomics. Nat Rev Genet 9:465. https://doi.org/10.1038/nrg2341
Takuno S, Ran J-H, Gaut BS (2016) Evolutionary patterns of genic DNA methylation vary across land plants. Nature Plants 2:15222. https://doi.org/10.1038/

nplants.2015.222, https://www.nature.com/articles/nplants2015222#supplementary-information
Tarakhovsky A (2010) Tools and landscapes of epigenetics. Nat Immunol 11(7):565–568
Thellier M, Lüttge U (2013) Plant memory: a tentative model. Plant Biol 15(1):1–12. https://doi.org/10.1111/j.1438-8677.2012.00674.x
Turck F, Coupland G (2014) Natural variation in epigenetic gene regulation and its effects on plant developmental traits. Evolution 68(3):620–631. https://doi.org/10.1111/evo.12286
Valdés A, Marteinsdottir B, Ehrlén J (2018) A natural heating experiment: phenotypic and genotypic responses of plant phenology to geothermal soil warming. Glob Chang Biol 0(ja). https://doi.org/10.1111/gcb.14525
Van Oosten MJ, Bressan RA, Zhu J-K, Bohnert HJ, Chinnusamy V (2014) The role of the epigenome in gene expression control and the epimark changes in response to the environment. Crit Rev Plant Sci 33 (1):64–87. https://doi.org/10.1080/07352689.2014.852920
Vazquez F, Hohn T (2013) Biogenesis and biological activity of secondary siRNAs in plants. Scientifica 2013:12. https://doi.org/10.1155/2013/783253
Vazquez F, Vaucheret H, Rajagopalan R, Lepers C, Gasciolli V, Mallory AC, Hilbert JL, Bartel DP, Crete P (2004) Endogenous trans-acting siRNAs regulate the accumulation of Arabidopsis mRNAs. Mol Cell 16(1):69–79. https://doi.org/10.1016/j.molcel.2004.09.028
Verhoeven KJF, Jansen JJ, van Dijk PJ, Biere A (2010) Stress-induced DNA methylation changes and their heritability in asexual dandelions. New Phytol 185 (4):1108–1118. https://doi.org/10.1111/j.1469-8137.2009.03121.x
Verhoeven KJF, vonHoldt BM, Sork VL (2016) Epigenetics in ecology and evolution: what we know and what we need to know. Mol Ecol 25(8):1631–1638. https://doi.org/10.1111/mec.13617
Vlot AC, Klessig DF, Park S-W (2008) Systemic acquired resistance: the elusive signal(s). Curr Opin Plant Biol 11(4):436–442. https://doi.org/10.1016/j.pbi.2008.05.003
Voinnet O (2009) Origin, biogenesis, and activity of plant MicroRNAs. Cell 136(4):669–687. https://doi.org/10.1016/j.cell.2009.01.046
Wang Y, Dasso M (2009) SUMOylation and deSUMOylation at a glance. J Cell Sci 122 (23):4249–4252. https://doi.org/10.1242/jcs.050542
Waterborg JH (2011) Plant histone acetylation: In the beginning…. Biochim Biophys Acta 1809(8):353–359. https://doi.org/10.1016/j.bbagrm.2011.02.005
Whittle CA, Otto SP, Johnston MO, Krochko JE (2009) Adaptive epigenetic memory of ancestral temperature regime in *Arabidopsis thaliana*. Botany 87(6):650–657. https://doi.org/10.1139/B09-030
Wigge PA (2013) Ambient temperature signalling in plants. Curr Opin Plant Biol 16(5):661–666. https://doi.org/10.1016/j.pbi.2013.08.004
Williams BP, Gehring M (2017) Stable transgenerational epigenetic inheritance requires a DNA methylation-sensing circuit. Nat Commun 8(1):2124. https://doi.org/10.1038/s41467-017-02219-3
Woo HR, Dittmer TA, Richards EJ (2008) Three SRA-domain methylcytosine-binding proteins cooperate to maintain global CpG methylation and epigenetic silencing in *Arabidopsis*. PLoS Genet 4(8): e1000156. https://doi.org/10.1371/journal.pgen.1000156
Wu G (2013) Plant microRNAs and development. J Genet Genomics 40(5):217–230. https://doi.org/10.1016/j.jgg.2013.04.002
Xia R, Xu J, Arikit S, Meyers BC (2015) Extensive families of miRNAs and PHAS Loci in Norway Spruce demonstrate the origins of complex phasiRNA networks in seed plants. Mol Biol Evol 32(11):2905–2918. https://doi.org/10.1093/molbev/msv164
Xie N, Zhou Y, Sun Q, Tang B (2018) Novel Epigenetic Techniques Provided by the CRISPR/Cas9 System. Stem Cells Int 2018:7834175. https://doi.org/10.1155/2018/7834175
Yadav CB, Pandey G, Muthamilarasan M, Prasad M (2018) Epigenetics and epigenomics of plants. In: Varshney RK, Pandey MK, Chitikineni A (eds) Plant genetics and molecular biology. Springer International Publishing, Cham, pp 237–261. https://doi.org/10.1007/10_2017_51
Yakovlev IA, Fossdal CG (2017) *In Silico* analysis of small RNAs suggest roles for novel and conserved miRNAs in the formation of epigenetic memory in somatic embryos of Norway Spruce. Front Physiol 8 (674):1–17. https://doi.org/10.3389/fphys.2017.00674
Yakovlev IA, Fossdal CG, Johnsen Ø (2010) MicroRNAs, the epigenetic memory and climatic adaptation in Norway spruce. New Phytol 187(4):1154–1169. https://doi.org/10.1111/j.1469-8137.2010.03341.x
Yakovlev IA, Lee Y, Rotter B, Olsen JE, Skrøppa T, Johnsen Ø, Fossdal CG (2014) Temperature-dependent differential transcriptomes during formation of an epigenetic memory in Norway spruce embryogenesis. Tree Genet Genom 10:355–366. https://doi.org/10.1007/s11295-013-0691-z
Yakovlev IA, Carneros E, Lee Y, Olsen JE, Fossdal CG (2016) Transcriptional profiling of epigenetic regulators in somatic embryos during temperature induced formation of an epigenetic memory in Norway spruce. Planta 243(5):1237–1249. https://doi.org/10.1007/s00425-016-2484-8
Yoshikawa M (2013) Biogenesis of *trans*-acting siRNAs, endogenous secondary siRNAs in plants. Genes Genet Syst 88(2):77–84
Yun M, Wu J, Workman JL, Li B (2011) Readers of histone modifications. Cell Res 21(4):564–578. https://doi.org/10.1038/cr.2011.42
Zhang Y (2003) Transcriptional regulation by histone ubiquitination and deubiquitination. Genes Dev 17 (22):2733–2740. https://doi.org/10.1101/gad.1156403
Zhang X, Yazaki J, Sundaresan A, Cokus S, Chan SWL, Chen H, Henderson IR, Shinn P, Pellegrini M,

Jacobsen SE, Ecker Joseph R (2006) Genome-wide high-resolution mapping and functional analysis of DNA methylation in arabidopsis. Cell 126(6):1189–1201

Zhang H, He X, Zhu J-K (2013a) RNA-directed DNA methylation in plants. RNA Biol 10(10):1593–1596. https://doi.org/10.4161/rna.26312

Zhang Y-Y, Fischer M, Colot V, Bossdorf O (2013b) Epigenetic variation creates potential for evolution of plant phenotypic plasticity. New Phytol 197(1):314–322. https://doi.org/10.1111/nph.12010

Zhang L, Yu H, Ma B, Liu G, Wang J, Wang J, Gao R, Li J, Liu J, Xu J, Zhang Y, Li Q, Huang X, Xu J, Li J, Qian Q, Han B, He Z, Li J (2017) A natural tandem array alleviates epigenetic repression of IPA1 and leads to superior yielding rice. Nat Commun 8:14789. https://doi.org/10.1038/ncomms14789

Zheng X, Pontes O, Zhu J, Miki D, Zhang F, Li W-X, Iida K, Kapoor A, Pikaard CS, Zhu J-K (2008) ROS3 is an RNA-binding protein required for DNA demethylation in Arabidopsis. Nature 455:1259. https://doi.org/10.1038/nature07305

Zilberman D, Coleman-Derr D, Ballinger T, Henikoff S (2008) Histone H2A.Z and DNA methylation are mutually antagonistic chromatin marks. Nature 456 (7218):125–129. https://doi.org/10.1038/nature07324

6 Comparative Genomics of Spruce and Other Gymnosperms

Amanda R. De la Torre

Abstract

In contrast to flowering plants, the genomes of gymnosperms are still poorly understood. Current knowledge on the genomic architecture, gene space, and macrostructure of gymnosperm genomes mainly comes from Pinaceae species, in which spruce species (genera with most sequenced genomes to date) had played a significant role. Whereas the accumulation of nuclear genomes' sequence information has followed a slow pace, the sequencing of chloroplast genomes skyrocketed in the last 10 years, with many spruce species sequenced so far. From the comparison of nuclear, mitochondrial, and chloroplast genomes, we can conclude that they all differ in their genomic macrostructure and rates of sequence evolution. The high structural conservation among species, genera, and families is characteristic of nuclear genomes, whereas chloroplast and mitochondrial genomes show less conserved structures due to genomic arrangements. Finally, *Picea* species follow trends observed in other gymnosperms and angiosperms, in which rates of sequence evolution are the highest in nuclear DNA, followed by chloroplast and then mitochondrial DNA.

6.1 Sequencing of Nuclear Genomes

The advent of new sequencing technologies and large-scale computational resources has allowed the development of genome-wide data for the large sizes (10–30 Gb) of conifer genomes (reviewed in De La Torre et al. 2014a; Neale et al. 2017). There has been significant interest in developing genomic resources for spruce species, due to their high economic importance in the Northern hemisphere. In fact, in the last 5 years, most of our knowledge on gymnosperm genomes has come from the genomes of spruce species, specifically *Picea glauca x engelmannii x sitchensis* (Birol et al. 2013), *Picea glauca* (Warren et al. 2015), and *Picea abies* (Nystedt et al. 2013). Reference genomes for these species were followed by efforts to do whole-genome resequencing analyses (*Picea abies*, Chap. 3 in this book), and whole-genome sequencing in other spruce species (*Picea engelmannii* in progress).

White spruce (*Picea glauca*) is the most widely planted conifer in Canada and one of the most economically important tree species worldwide. The genotype selected for the first spruce genome sequencing was PG29, an important genotype in British Columbia's tree-breeding

A. R. De la Torre (✉)
School of Forestry, Northern Arizona University, 200 E. Pine Knoll, Flagstaff, AZ 86001, USA
e-mail: Amanda.de-la-torre@nau.edu

I. M. Porth and A. R. De la Torre (eds.), *The Spruce Genome*, Compendium of Plant Genomes,
https://doi.org/10.1007/978-3-030-21001-4_6

program for its resistance to the insect pest *Pissodes strobi*. Since 2013, several updated versions of the sequence assembly have been improved with increased scaffold contiguity (see Chap. 2 in this book). Transcriptome re-scaffolding and the use of large fragment mate pair sequences significantly increased the contiguity of the PG29 assembly, resulting in a ten-fold increase in scaffold length (N50), when compared to the draft version of Norway spruce (assembly version 1, v1). Besides increases in contiguity, all spruce (and conifers) reference genomes to date are still highly fragmented due to the challenges associated with sequencing and assembling highly repetitive, complex and very large genome sizes. Later studies (De La Torre et al. 2014b) revealed PG29 to be an admixed individual containing ancestry of *P. glauca*, *P. engelmannii,* and *P. sitchensis*. For this reason, a second non-admixed genotype (WS77111) was sequenced (Warren et al. 2015). High levels of synteny were found between PG29 and WS77111, despite introgression in PG29 (Warren et al. 2015).

Norway spruce (*Picea abies*) is one of the most widespread, ecologically and economically important tree species in Europe (Nystedt et al. 2013). The draft genome sequence (19.6 Gb) of the species came from an individual (Z4006) extensively used in Swedish breeding programs. Haploid and diploid whole-genome shotgun (WGS) data, together with RNA sequencing of several tissues was used to generate the draft genome (Nystedt et al. 2013). Although the assembly showed high levels of fragmentation (>10 million scaffolds), 28k high-confidence genes were identified. Two main findings derived from this study. The first one suggests the same type of long-terminal repeats (LTR) are largely shared among extant gymnosperms; and the second, that the accumulation of these LTRs may explain the large genome sizes in conifers and other gymnosperms (Nystedt et al. 2013). In any case, the current knowledge of the role of transposable elements in the genomes of conifers is still scarce and will further benefit from the sequencing of additional gymnosperm genomes.

6.1.1 Genetic Linkage Maps and Comparative Mapping

Due to the high levels of fragmentation and the absence of chromosome-scale reference genomes in gymnosperms, genetic linkage maps have proven to be useful for anchoring and ordering scaffolds into linkage groups improving the quality of the genome assemblies. A linkage map locating 1.1k markers over 12 linkage groups was developed for black spruce (*Picea mariana*; Kang et al. 2010). The most recent white spruce linkage map (Pavy et al. 2017) was able to locate the position of 8.8k genes across linkage groups, with a significant increase in gene density in comparison to previous linkage maps in the species (Pavy et al. 2008; Pelgas et al. 2011; Pavy et al. 2012). More recently, an ultra-dense linkage map built with 21k SNP markers located 17k genes in the genome of Norway spruce (Bernhardsson et al. 2019). Ultra-dense linkage maps have only been developed for two other gymnosperms to date, *Pinus taeda* (De La Torre et al. 2019) and *Ginkgo biloba* (Liu et al. 2017). In addition to assisting in genome assemblies, linkage maps in spruce have also been developed in conjunction with quantitative trait loci (QTL) mapping to identify QTLs associated with traits of importance such as disease (Lind et al. 2014) and phenology (Pelgas et al. 2011).

Linkage maps have revealed highly conserved genomic macrostructure, with high levels of synteny and co-linearity between spruces, and among spruces and pines (Pavy et al. 2012; De Miguel et al. 2015; Pavy et al. 2017; Bernhardsson et al. 2019). High correlation of marker order was found between the Pavy et al. 2017 white spruce map and the Norway spruce map (Bernhardsson et al. 2019). Similarly, De Miguel et al. (2015) found 513 bridges (orthologous genes based on sequence similarity) between white spruce and a pine composite map (*Pinus pinaster* and *Pinus taeda*), whereas Pavy et al. (2017) found 1,851 *Picea*–*Pinus* bridges and no evidence for large-scale genomic rearrangements between the genera. Considering that *Picea* and

Pinus diverged well over 100 million years ago, this level of macroscale synteny is unprecedented in other seed plant genera. Chromosomal rearrangements, although rare in gymnosperms, have been found between Pinaceae and Cuppresaceae, taxa that diverged 200 million years ago (De Miguel et al. 2015). Ultra-dense linkage maps have greatly contributed to our understanding of the macrostructure of gymnosperm genomes; however, their genetic information still accounts for a small percentage of the very large genome sizes of gymnosperms. Limitations include the scarcity of large-scale full-sib pedigrees in long-generation conifers (although see Bernhardsson et al. 2019 for the use of megagametophytes as an alternative), sequencing costs, computational issues, and the fact that published linkage map studies have greatly focused on the gene space, which represents a very small fraction of gymnosperm genomes. New technologies involving extra-long read sequencing and the use of proximity ligation (Hi-C technology), in combination with ultra-dense linkage maps, will significantly increase the accuracy of reference genomes.

6.1.2 Rates of Sequence Evolution

The high conservation of genomic macrostructure, low morphological diversity, and levels of speciation in conifers and gymnosperms in general, have led scientists to believe conifers are slow-evolving plants (Buschiazzo et al. 2012; Pavy et al. 2012; De la Torre et al. 2014a; De La Torre et al. 2017). Early studies based on the comparison of a few Pinaceae against arabidopsis and poplar found that rates of sequence evolution were slower in Pinaceae in comparison to flowering plants (Buschiazzo et al. 2012; Chen et al. 2012). More recently, using a broad taxonomic sampling that included representatives of all gymnosperms (with exception of Araucariaceae) and all angiosperms with a sequenced genome, De La Torre et al. 2017 estimated absolute rates of silent-site divergence per year (μ) for each taxonomic family and genera. This study found that although μ was much lower in gymnosperms than angiosperms, the ratio of non-synonymous to synonymous substitution rates (d*N*/d*S* or ω) was higher in gymnosperms (De La Torre et al. 2017). A more recent study investigating somatic mutations in *Picea sitchensis* found even lower μ measures per year than those found by De La Torre et al. 2017; however, found a higher μ per generation when comparing *P. sitchensis* to other plants and animals (Hanlon et al. 2019). Large genome sizes and long generation times may partly explain the slow rates of sequence evolution in gymnosperms (De La Torre et al. 2017).

In strikingly contrast to the slow rates of sequence evolution (site/year), Pinaceae species (including *Picea spp.*) are often well adapted to local climates across their usually wide geographical distributions, and have evolved resistance to biotic and abiotic stresses. Several hypotheses have been suggested to explain this discrepancy. In De La Torre et al. 2017, several molecular evolution models (site and branch codon-based likelihood) suggested a higher number of sites under diversifying selection in *Picea* and *Pinus* in relation to other gymnosperms and angiosperms, suggesting relatively recent evolutionary radiations in these genera might have resulted in lower numbers of synonymous substitutions (d*S*) and higher d*N*/d*S* than in other gymnosperms (Palme et al. 2009; De La Torre et al. 2017). Hanlon et al. (2019), however, suggests that the high somatic base substitution rate per base per generation in *Picea sitchensis* might generate a significant amount of the genetic variation that creates and maintains local adaptation in the species. It is important to mention, however, that the number of somatic mutations that are passed on to the next generation is unknown in gymnosperms. Other studies have suggested copy number variation (CNV) has a direct role in climate adaptation and insect resistance in *Picea spp.* such as white spruce and *P. sitchensis* (Hall et al. 2011; Prunier et al. 2017a, b). Finally, besides CNVs, epigenetic variation affecting the regulation of gene expression also represent a major mechanism of genetic variation from which phenotypic variation and local adaptation may evolve (Prunier et al. 2016). The role of epigenetic variation in adaptation to climate in early embryo development has been extensively reported in *Picea abies* (Yakovlev et al. 2011, 2016). The role of gene expression variation in phenotypic plasticity and adaptive evolution in gymnosperms also deserves more attention (Prunier et al. 2016). In any case, the molecular basis of local adaptation in long-generation conifers probably results from a combination of several of the above or other complex processes that are yet to be elucidated.

6.2 Sequencing of Mitochondrial and Chloroplast Genomes

In contrast to the nuclear reference genomes which represent only a handful of gymnosperm species, organelle genomes have skyrocketed in numbers in the last 10 years. Chloroplast genomes of 80 species representing 10 of the 12 extant gymnosperm families (only excluding Zamiaceae and Sciadopityaceae) have been published to date (Fig. 6.1). Half of these studies are in Pinaceae species, and from these, six are in *Picea spp* (Nystedt et al. 2013; Jackman et al. 2015; Coombe et al. 2016; Yang et al. 2016; Li et al. 2019; Lin et al. 2019). Chloroplast genomes varied in size from 109 to 163 kb across gymnosperm families, with Cycadaceae and Ginkgoaceae being the largest (Fig. 6.1). Mitochondrial genomes are substantially larger than chloroplast genomes ranging from 347 kb (*Ginkgo biloba*) to 5900 kb (*Picea glauca*) (Jackman et al. 2015; Guo et al. 2016). All *Picea spp.* (*abies*, *glauca,* and *sitchensis*) mitochondrial genomes are larger than the rest of gymnosperms; however, only *Ginkgo biloba*, *Cycas taitungensis*, *Welwitschia mirabillis*, *Pinus taeda,* and *Larix spp.* have been sequenced to date (Fig. 6.1; Chaw et al. 2008; Guo et al. 2016; Jackman et al. 2015, 2020; Sullivan et al. 2019; Zimmerman et al. 2019).

In contrast to most seed plants in which cytoplasmic genomes are usually maternally inherited, Pinaceae have different forms of inheritance for chloroplast and mitochondrial DNA. Chloroplast DNA is paternally inherited and mitochondrial DNA is maternally inherited in Pinaceae species (Neale and Sederoff 1989). This provides a unique opportunity to differentiate gene flow coming from seeds or pollen, create markers to trace lineages over time, and identify episodes of species' geographic range expansions or contractions due to changes in the climate. Organelle genomes have been sequenced as tools for phylogenomic and phylogeographic analyses (Chaw et al. 2008; Chen et al. 2015; Hao et al. 2016; Asaf et al. 2018; Zang et al. 2019), for conservation purposes in endemic or threatened species (Chen et al. 2018; Miao et al. 2019); and for comparative or molecular evolution analyses (Tomonori et al. 2008; Lin et al. 2010; Wu et al. 2011a, b; Smith 2015; Guo et al. 2016; Wu and Chaw 2016).

6.2.1 Comparative Mapping–Chromosomal Rearrangements

Chloroplast genomes in angiosperms are composed by two duplicate inverted repeats (IRs) separated by a large single-copy (LSC) and small single-copy (SSC) regions that are relatively conserved in gene order and content (Sugiura 1992). In contrast, the chloroplast genomes of the five extant gymnosperm groups show different structural patterns. For example, those of gnetophytes present multiple inversions and have undergone multiple contractions or expansions (Chaw et al. 2018). In contrast, Cycads and Ginkgo present higher levels of conservation and larger IRs than those in Pinaceae (Palmer and Stein 1986; Tsudzuki et al. 1992; Wu et al. 2007; Lin et al. 2012). Pinaceae and Cuppresaceae present highly rearranged chloroplast genomes that lack canonical IRs and contain lineage-specific repeats (Strauss et al. 1988; Hirao et al. 2008; Wu et al. 2011a, b; Chaw et al. 2018). Major rearrangements were found among spruces species *Picea sitchensis*, *Picea morrisonicola,* and *Picea abies*. Whereas a 46-kbp inversion differentiates the cpDNAs of *P. abies* and *P. morrisonicola*, a 23k-bp inversion in addition to a 23-kbp translocation

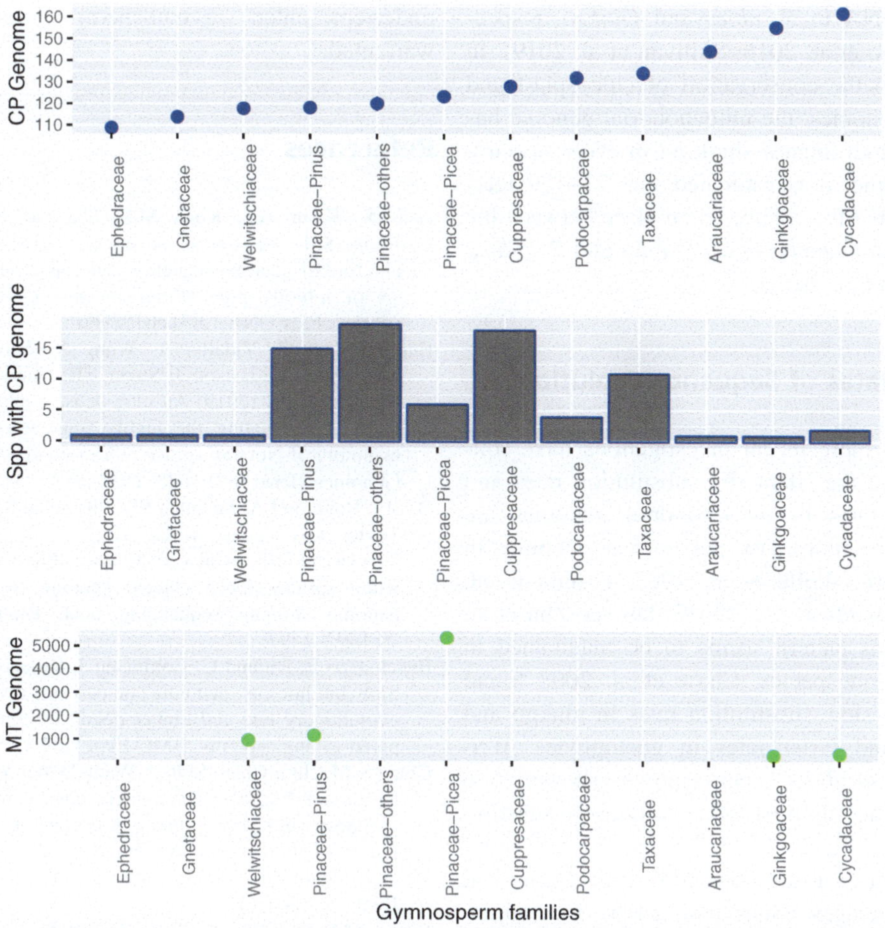

Fig. 6.1 Comparison among published organelle genomes for gymnosperm families. Top graph shows the different genome sizes for chloroplast genomes in kilo base pairs (kb); Middle graph shows the number of species with a sequenced chloroplast genome to date; Bottom graph shows the different genome sizes for mitochondrial genomes in kb. The Pinaceae family is divided in three groups: *Picea*, *Pinus,* and Others (*Abies*, *Cathaya*, *Cedrus*, *Larix*, *Pseudotsuga*, and *Tsuga*) for graphic purposes

differentiate the cpDNAs of *P. sitchensis* and *P. abies* (Nystedt et al. 2013). No structural rearrangements were found between the cpDNAs of *P. glauca* and *P. abies* despite a species divergence of over 10 million years (Jackman et al. 2015). Consistently, *P. glauca*, *P. abies,* and *P. sitchensis* have all their 114 chloroplast genes in the same order and copy number (Coombe et al. 2016).

Mitochondria have larger and more variable genomes sizes than chloroplasts in gymnosperms. Although spruces have more repetitive content which could account for larger genome sizes than other gymnosperms, the relationship between repetitive content and genome size is inconclusive (Guo et al. 2016). The mtDNAs of Pinaceae (*Picea abies*, *Picea glauca,* and *Pinus taeda*) showed conservation in relation to the number of protein-coding genes inferred to be present in the last common ancestor of angiosperms; and *P. abies* and *P. glauca* had similar intron and tRNA content (Jackman et al. 2015; Sullivan et al. 2019). In addition, some mtDNA is believed to be transferred from cpDNA and nuclear DNA in *P. abies* and *Ginkgo*, respectively (Guo et al. 2016; Sullivan et al. 2019). In contrast to the conservation of chloroplast and nuclear genomes, the mitochondrial genome has gone through intense shuffling in *Picea* species. For example, it is estimated that 71–129 rearrangements Mya^{-1} have taken place between the mitochondrial genomes of *P. abies* and *P. glauca* (Sullivan et al. 2019).

6.2.2 Rates of Sequence Evolution

Early and more recent investigations have concluded that the silent site substitution rates and mutation rates in mitochondrial genomes are lower than chloroplast and nuclear genomes in land plants (Wolfe et al. 1987, Drouin et al. 2008; Richardson et al. 2013; but see Zhu et al. 2014). This pattern seems to be maintained in gymnosperm species. Chaw et al. 2008 found extremely slow substitution rates in ginkgo and cycads, and faster rates in gnetophytes. *Welwitschia mirabilis*, a gnetophyte, showed 5–6 times higher number of synonymous substitutions (d*S*) and seven times higher non-synonymous substitutions (d*N*) than *Cycas* and *Ginkgo* lineages (Guo et al. 2016). In comparison, Pinaceae species seem to have faster substitution rates than those in *Cycas* and *Ginkgo*, but slower than gnetophytes (Guo et al. 2016; Chaw et al. 2018; Sullivan et al. 2019). *Picea abies'* number of synonymous substitutions (dS) was estimated as 1.53×10^{-10} and the number of non-synonymous substitutions (dN) was 0.8×10^{-10} (Sullivan et al. 2019). These rates place spruce species in the lower end of the absolute rates observed in angiosperms, but above the rates observed for *Pinus* species (Wang and Wang 2014). As an example, dS were 0.12×10^{-10} and 0.63×10^{-10}, whereas dN were 0.06×10^{-10} and 0.34×10^{-10} site/year in *Pinus taeda* and *Pinus strobus*, respectively (Richardson et al. 2013; Sullivan et al. 2019). It would be interesting to see whether the patterns observed to date are maintained when more mitochondrial genomes are sequenced. A wider representation of Pinaceae species and other underrepresented gymnosperm families is warranted.

References

Asaf S, Khan AL, Khan MA, Shahzad R, Lubna, Kang SM, Al-Harrasi A et al (2018) Complete chloroplast genome sequence and comparative analysis of loblolly pine (*Pinus taeda L.*) with related species. Plos ONE 13(3):e0192966

Bernhardsson C, Vidallis A, Wang X, Scofield DG, Schiffthaler B, Baison J, Street NR, Garcia-Gil MR, Ingvarsson PK (2019) An ultra-dense haploid genetic map for evaluating the highly fragmented genome assembly of Norway spruce *(Picea abies)*. G3: Genes Genomes Genet 9(5):1623–1632

Birol I, Raymond A, Jackman SD, Pleasance S, Coope R, Taylor GA, Yuen MMS, Keeling CI, Brand D, Vandervalk BP, et al (2013) Assembling the 20 Gb white spruce (Picea glauca) genome from whole-genome shotgun sequencing data. Bioinformatics 29:1492–1497

Buschiazzo E, Ritland C, Bohlmann J, Ritland K (2012) Slow but not low: genomic comparisons reveal slower evolutionary rate and higher dN/dS in conifers compared to angiosperms. BMC Evol Biol 12:8

Chaw S-M Chun-Chieh Shih A Wang D Wu Y-W Liu S-M Chou T-Y (2008) The mitochondrial genome of the gymnosperm *Cycas taitungensis* contains a novel family of short interspersed elements, *Bpu* sequences, and abundant RNA editing sites. Mol Biol Evol 25:603–615

Chaw SM, Wu SC, Sudianto E (2018) Evolution of gymnosperm plastid genomes. Adv Bot Res 85:195–222

Chen J, Uebbing S, Gyllenstrand N, Lagercrantz U, Lascoux M, Kallman T (2012) Sequencing of the needle transcriptome from Norway spruce (Picea abies Karst L.) reveals lower substitution rates, but similar selective constraints in gymnosperms and angiosperms. BMC Genomics 13:589

Chen J, Hao Z, Xu H, Yang L, Liu G, Shen Y, Zheng C, Zheng W, Cheng T, Shi J (2015) The complete chloroplast genome sequence of the relict woody plant *Metasequoia glyptostroboides* Hu et Cheng. Front Plant Sci 6:447

Chen L, Li L, Yang G, Qian H, Li M (2018) Characterization of the complete chloroplast genome sequence of *Tsuga longibracteata* W.C. Cheng (Pinaceae). Conserv Genet Resour. https://doi.org/10.1007/s12686-017-0974-z

Coombe L, Warren RL, Jackman SD, Yang C, Vandervalk BP, Moore RA, Pleasance S, Coope RJ, Bohlmann J, Holt RA, Jones SJM, Birol I (2016) Assembly of the complete sitka spruce chloroplast genome using 10X genomics Gemcode sequencing data. PLos One 11(9):e0163

De La Torre AR, Birol I, Bousquet J, Ingvarsson PK, Jansson S, Jones SJM, Keeling CI, MacKay J, Nilsson O, Ritland K, Street N, Yanchuk A, Zerbe P, Bohlmann J (2014a) Insights into Conifer Giga-genomes. Plant Physiol 166: 1–9

De La Torre AR, Roberts D, Aitken SN (2014b) Genome-wide admixture and ecological niche modeling reveal the maintenance of species boundaries despite long history of interspecific gene flow. Mol Ecol 23:2046–2059

De La Torre AR, Li Z, Van de Peer Y, Ingvarsson PK (2017) Contrasting rates of molecular evolution and patterns of selection among gymnosperms and flowering plants. Mol Biol Evol 34(6):1363–1377

De La Torre AR, Wilhite B, DB Neale (2019) Environmental genome-wide association reveals climate adaptation is shaped by subtle to moderate allele frequency shifts in loblolly pine. Genome Biol Evol (In Press). https://doi.org/10.1093/gbe/evz220

De Miguel M, Bartholomé J, Ehrenmann F, Murat F, Moriguchi Y et al (2015) Evidence of intense chromosomal shuffling during conifer evolution. Genome Biol Evol 7:2799–2809. https://doi.org/10.1093/gbe/evv185

Drouin G, Daoud H, Xia J (2008) Relative rates of synonymous substitutions in the mitochondrial, chloroplast and nuclear genomes of seed plants. Mol Phylogenet Evol 49:827–831

Guo W, Grewe F, Fan W, Young GJ, Knoop V, Palmer JD, Mower JP (2016) *Ginkgo* and *Welwitschia* mitogenomes reveal extreme contrasts in gymnosperm mitochondrial evolution. Mol Biol Evol 33(6):1448–1460. https://doi.org/10.1093/molbev/msw024

Hall DE, Robert JA, Keeling CI, Domanski D, Quesada AL, Jancsik S, Kuzyk MA, Hamberger B, Borchers CH, Bohlmann J (2011) An integrated genomic, proteomic and biochemical analysis of (+)-3-carene biosynthesis in Sitka spruce (*Picea sitchensis*) genotypes that are resistant or susceptible to white pine weevil. Plant J 65:936–948

Hanlon VCT, Otto SP, Aitken SN (2019) Somatic mutations substantially increase the per-generation mutation rate in the conifer *Picea sitchensis*. Evol. Lett. 3–4:348–358

Hao Z, Cheng T, Zheng R, Xu H, Zhou Y, Li M, Lu F, Dong Y, Chen J, Shi J (2016) The complete chloroplast genome sequence of a relict conifer *Glyptostrobus pensilis*: comparative analysis and insights into dynamics of chloroplast genome rearrangement in Cupressophytes and Pinaceae. PLoS One 11(8): e0161809

Hirao, T; Watanabe A; Kurita M; Kondo T; Takata K (2008) Complete nucleotide sequence of the *Cryptomeria japonica* D. Don. chloroplast genome and comparative chloroplast genomics: diversified genomic structure of coniferous species. BMC Plant Biol 8 (70)

Jackman SD, Warren RL, Gibb EA, Vandervalk BP, Mohamadi H, Chu J, Raymond A, Pleasance S, Coope R et al (2015) Organellar genomes of white spruce (*Picea glauca*): assembly and annotation. Genome Biol Evol 8:29–41. https://doi.org/10.1093/gbe/evv244

Jackman SD, Coombe L, Warren RL, Kirk H, Trinh E, MacLeod T, Pleasance S, Pandoh P, Zhao Y, Coope RJ, Bousquet J, Bohlmann J, Jones SJM, Birol I (2020) Complete mitochondrial genome of a gymnosperm, Sitka spruce (*Picea sitchensis*), indicates a complex physical structure. Genome Biol Evol evaa108. https://doi.org/10.1093/gbe/evaa108

Kang BY, Mann IK, Major JE, Rajora OP (2010) Near-saturated and complete genetic linkage map of black spruce (*Picea mariana*). BMC Genomics 11:515

Li Y, Wang J, Li P, Fan W, Chen Q (2019) The complete chloroplast genome sequence of *Picea neoveitchii* (Pinaceae: Piceoideae). Mitochondrial DNA Part B 4:1, 448–449. https://doi.org/10.1080/23802359.2018.1553528

Lin CP, Huang JP, Wu CS, Hsu CY, Chaw SM (2010) Comparative chloroplast genomics reveals the evolution of Pinaceae genera and subfamilies. Genome Biol Evol 2:504–517

Lin CP, Wu CS, Huang YY, Chaw SM (2012) The complete chloroplast genome of *Ginkgo biloba* reveals the mechanism of inverted repeat contraction. Genome Biol Evol 4(3):374–381. https://doi.org/10.1093/gbe/evs021

Lin D, Coombe L, Jackman SD, Gagalova KK, Warren RL et al (2019) Complete chloroplast genome sequence of an Engelmann spruce (*Picea engelmannii*, genotype Se404-851) from western Canada. Microbiol Resour Announcements 8(24):e00382-19

Lind M, Kallman T, Chen J, Ma XF, Bousquet J, Morgante M, Zaina G, Karlsson B, Elfstrand M, Lascoux M, Stenlid J (2014) A Picea abies linkage map based on SNP markers identifies QTLs for four aspects of resistance to Heterobasidion parviporum infection. PLos one 9(7):e101049

Liu H, Cao F, Yin T, Chen Y (2017) A highly dense genetic map for *Ginkgo biloba* constructed using sequence-based markers. Front Plant Sci 8:1041

Miao J, Tso S, Li J, Xie S, Mao K (2019) The complete chloroplast genome of Juniperus tibetica (Cupressaceae), the conifer that occupies the highest known treeline in the Northern Hemisphere, Mitochondrial DNA Part B, 4(1):609–611. https://doi.org/10.1080/23802359.2018.1561229

Neale DB, Sederoff RR (1989) Paternal inheritance of chloroplast DNA and maternal inheritance of mitochondrial DNA in loblolly pine. Theoret Appl Genet 77:212. https://doi.org/10.1007/BF00266189

Neale DB, Martinez-Garcia PJ, De La Torre AR, Montanari S, Wei X (2017) Novel insights into tree biology and genome evolution as revealed through genomics. Annu Rev Plant Biol 68:13.1–13.27

Nystedt B, Street NR, Wetterbom A, Zuccolo A, Lin YC, Scofield DG, Vezzi F, Delhomme N, Giacomello S, Alexeyenko A et al (2013) The Norway spruce genome sequence and conifer genome evolution. Nature 497:579–584

Palme AE, Pyhajarvi T, Wachowiak W, Savolainen O (2009) Selection on nuclear genes in a Pinus phylogeny. Mol Biol Evol 26(4):893–905

Palmer JD, Stein DB (1986) Conservation of chloroplast genome structure among vascular plants. Curr Genet 10(11):823–833

Pavy N, Pelgas B, Beauseigle S et al (2008) Enhancing genetic mapping of complex genomes through the design of highly-multiplexed SNP arrays: application to the large and unsequenced genomes of white spruce and black spruce. BMC Genomics 9:21

Pavy N, Pelgas B, Laroche J, Rigault P, Isabel N, Bousquet J (2012) A spruce gene map infers ancient plant genome reshuffling and subsequent slow evolution in the gymnosperm lineage leading to extant conifers. BMC Biol 10:84

Pavy N, Lamothe M, Pelgas B, Gagnon F, Birol I, Bohlman J, Mackay J, Isabel N, Bousquet J (2017) A high-resolution reference genetic map positioning 8.8 K genes for the conifer white spruce: structural genomics implications and correspondence with physical distance. Plant J 90:189–213

Pelgas B, Bousquet J, Meirmans PG, Ritland K, Isabel N (2011) QTL mapping in white spruce: gene maps and genomic regions underly- ing adaptive traits across pedigrees, years and environments. BMC Genomics 12:145

Prunier J, Verta JP, MacKay JJ (2016) Conifer genomics and adaptation: at the crossroads of genetic diversity and genome function. New Phytol 209:44–62

Prunier J, Caron S, MacKay JJ (2017a) CNVs into the wild: screening the genomes of conifer trees (*Picea spp.*) reveals fewer gene copy number variations in hybrids and links to adaptation. BMC Genomics 18:97

Prunier J, Caron S, Lamothe M, Blais S, Bousquet J, Isabel N, MacKay J (2017b) Gene copy number variations in adaptive evolution: the genomic distribution of gen copy number variation revealed by genetic mapping and their adaptive role in an undomesticated species, white spruce (*Picea glauca*). Mol. Ecol. 26:5989–6001

Richardson AO, Rice DW, Young GJ, Alverson AJ, Palmer JD (2013) The "fossilized" mitochondrial genome of *Liriodendron tulipifera*: ancestral gene content and order, ancestral editing sites, and extraordinarily low mutation rate. BMC Biol 11:29

Smith DR (2015) Mutation rates in plastid genomes: they are lower than you might think. Genome Biology and Evol 7(5):1227–1234. https://doi.org/10.1093/gbe/evv069

Strauss SH, Palmer JD, Howe GT, Doerksen AH (1988) Chloroplast genomes of two conifers lack a large inverted repeat and are extensively rearranged. Proc Natl Acad Sci 85(11):3898–3902. https://doi.org/10.1073/pnas.85.11.3898

Sugiura M (1992) The chloroplast genome. Plant Mol Biol 19(1):149–168. Epub 1992/05/01. PMID: 1600166

Sullivan AR, Eldfjell Y, Schiffthaler B, Delhomme N, Asp T, Hebelstrup KH, Keech O, Oberg L, Moller IM, Arvestad L, Street NR, Wang XR (2019) The mitogenome of Norway spruce and a reappraisal of mitochondrial recombination in plants. http://dx.doi.org/10.1101/682732

Tomonori H, Watanabe A, Kurita M, Kondo T, Takata K (2008) Complete nucleotide sequence of the Cryptomeria japonica D. Don. Chloroplast genome and comparative chloroplast genomics: diversified genomic structure of coniferous species. BMC Plant Biol 8:70

Tsudzuki J, Nakashima K, Tsudzuki T, Hiratsuka J, Shibata M, Wakasugi T, et al (1992) Chloroplast DNA of black pine retains a residual inverted repeat lacking rRNA genes: nucleotide sequences of trnQ, trnK, psbA, trnI and trnH and the absence of rps16. Mol Gen Genet 232(2):206–214. Epub 1992/03/ 01. PMID: 1557027

Yakovlev IA, Asante DKA, Fossdal CG, Junttila O, Johnsen Ø (2011) Differential gene expression related to an epigenetic memory affecting climatic adaptation in Norway spruce. Plant Sci 180:132–139

Yakovlev IA, Carneros E, Lee YK, Olsen JE, Fossdal CG (2016) Transcriptional profiling of epigenetic regulators in somatic embryos during temperature induced formation of an epigenetic memory in Norway spruce. Planta 243:1237–1249

Wang B, Wang X-R (2014) Mitochondrial DNA capture and divergence in Pinus provide new insights into the evolution of the genus. Mol Phylogenet Evol 80:20–30, 783. https://doi.org/10.1016/j.ympev.2014.07.014

Warren RL, Keeling CI, Yuen MMS, Raymond A, Taylor GA, Vandervalk BP et al (2015) The improved white spruce (*Picea glauca*) genome assemblies and annotation of large gene families of conifer terpenoid and phenolic defense metabolism. Plant J 83:189–212

Wolfe KH, Li WH, Sharp PM (1987) Rates of nucleotide substitution vary greatly among plant mitochondrial, chloroplast, and nuclear DNAs. Proc Natl Acad Sci USA 84:9054–9058

Wu CS, Wang YN, Liu SM, Chaw SM (2007) Chloroplast genome (cpDNA) of *Cycas taitungensis* and 56 cp protein-coding genes of *Gnetum parvifolium*: insights into cp DNA evolution and phylogeny of extant seed plants. Mol Biol Evol 24:1366–1379. https://doi.org/10.1093/molbev/msm059

Wu CS, Wang YN, Hsu CY, Lin CP, Chaw SM (2011a) Loss of different inverted repeat copies from the chloroplast genomes of Pinaceae and Cupressophytes and influence of heterotachy on the evaluation of gymnosperm phylogeny. Genome Biol Evol 3:1284–1295. https://doi.org/10.1093/gbe/evr095

Wu CS, Lin CP, Hsu CY, Wang RJ, Chaw SM (2011b) Comparative chloroplast genomes of pinaceae: insights into the mechanism of diversified genomic organizations. Genome Biol Evol 3: 309–319. https://doi.org/10.1093/gbe/evr026

Wu CS, Chaw SM (2016) Large-scale comparative analysis reveals the mechanisms driving plastomic compaction, reduction, and inversions in conifers II (*Cupressophytes*). Genome Biology and Evolution 8 (12):3740–3750

Yang JC, Joo M, So S, Yi DK et al (2016) The complete plastid genome sequence of Picea jezoensis (*Pinaceae: Piceoideae*). Mitochondrial DNA Part A 27 (5):3761–3763

Zang M, Su Q, Weng Y, Lu L, Zheng X, Ye D, Zheng R et al (2019) Complete chloroplast genome of Fokienia hodginsii (Dunn) Henry et Thomas: Insghts into repeat regions variation and phylogenetic relationships in Cupressophyta. Forests 10:528

Zhu A, Guo W, Jain K, Mower JP (2014) Unprecedented heterogeneity in the synonymous substitution rate within a plant genome. Mol Biol Evol 31:1228–1236

Zimmerman HH, Harms L, Epp LS, Mewes N, Bernhardt S et al (2019) Chloroplast and mitochondrial genetic variation of larches at the Siberian tundra-taiga ecotone revealed by de novo assembly. PLoS One 14 (7):e0216966

Back to BACs: Conifer Genome Exploration with Bacterial Artificial Chromosomes

7

Kermit Ritland, Nima Farzaneh, Claire Cullis, Agnes Yuen, Michelle Tang, Joël Fillon, Sarah Chao, Daniel G. Peterson, and Carol Ritland

Abstract

We sought to ascertain the structure of spruce and pine genomes via sequencing of randomly selected bacterial artificial chromosomes (BACs) from both white spruce and loblolly pine. In the first genomic comparison of spruce and pine for its time, using 454 sequencing, we compared 96 random BACs from each species. Thirty-three (spruce) and 24 (pine) genes were found in the BACs that spanned 8.7 MB (spruce) and 10.3 MB (pine), respectively; greater numbers of pseudogenes were also found. The gene numbers found in BACs gives conflicting estimates about the true numbers of genes in the genome (46K in pine, 75K in spruce). There were ten times as many retrotransposons as genes, and clustering of inverted repeats showed transposition may have peaked in both species about 50 million years ago. Dotplots of BACs show interesting repeat structures in these conifers. A phylogeny of retrotransposons shows recent periods of expansion in both species, and a general peak of activity ca. 50 million years ago. Besides providing a snapshot of conifer genome structure and evolution, BACs can aid in the assembly of conifer genomes.

7.1 Introduction

BACs are "bacterial artificial chromosomes," and they contain inserts of roughly 150,000 base pairs of the genome of interest. They can be easily shotgun sequenced due to their relatively small size, and provide a snapshot about the nature of a genome. They have been of particular use for genomes of large size which, until the advent of next-generation sequencing, have been difficult to study. Regardless of such technical advances, we can look backward to existing data and recover valuable insights about genomes.

Earlier examples of using BACs for large genomes include corn with the analysis of the 74 random maize BACs; maize gene number was estimated at 37,000 but 30% of BACs contained no genes, and also the gene density varied from 0 to 7.9 per 100 kb (Liu et al. 2007). Earlier, Haberer et al. (2005) characterized a sample of four maize BACs. By comparing the two ends of a retrotransposon (which are identical at insertion), the date of transposition can be inferred by the sequence divergence. They inferred that two major bursts of retrotransposon activity in the last

K. Ritland (✉) · N. Farzaneh · C. Cullis · A. Yuen · M. Tang · J. Fillon · S. Chao · C. Ritland
Department of Forest and Conservation Sciences, University of British Columbia, Vancouver, BC, Canada
e-mail: kermit.ritland@ubc.ca

D. G. Peterson
Institute for Genomics, Biocomputing & Biotechnology, Mississippi State University, Starkville, MS, USA

I. M. Porth and A. R. De la Torre (eds.), *The Spruce Genome*, Compendium of Plant Genomes,
https://doi.org/10.1007/978-3-030-21001-4_7

2 million years are responsible for the large maize genome size. These results also show that sequencing and annotation of a few randomly selected BACs can provide a robust description of a complex plant genome.

The most abundant subclass of transposable elements in plants are Long Terminal Repeat (LTR) retrotransposons and two superfamilies are present in plants: copia and gypsy (Zuccolo et al. 2015). TEs were long dismissed as junk DNA but have since been recognized as important players in shaping eukaryotic genome structure and evolution, and can accumulate by waves of sustained activity over sometimes short evolutionary times, as in rice and the maize example above.

The "pooling" of BACs has also been thought to be one route to genome sequencing. This was discussed in a special symposium, "International Conifer Genome Workshop" at Cold Springs Harbor Laboratories in March 2007. One justification for our test of pooled BAC sequencing was that sequencing of pooled BACs might be a better way to obtain reads that are easily assembled, as the sequence length in any pool of BACs is much shorter than the genome sequence. However, as shown by the recently published spruce genomes for white and Norway spruces, pure shotgun sequencing across numerous libraries, together with new bioinformatics approaches for genome assembly, is the route for genome sequencing, as shown by publication of both the Norway (Nystedt et al. 2013) and white (Birol et al. 2013) spruce sequences. Nevertheless, BAC comparisons are still revealing in conifers (Kovach et al. 2010; Perera et al. 2018).

7.2 BAC DNA Isolation

In both white spruce and loblolly pine, there were available BAC libraries of sizes 1.2 million clones for spruce (Hamberger et al. 2009) and 1.8 million clones for pine (Magbanua et al. 2011). To obtain 96 BACs separately for loblolly pine and white spruce, in the size range of 97–120 KB, from these libraries we screened a total of 502 clones in spruce and 408 clones in pine. We then either grew BAC clones individually (in 3-ml culture tubes) or pooled several clones (in 40-ml culture tubes), combined 4–10 single colonies for DNA isolation using either Sigma Phaseprep BAC DNA kit/Plasmid-Safe ATP-Dependent DNase from Epicentre or QIAGEN Large Construct Kit (the latter giving higher yield and quality).

7.3 Random BAC Sequencing

Initially, we proposed using Sanger sequencing to sequence 64 randomly selected BACs to about 4X coverage. When the proposal for sequencing BACs was submitted, 454 sequencing technology yielded 200-bp fragments, but with current levels of 400 bp and higher, it was reasonable for de novo shotgun sequencing at much higher coverage. With the 454, we performed both single reads and paired-end reads, depending on the preparation of the pool templates for sequencing.

Table 7.1 shows the various pool sizes, read directions, and coverage of the pooled BACs that we sequenced. Pool sizes were initially small (4–16) for testing reasons, then pools increased in size to 40 and even 48. Initially, single reads were performed, but then paired-end reads were obtained. Assemblies were performed by Mira for unpaired-end reads, and by Newbler for paired-end reads. Genes were identified by highly significant BLASTs to known genes in plants, and pseudogenes were identified by the presence of a stop codon within the gene.

Table 7.2 gives the numbers of genes, pseudogenes, and retroelements identified among the 10 Mb of BAC sequence data generated for each species. We, respectively, found 24 (pine) and 33 (spruce) functional genes, based upon highly significant protein BLASTs to Genbank. This contrasts to the ten genes (all targeted) with high support in 100 BACs found by Perera et al. (2018); this is a minimum estimate as their clones were not entirely random (some were targeted for specific genes, others were screened for fewer repeats), and some of the 144 other candidate genes they identified may be protein-coding genes, and not pseudogenes.

While Table 7.2 seems to indicate the pine genome might have more coding genes than

Table 7.1 Numbers of pooled BACs, average contig size, largest contig size, and fold coverage of the 454 sequencing

	No. pooled BACs	Average size	Largest size	Coverage
Initial attempts, single-end reads				
Pine	16	29,635	109,089	18X
Pine	8	37,670	105,604	72X
Spruce	4	31,345	73,254	56X
Spruce	8	26,408	79,831	34X
Production runs, unpaired-end reads				
Pine	40	21,052	83,255	25X
Spruce	40	23,768	87,123	24X
Production runs, paired-end reads				
Pine	40	39,613	12,112	42X
Spruce	48	27,825	95,292	34X

Table 7.2 Numbers of genes, pseudogenes, and retroelements found in the pine and spruce BACs, and their density, in the 454 sequencing

	Loblolly pine	White spruce
Number of bases sequenced	10,350,783	8,743,487
Identified genes	72	69
Functional genes	24	33
Pseudogenes	48	36
Retroelements	860	689
Nucleotides/gene	431,283	264,954
Nucleotides/retroelement	12,036	12,690

spruce, the difference of 33 versus 24 is confounded by a number of sampling biases and errors, and their values are only first indicators of genome number, and mainly illustrate the limitations of BAC comparisons. Likewise, the estimates of number of pseudogenes in each species are subject to unknown biases and errors. The presence of a pseudogene was inferred by a stop codon, which is standard in the literature, but genome annotation errors may also contribute an overestimation of pseudogene number. Again, this BAC data are limited, and any apparent differences of pseudogene number are more likely due to these sources of error.

In Table 7.2, the number of pseudogenes was about equal to the number of functional coding genes (pseudogenes were identified by the presence of a stop codon within the gene). The number of retroelements (elements were found with the software ltr_finder) was very large, ca. 700–800, ten times the number of other genes. From Table 7.2, the number of nucleotides per gene (e.g., number of bases on average between genes) was 431K in pine, and 265K in spruce. We can estimate the gene number in spruce and pine as 20 Gb/431K = 46,500 in pine and 20 Gb/265K = 75,500 in spruce. These are probably overestimates, but suggest, at face value, that spruce has a more complex genome than pine. There is no reason to expect this, so as above, we defer any conclusions to the biases and errors incurred by BAC sampling and sequencing.

Figure 7.1 gives the distribution of contigs of size 10 Kb and larger. Contigs of at least 10 kb in size would be expected to give reliable annotations of genes as most genes and transposons are much shorter than 10 Kb. Contig sizes were almost double the size for pair-end reads, as compared to single-end reads. With pair-end reads, we also achieved greater numbers with white spruce, as compared to loblolly pine.

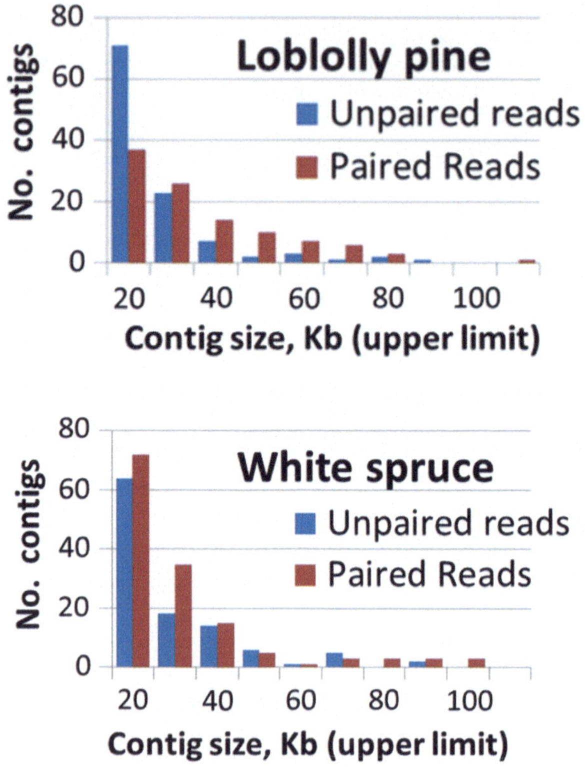

Fig. 7.1 Distribution of contig sizes in assembled BACs, for (1) loblolly pine versus white spruce and for (2) unpaired reads versus paired reads

Figure 7.2 gives two examples of dotplots of BACs with themselves. The first dotplot is rather typical for the spruce and pine BACs; there are moderate numbers of tandem repeats (segments parallel to the diagonal line) and also an inverted repeat (segment perpendicular to the diagonal); moderate and lowly repetitive DNA can be seen as a cloud of stars and galaxies like the night sky. The second dotplot is an extreme example of the repetitive DNA that can exist (here, probably at the centromere). Such bizarre patterns were sometimes observed at a smaller scale in the other BACs.

7.4 Transposable Element Activity

Figure 7.3 gives the distribution of inverted repeat similarity, separately for pine versus spruce species, and separately for copia versus gypsy elements. The frequency of similarity is plotted downward from the value of 1.0 (complete similarity) as this corresponds to a repeat of most recent origin. Through time, mutations accumulate within an inverted repeat, and the similarity declines. Hence, the numbers of transpositions at a given similarity level, corresponds to the number of transposons that originated during a certain time, which can be estimated with a molecular clock.

In Fig. 7.3, as we move to the right, it is apparent that for both spruce and pine, that there was a peak of transposition activity, corresponding to about 5% divergence (95% identity) for the inverted repeats. The peak for both species is about the same, and also, for both species the frequency of transposition has decreased since that peak. In the other direction of time, there is also a reduction in the frequency of inferred transposition to the right of that peak, but this is probably a statistical bias caused by omission of more non-significant similarities as overall similarity decreases.

At least we can attempt to date this peak of transposition activity. We can date the timing of transposition events assuming a molecular clock

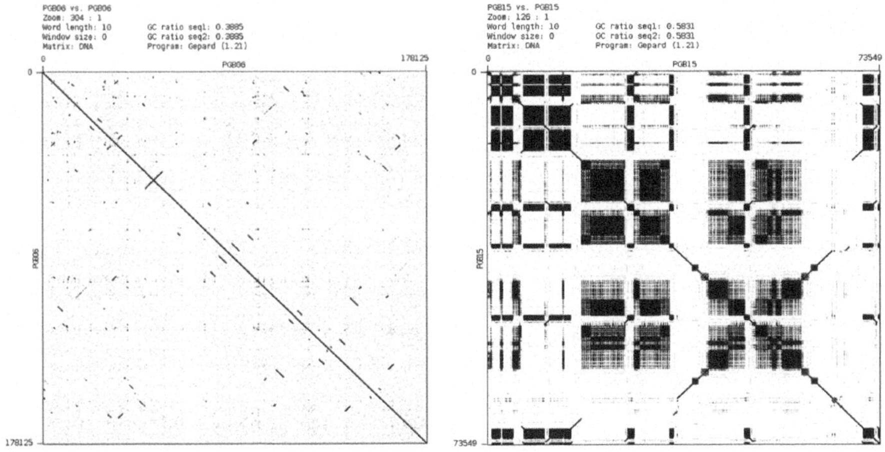

Fig. 7.2 Dotplots of two BACs, each with itself

Fig. 7.3 Distributions of inverted repeat similarities, separately for loblolly pine and white spruce. More dissimilar repeats to the right represent more ancient transpositions, and 95% similarity corresponds to about 50 million years ago

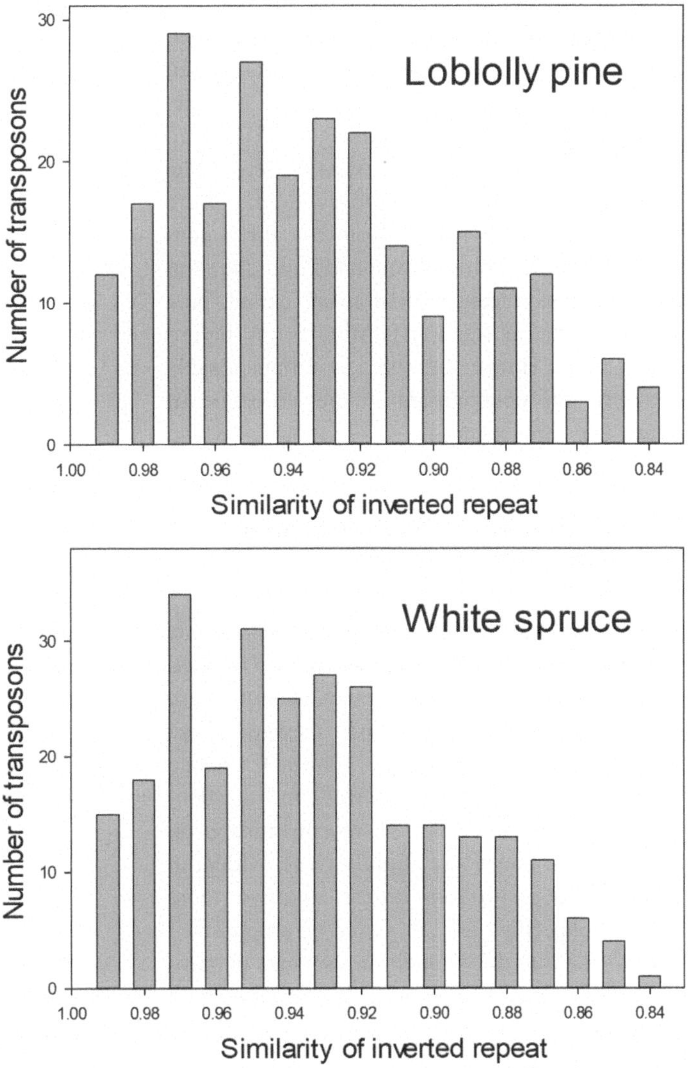

(the inverted repeats at the two ends are identical at the time of transposition). If the observed fraction of divergence between two sequences is d, then at a base substitution rate of r per site per year, then for low total divergence when multiple substitutions can be neglected. $d = rt$, where t is the number of years. As discussed in Buschiazzo, Ritland, Bohlmann, and Ritland (2012), a good estimate for mutation rate for TE elements is 1.0×10^{-9}, so this gives an estimate of $t = d/r = (0.05)(10^9) = 50$ million years at the point of 5% divergence. The divergence of spruce from pine was about 140 million years ago (Ran et al. 2006), or even just 128 MYA (Eckert and Hall 2006). This would correspond to the 14–15% divergence (0.86 on far right), so most of the two histograms document evolution in independent lineages.

Interestingly, this is near the Paleocene–Eocene Thermal Maximum (55.8 MYA). Based on the information on plate tectonics, climate, fossils, and biogeography, Millar (1993) argued that the Eocene (56–35 MYA) was one of the most important phases in pine evolution. More recent molecular evidence from the three genomes (nuclear, chloroplast, and mitochondia) also indicates recent radiation of extant pine sections later, in the Early Miocene (Willyard et al. 2007). Hence the spread of transposable elements was concomitant in the evolution of spruce and pine species.

7.5 Phylogeny of Transposable Elements

Figure 7.4 shows the phylogeny of transposable elements, separately for spruce (with 81 transposons) and pine (with 94 transposons). These were elements whose paired ends showed at least 80% similarity. In this figure, ClustalV created the dendrogram. The larger arrows in both dendrograms indicate bouts of recent transposon activity. The "bouts" are slightly more evident in loblolly pine, with four recent and two more ancient bouts (Fig. 7.4). It should be emphasized that this is quite different from a species dendrogram. Here, the clusters indicate multiplication of taxa, almost like an infectious disease, as opposed to the splitting of taxa, which are like branches falling off a tree.

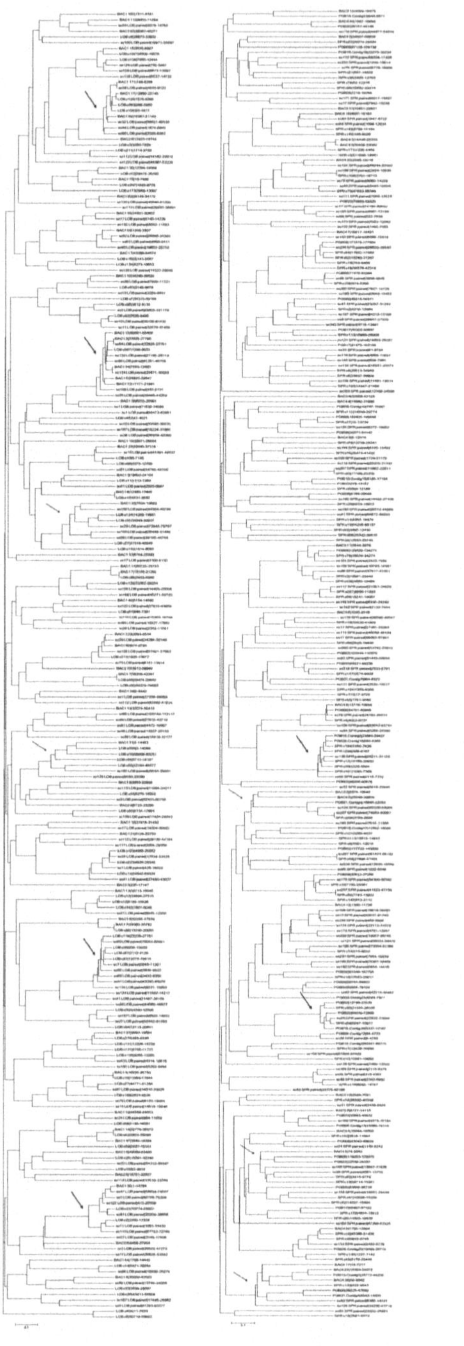

Fig. 7.4 Phylogenies of transposons within loblolly pine (left) and white spruce (right). Arrows indicate some cases of transposon activity (larger arrows are more recent)

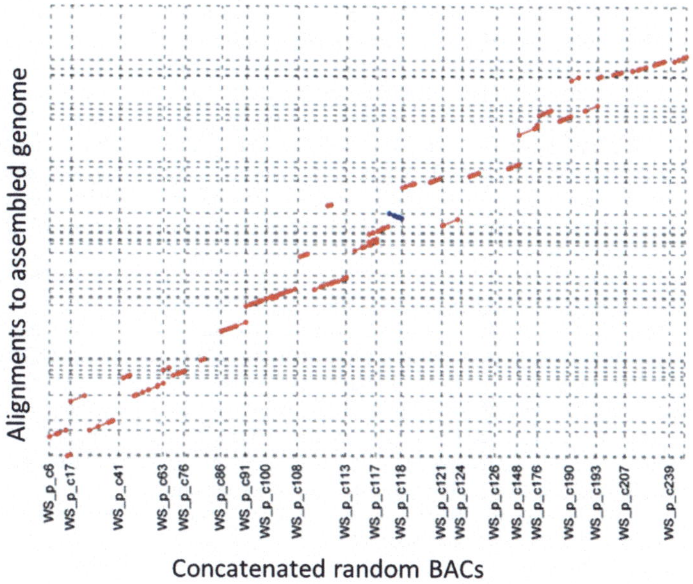

Fig. 7.5 MUMMER alignment of a concatenation of 21 BACs against the assembled spruce genome

7.6 Alignment of BACs with the Spruce Genome Sequence

Figure 7.5 shows a MUMMER alignment of a concatenation of 21 BACs against the assembled spruce genome (Birol et al. 2013). A concatenation is when BACs are arrayed end-at-end without regard to order. The alignment is excellent, the contrasting blue line is an inversion, ordering is intact in the rest of the alignment except for gaps. In general, these BAC sequences can serve as "quality controls" for shotgun assemblies of spruce and pine.

Acknowledgments This work was primarily funded by a Science Opportunity Fund grant "Exploration of conifer genomes" from Genome British Columbia to KR. Our co-investigators from Arborea (Genome Quebec) and Mississippi Genome Exploration Laboratory (MGEL) assisted with our objectives. The Genome Quebec project, "Pilot studies and resource development for Conifer Genome Sequencing; genome characterization and gene space," served as our co-funding and their study had complementary activities.

References

Birol I, Raymond A, Jackman S, Pleasance S, Coope R, Taylor G, Yuen MMS, Keeling CI, Brand D, Vandervalk BP, Kirk H, Pandoh P, Moore RA, Zhao Y, Mungall AJ, Jaquish B, Yanchuk A, Ritland C, Boyle B, Bousquet J, Ritland K, Mackay J, Bohlmann J, Jones SJM (2013) Assembling the 20 Gb white spruce (Picea glauca) genome from whole-genome shotgun sequencing data, vol 29

Buschiazzo E, Ritland C, Bohlmann J, Ritland K (2012) Slow but not low: genomic comparisons reveal slower evolutionary rate and higher dN/dS in conifers compared to angiosperms. BMC Evol Biol 12:8. https://doi.org/10.1186/1471-2148-12-8

Eckert AJ, Hall BD (2006) Phylogeny, historical biogeography, and patterns of diversification for Pinus (Pinaceae): phylogenetic tests of fossil-based hypotheses. Mol Phylogenet Evol 40(1):166–182. https://doi.org/10.1016/j.ympev.2006.03.009

Haberer G, Young S, Bharti AK, Gundlach H, Raymond C, Fuks G, Butler E, Wing RA, Rounsley S, Birren B, Nusbaum C, Mayer KFX, Messing J (2005) Structure and architecture of the maize genome. Plant Physiol 139(4):1612–1624. https://doi.org/10.1104/pp.105.068718

Hamberger B, Hall D, Yuen M, Oddy C, Hamberger B, Keeling CI, Ritland C, Ritland K, Bohlmann J (2009) Targeted isolation, sequence assembly and characterization of two white spruce (Picea glauca) BAC clones for terpenoid synthase and cytochrome P450 genes involved in conifer defence reveal insights into a conifer genome. BMC Plant Biol 9:106. https://doi.org/10.1186/1471-2229-9-106

Kovach A, Wegrzyn JL, Parra G, Holt C, Bruening GE, Loopstra CA, Hartigan J, Yandell M, Langley CH, Korf I, Neale DB (2010) The Pinus taeda genome is characterized by diverse and highly diverged repetitive sequences. BMC Genomics 11(1):420. https://doi.org/10.1186/1471-2164-11-420

Liu R, Vitte C, Ma J, Mahama AA, Dhliwayo T, Lee M, Bennetzen JL (2007) A GeneTrek analysis of the maize genome. Proc Natl Acad Sci USA 104(28):11844–11849. https://doi.org/10.1073/pnas.0704258104

Magbanua ZV, Ozkan S, Bartlett BD, Chouvarine P, Saski CA, Liston A, Cronn RC, Nelson CD, Peterson DG (2011) Adventures in the enormous: a 1.8 Million Clone BAC Library for the 21.7 Gb Genome of Loblolly Pine. PLOS ONE 6(1):e16214. https://doi.org/10.1371/journal.pone.0016214

Millar CI (1993) Impact of the eocene on the evolution of Pinus L. Ann Mo Bot Gard 80(2):471–498. https://doi.org/10.2307/2399795

Nystedt B, Street N, Wetterbom A, Zuccolo A, Lin Y-C, Scofield DG, Vezzi F, Delhomme N, Giacomello S, Alexeyenko A, Vicedomini R, Sahlin K, Sherwood E, Elfstrand M, Gramzow L, Holmberg K, Hällman J, Keech O, Klasson L, Koriabine M, Kucukoglu M, Käller M, Luthman J, Lysholm F, Niittylä T, Olson A, Rilakovic N, Ritland C, Rosselló JA, Sena J, Svensson T, Talavera-López C, Theißen G, Tuominen H, Vanneste K, Wu Z-Q, Zhang B, Zerbe P, Arvestad L, Bhalerao R, Bohlmann J, Bousquet J, Gil RG, Hvidsten TR, de Jong P, MacKay J, Morgante M, Ritland K, Sundberg B, Lee Thompson S, Van de Peer Y, Andersson B, Nilsson O, Ingvarsson PK, Lundeberg J, Jansson S (2013). The Norway spruce genome sequence and conifer genome evolution. Nature 497

Perera D, Magbanua ZV, Thummasuwan S, Mukherjee D, Arick II M, Chouvarine P, Nairn CJ, Schmutz J, Grimwood J, Dean JF, Peterson DG (2018) Exploring the loblolly pine (Pinus taeda L.) genome by BAC sequencing and Cot analysis. Gene 663:165–177. https://doi.org/10.1016/j.gene.2018.04.024

Ran J-H, Wei X-X, Wang X-Q (2006) Molecular phylogeny and biogeography of Picea (Pinaceae): implications for phylogeographical studies using cytoplasmic haplotypes. Mol Phylogenet Evol 41(2):405–419. https://doi.org/10.1016/j.ympev.2006.05.039

Willyard A, Syring J, Gernandt DS, Liston A, Cronn R (2007) Fossil calibration of molecular divergence infers a moderate mutation rate and recent radiations for Pinus. Mol Biol Evol 24(1):90–101. https://doi.org/10.1093/molbev/msl131

Zuccolo A, Scofield DG, De Paoli E, Morgante M (2015) The Ty1-copia LTR retroelement family PARTC is highly conserved in conifers over 200 MY of evolution. Gene 568(1):89–99. https://doi.org/10.1016/j.gene.2015.05.028

Genomic Selection in Canadian Spruces

8

Yousry A. El-Kassaby, Blaise Ratcliffe, Omnia Gamal El-Dien, Shuzhen Sun, Charles Chen, Eduardo P. Cappa, and Ilga M. Porth

Abstract

The genetic gain of spruce (*Picea spp.*) breeding programs is impeded by long recurrent selection cycles stemming from biological constraints such as late expression of traits, weak juvenile mature correlations, and late onset of sexual maturity. Genomic selection (GS) is capable of addressing these barriers to improving the rate of genetic gain via early prediction of phenotypes using dense genetic marker arrays. Results from GS studies focused on spruce species in Canada thus far have produced encouraging results to capture additional genetic gain for wood quality, growth, and insect resistance traits either through the re-analysis of existing progeny trials with genomic information or via prediction of phenotypes for untested candidate trees. With the continual improvement of phenotyping technologies and spruce genomic resources, we expect the capability of GS to capture genetic gain to greatly exceed that of traditional pedigree-based selection methods in the future.

Y. A. El-Kassaby (✉) · B. Ratcliffe · O. G. El-Dien · S. Sun
Department of Forest and Conservation Sciences, Faculty of Forestry, The University of British Columbia, 2424 Main Mall, Vancouver, BC V6T 1Z4, Canada
e-mail: y.el-kassaby@ubc.ca

B. Ratcliffe
e-mail: blaise.ratcliffe@ubc.ca

O. G. El-Dien
e-mail: omnia.gamal@alumni.ubc.ca

S. Sun · C. Chen
Department of Biochemistry and Molecular Biology, Oklahoma State University, Stillwater, OK 74078-3035, USA
e-mail: charles.chen@okstate.edu

E. P. Cappa
Instituto de Recursos Biologicos, Centro de Investigacion en Resursos Naturales, Instituto Nacional de Tecnologis Agropecuaria (IINTA), De Los Reseros y, Buenos Aires, Argentina
e-mail: cappa.eduardo@inta.gob.ar

I. M. Porth
Départment des Sciences du Bois et de la Forêt, Faculté de Foresterie, de Géographie et Géomatique, Université Laval, Quebec City, QC G1V 0A6, Canada
e-mail: ilga.porth@sbf.ulaval.ca

8.1 Introduction

Spruce tree breeding programs, by their nature, are long-term enterprises demanding continuous financial, technical, and logistical support. Most of these programs' traits of interest (i.e., economically important such as wood quality and tree volume) are challenging due to their late phenotypic and their polygenic architecture. A common feature among tree breeding programs is the adoption of the recurrent selection scheme which consists of repeated cycles of

I. M. Porth and A. R. De la Torre (eds.), *The Spruce Genome*, Compendium of Plant Genomes,
https://doi.org/10.1007/978-3-030-21001-4_8

selection, breeding, and testing (El-Kassaby 1995; White et al. 2007). The time span needed for the successful completion of one recurrent selection cycle is exceedingly long and could extend to multiple decades as each phase encounters its own specific challenges (El-Kassaby 1995; El-Kassaby and Lstiburek 2009). For example, the initial selection phase covers large breeding zone's geography (Ying and Yanchuk 2006), the breeding phase usually encounters issues related to sexual maturity (e.g., time to reproduction, fertility variation, and reproductive asynchrony Funda and El-Kassaby 2012), and finally, the testing phase involves thousands of individuals planted over multiple locations, covering vast geographic territories. Therefore, the development of shortcuts to circumvent these protracted breeding cycle's phases are of great practical and economical values.

The substantial advances of DNA fingerprinting, including sequencing technologies, has provided tree breeders with the genetic markers needed to augment the recurrent selection cycle. These include: quantitative trait loci (QTL) identification (El-Kassaby 1982; Sewell et al. 2000, 2002; Brown et al. 2003), pedigree reconstruction (El-Kassaby and Lstiburek 2009; Doerksen and Herbinger 2010; El-Kassaby et al. 2011), genome-wide associations (Porth et al. 2013; La Mantia et al. 2013; McKown et al. 2014, 2018), and, genomic selection (Meuwissen et al. 2001). Regardless of how successful and/or practical the former methods have been, genomic selection (GS) by far has been prominent in revolutionizing the genetic improvement of the dairy industry (VanRaden et al. 2009), thus attracting the attention of virtually all plant and animal improvement programs. It is noteworthy that the success of GS in dairy cattle became possible due to the organized cooperative structure of this industry, features difficult to match in other phenotypic improvement programs. All dairy genetic improvement programs are dependent on a small number of breeding bulls as the unit of selection, based on artificial insemination for siring as many heifers as the farmers allow, provided unique improvement circumstances. Additionally, the dairy industry is known for its meticulous record-keeping including pedigree and production. This situation is unique and cannot be generalized to either other livestock or plant breeding programs, as these programs are not centered on few genetically elite males (i.e., small effective population size). Additionally, in plants their development requires testing over multiple environments, introducing further complications (e.g., genotype × environment interaction).

Unlike traditional genetic evaluation methods that categorically rely on either: (1) phenotypic data and pedigree information for determining breeding values (Henderson 1976) or (2) marker-QTL associations identifying "major" gene effects (Buerstmayr et al. 2009), GS provides simultaneous estimation of the collective effect of thousands of loci throughout the genome, irrespective of effect size, for the trait of interest. Additionally, these many markers can precisely measure the degree of genetic similarity among individuals (Meuwissen 2007), and thus, introduces a new paradigm as the model unit of quantitative genetics evaluation has shifted from the line of decent (i.e., pedigree) to the presence or absence of specific alleles. The presence or absence of specific alleles and their subsequent effect on individuals' breeding values have a major appealing feature as it shifts the focus from phenotyping (testing) to genotyping (prediction). This drastic shift, if real and reproducible, would permit bypassing of progeny testing, a major component of the recurrent selection cycle. Thus, the time needed for selection is reduced, accelerating the recurrent selection cycle's turnover, resulting in maximizing genetic gain per unit time and cost.

Genomic selection, while applying the same genetic principles, encompasses two major aspects: namely, components of genetic variance estimation and genetic prediction. In this chapter, we will focus on these two aspects using studies conducted on four spruces grown commercially in Canada: black (*Picea mariana*) (Lenz et al. 2017), white (*Picea glauca*) (Beaulieu et al. 2014a, b, Gamal El-Dien et al. 2016; Ratcliffe et al. 2017), interior (*Picea engelmannii x glauca*) (Gamal El-Dien et al. 2015, 2018; Ratcliffe et al. 2015), and Norway spruce (*Picea abies*) (Lenz et al. 2019). These studies utilized different genotyping platforms for accessing SNP information including, high-density SNP arrays (white and Norway spruce: Pavy et al. 2016; Lenz et al. 2019), exome capture (black spruce: Lenz et al. 2017), and genotyping-by-sequencing (GBS) (interior spruce: Gamal El-Dien et al. 2015). Due to the unordered and random nature of GBS and the commonly observed missing data, a specific section on missing data imputation is included in this chapter.

The principle behind GS hinges on the assumption that the SNP markers are in linkage disequilibrium (LD) with loci controlling the trait of interest; thus, the extent of LD is of major importance as it is dependent on many factors specific to each species' breeding system and the breeding populations' generation numbers and mating scheme. This brings the issue of genome size complexity (De La Torre et al. 2019) to the forefront of GS's success in spruces, and conifers in general, and the question of how many molecular markers are needed to successfully capture LD with a trait's causal genes (Thistlethwaite et al. 2017, 2019).

Wood quality attributes are ideal candidates for GS due to their late expression and the fact that they are costly and laborious to determine, so finding indirect determination methods are of great value. So far, GS is the only viable option for this purpose as methods relying on the classical correlated response between wood quality attributes and early, easy-to-measure traits have not been developed (Falconer and MacKay 1969). Wood density (g cm^{-3}) is the most important intrinsic characteristic of wood quality as it directly influences flexural wood properties (Evans et al. 2000; Andrews 2002), including bending strength and stiffness. Therefore, wood density has been one of the main selection criteria in most tree-breeding programs (Zobel and Talbert 1984). For spruces, wood density was determined using X-ray densitometry analysis (interior, black, and Norway spruces: Gamal El-Dien et al. 2015; Lenz et al. 2017, 2019) and the SilviScan Technology (white spruce: Beaulieu et al. 2014a).

8.2 Genetic Components of Variance Estimation

The genetic evaluation and ranking of parents and offspring, in virtually all tree improvement programs, are based on the intraclass correlation defined by a structured pedigree (Falconer and Mackay 1969). Forest trees' structured pedigree ranges between the simple open-pollinated families and the exceedingly complicated multigenerational breeding (Askew and El-Kassaby 1994). These genealogical relationships among parents and offspring (i.e., pedigree-based relationship (***A***-matrix: Wright 1922) form the backbone of quantitative genetic analyses (Xu 2003). The best linear unbiased predictor (BLUP: Henderson 1976) using the ***A***-matrix is the analysis of choice in these cases, known as ABLUP. However, the ABLUP method has its own limitations even under the simplest pedigree family structures (e.g., half- and full-sib), as it fails to accurately differentiate within family genetic variation. Under the ABLUP approach, all members of a half-sib or a full-sib family are given a constant expected relationship estimate (average numerator relationship) without any consideration to the random segregation of alleles from the parents, known as the Mendelian sampling term (Hill and Weir 2011). Furthermore, the use of the average numerator relationships, specifically for the relatively undomesticated nature of forest tree breeding programs, is incapable of detecting unknown genetic population structure relationships and/or cryptic inbreeding, resulting in an overestimation of the additive genetic variance and heritability estimates, leading to exaggerated estimates of genetic gain (Askew and El-Kassaby 1994). With the availability of genomic data, a realized genomic relationship (***G***-matrix; VanRaden 2008) between individuals of any group can be estimated to a degree of great precision.

The realized relationship is based on the actual proportion of alleles shared among individuals, and thus both unknown genetic population structure and cryptic inbreeding can be observed, whereas these relationships remain hidden in the ABLUP analysis. This development (i.e., independence from "structured pedigree") paved the way for the introduction of "pedigree-free-models" (Frentiu et al. 2008) for estimating unconfounded genetic variance components, without the reliance on the creation of a pedigree via structured mating designs. The development of the ***G***-matrix and its use in BLUP analysis (i.e., GBLUP) is a quantitative genetics landmark as most, if not all, the shortcomings of the ABLUP analysis was overcome in one single step. These shortcomings include: (1) dependence of structured pedigree, (2) detection of hidden relationships among individuals, (3) detection of cryptic inbreeding, and (4) production of precise and accurate genetic covariance among relatives (i.e., accounting for the Mendelian sampling term Visscher et al. 2006), and (5) providing opportunities for capturing alleles in LD with QTL (Lippert et al. 2013).

Forest geneticists have taken full advantages of the ***G***-matrix in their quantitative genetic analyses and were successful in estimating: (1) the genetic variance components under "no pedigree" scenario (El-Kassaby et al. 2012; Porth et al. 2013), (2) uncoupling additive and non-additive effects from existing pedigreed progeny test trials (Zapata-Valenzuela et al. 2013; Muñoz et al. 2014; Gamal El-Dien et al. 2016, 2018), and (3) accounting for hidden relationships in a simple open-pollinated progeny testing design (Gamal El-Dien et al. 2016, 2018). Gamal El-Dien et al. (2016, 2018) studies on white and interior spruce's wood density assumed "half-sib" family structure and thus permitted the genetic variance decomposition and the estimation of: (1) more accurate additive genetic variance, (2) non-additive genetic effect estimates: (a) dominance genetic variance and (b) epistatic genetic variance (additive × additive, additive × dominance), and (3) genotype × environment interactions (Table 8.1). Using this approach, realistic and accurate wood density heritability estimates were obtained for white (0.609 vs. 0.179; ABLUP vs. GBLUP) and interior (0.260 vs. 0.240) spruce, resulting in increased reliability of individual tree rankings and genetic gain estimates.

While the introduction of the ***G***-matrix to forest trees' quantitative genetic analyses was exceedingly successful in unravelling hidden relationships, inbreeding, and addressing the shortcomings of the ***A***-matrix, the implementation of this approach is somewhat restrictive, as it requires genotype data for all individuals of a breeding population. However, and without exception, all progeny testing trials involve thousands of genetically related individuals, thus genotyping the entire population for determining its genomic-based relationships is virtually impossible both financially and logistically. An innovative approach, called the single-step

Table 8.1 Illustration of possible estimates of genetic variance components under ***A***-matrix and ***G***-matrix for single (white spruce) and multiple (interior spruce) environments[a]

Single environment			Multiple environments		
S.O.V.	Variance component		S.O.V.	Variance component	
	A-matrix	***G***-matrix		***A***-matrix	***G***-matrix
σ^2_A	X	X	σ^2_A	X	X
σ^2_D		X	σ^2_D		X
σ^2_{AxA}		X	σ^2_{AxA}		X
σ^2_{DxD}		X	σ^2_{DxD}		X
σ^2_{AxD}		X	σ^2_{AxD}		X
			σ^2_{Axe}	X	X
			σ^2_{Dxe}		X
σ^2_e			σ^2_e		
h^2	X	X	h^2	X	X

[a]see Gamal El-Dien et al. (2016, 2018) for exact estimates

method (***H***-matrix), where the genomic information (genotypes) of a subset of individuals can be incorporated with a larger subset of un-genotyped individuals with known pedigree for facilitating the genetic evaluation of progeny testing trials, has been recently proposed (Misztal et al. 2009; Legarra et al. 2009; Christensen and Lund 2010). This approach is very appealing as it permits analysing progeny tests with their very large data at a reasonable cost as well as increased accuracy. The method involves constructing two matrices, an ***A***-matrix with all genotyped and non-genotyped individuals and the ***G***-matrix of the genotyped subset, and combining them to form a blended (***H***-matrix) for use in quantitative genetics analysis (HBLUP) of the entire progeny test. The added advantage of the single-step method is visually depicted (Fig. 8.1) using network representation of the white spruce open-pollinated progeny test data of Beaulieu et al. (2014a) as follows: (1) ***A***-matrix (Fig. 8.1) illustrates the pedigree clustering of every half-sib family member and the lack of connectedness among the various families, (2) ***G***-matrix (Fig. 8.1) illustrates the genomic relationship among genotyped half-sib families with within-family clustering as well as connectedness among related families, and (3) ***H***-matrix (Fig. 8.1) illustrates merging the ***A***- and ***G***-matrices showing within half-sib family clustering as well as increasingly obvious connectedness among members of all genotyped and non-genotyped half-sib families' members. This increased connectedness of the ***H***-matrix (Fig. 8.1) is the essence of the single-step approach as increased connectedness results in the production of a more informative relationship matrix. The increase in genetic information is also well illustrated in Legarra et al. (2009), their Tables 1, 2 & 3, showing the progression of relatedness from ***A***- to ***G***- to ***H***-matrices.

The work of Ratcliffe et al. (2017) using an open-pollinated white spruce population (Beaulieu 2014a, b) was aimed at determining the genotyped to non-genotyped proportional threshold, so the genotyping and phenotyping costs and efforts could be minimized. In this study, genotyping efforts of 0, 25, 50, 75, and

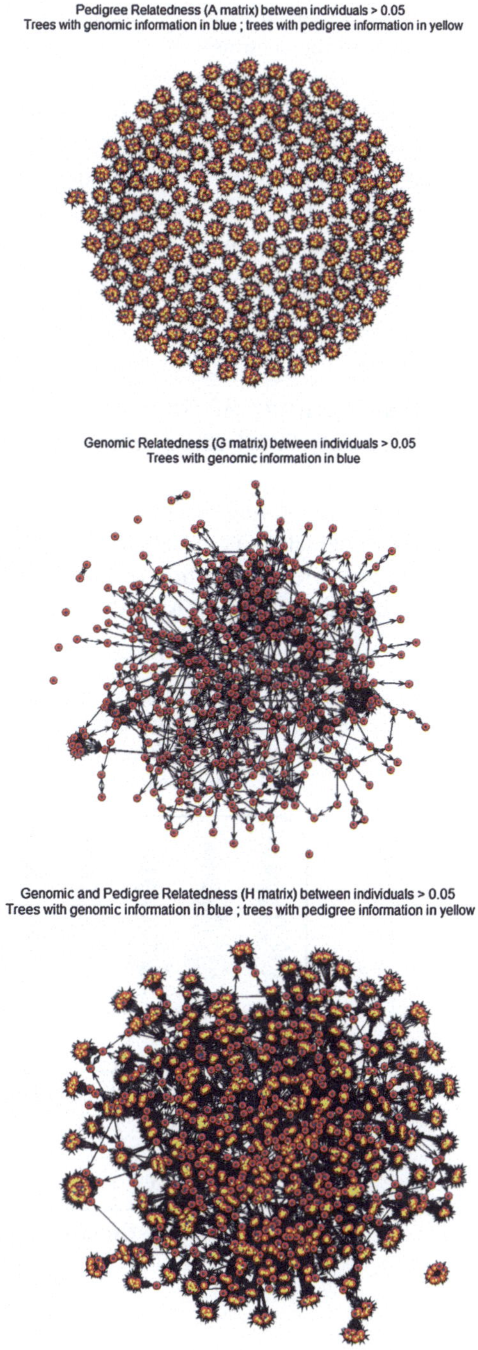

Fig. 8.1 Network representation of pedigree (***A***-matrix: top), genomic (***G***-matrix: middle), and blended pedigree and genomic (***H***-matrix: bottom) relatedness among 214 half-sib white spruce families (Ratcliffe et al. 2017). Clusters and lines represent half-sib families and connections among individuals, respectively (relatedness among individuals is > 0.05). Yellow, and blue dots represent individuals with genomic and pedigree information, respectively

100% were tested and the results indicated improvement in the accuracy of genetic variance components at even the lowest genotyping effort (25%) over the pedigree based analysis (0%). Ratcliffe et al. (2017) thus highlighted the importance of optimization of genotyping efforts for this approach. In other forest trees, the single-step approach has been used in single- (Cappa et al. 2017) and multi-traits (Cappa et al. 2018) evaluation of *Eucalyptus* progeny testing trials.

8.3 Development of Genomic Selection Predictive Models

Before describing the main steps for developing GS predictive models, it is important to highlight the role of the correct estimation of SNP marker effects on the trait of interest. An individual SNP marker effect is influenced by many factors including: (1) genotype × environment interaction (G × E), specifically if testing is conducted on multiple environments, (2) epistasis which cannot be discerned unless full genotypic and phenotypic complements are available, and (3) rare alleles that tend to be excluded from the analysis [SNPs are included in the analyses only if they meet the specific population minor allele frequency (MAF) threshold, usually 0.10]. All these factors are virtually unknown for GS studies conducted on trees including spruces (see Grattapaglia et al. 2018, for review); thus, tree species GS predictive accuracies should be interpreted with caution as they still represent coarse estimates.

The development of GS predictive models involves the two main types of populations; namely, the training and the selection populations (Fig. 8.2). The relationship between the training

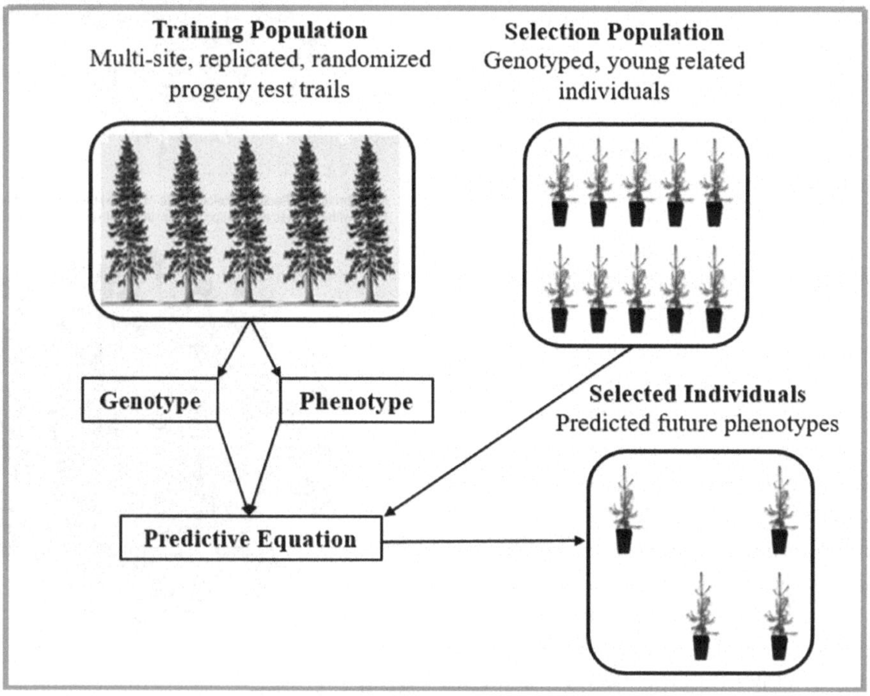

Fig. 8.2 Illustration of genomic selection (GS) steps starting from the training population with genotyped and phenotyped individuals, the development of GS predictive model using the training population information, implementation of the GS predictive model in the selection population for the identification of superior individuals and their selection based on their estimated genomic breeding values

and selection populations determines the degree of GS success, and the higher the degree of genetic similarity between them, the better the GS predictions (i.e., common LD between causal genes and SNPs). In this respect, a study using an open-pollinated white spruce breeding population demonstrated that when relatedness is minimized between the training and selection populations, the prediction accuracy is greatly reduced, or nonexistent, due to the lack of LD (Beaulieu et al. 2014a).

Additionally, the size of the training population is one of the main factors determining GS success (Grattapaglia and Resende 2010). Spruces' training population sizes ranged between 734 (Lenz et al. 2017) and 1,748 (Beaulieu et al. 2014b) individuals for black and white spruces, respectively. Individuals in the training population are genotyped and phenotyped for predictive model development (Heslot et al. 2011). The predictive accuracies for the spruces' GS models varied depending on the species, trait heritability, cross-validation method, and statistical model used, and the size of the training population.

Spatial and temporal GS studies are of great interest as they present scenarios where predictive models' transferability is evaluated (Beaulieu et al. 2014a, Gamal El-Dien et al. 2015; Ratcliffe et al. 2015; Lenz et al. 2017). The number of environments (sites) used in spruce GS studies have ranged between one (Beaulieu et al. 2014a) and three (Gamal El-Dien et al. 2015) for white and interior spruces, respectively. Generally, GS prediction accuracy of a specific environment is not transferable to other environments (i.e., test population(s)), highlighting the role of G × E (Beaulieu et al. 2014a, Gamal El-Dien et al. 2015; Lenz et al. 2017). However, when data from multiple environments are used to develop a 'universal' predictive model, higher predictive accuracies can be achieved as the combined model accounted for the spatial differences among environments (Gamal El-Dien et al. 2015). Thus, it is preferable to have the training population planted over multiple environments, so the effect of G × E is accounted for. The ideal situation is using clonally propagated material replicated over multiple environments as has been done by Resende et al. (2012). Using clonally propagated material over multiple environments results in estimating G E with high fidelity, a situation not comparable for species that are difficult to clonally propagate.

Temporal GS predictive accuracy has been reported for Interior spruce tree height and it was well correlated with the age–age genetic correlation and showed a substantial decreasing trend with increasing age difference between the training and the validation populations (Ratcliffe et al. 2015). The widespread lack of wood quality temporal data is expected as traits change over time (Franceschini et al. 2013) as well as it is related to the difficulties associated with the added costs and sampling logistics; nonetheless, temporal GS models' predictive accuracies are of great value as they determine the time needed for obtaining reliable predictions.

It is noteworthy to state that so far, since all GS studies conducted on spruces within Canada used within-generation cross-validation for testing the accuracy of their developed predictive models, all lacked the use of an independent validation population (i.e., the progeny generation), yet an essential component of GS implementation in breeding programs. After the development of a reliable (tested/validated) GS predictive model, this can be used to predict the genomic estimated breeding value (GEBV) and rank of all members of the selection population (Fig. 8.2). Thus, the untested progeny of this training population can be selected for further breeding and production without the traditional progeny evaluation. At this stage, it should be stated that all tree species' GS studies have to be considered as proof-of-concept as no successful and proven GS implementation in spruce breeding programs has been documented so far.

8.4 Innovative Genomic Selection Approach

GS is particularly beneficial to select for labor intensive and expensive to measure traits such as wood quality characteristics. One of the strategies of GS that allows increasing the accuracy of prediction of breeding values of these traits is to use the multiple-trait GBLUP approach. This methodology allows predicting the genetic values of trees for a difficult to measure trait using one or several indicator traits. This indicator trait (s) should be easily measurable over the entire population with low assessed cost, should be heritable and, most importantly, genetically correlated with the trait of interest (Pszczola et al. 2013). The multiple-trait GBLUP approach was successfully used to increase the prediction accuracy of poorly measured traits in animals (Pszczola et al. 2013; Calus and Veerkamp 2011) and agricultural crops (Schulthess et al. 2016; Fernandes et al. 2018) using simulated and empirical datasets. In the tree breeding context, many phenotypic observations already exist for routinely recorded traits, for example, height and diameter, and these yield traits could be used as predictor trait(s) to predict correlated wood quality traits. However, until recently, the multiple-trait GS approach has not been extensively applied in tree breeding (Cappa et al. 2018). Lenz et al. (2019) explicitly tested this idea for Norway spruce and demonstrated that when full phenotypic data was not available for weevil insect resistance, the inclusion of genetically correlated growth traits boosted prediction accuracy above that of a single trait model. We expect that the simulations and empirical studies in animal and crop breeding as well as recent findings in tree breeding using multiple-trait improvement by means of GS opens the door to future studies of this approach in difficult to measure traits such as wood quality attributes.

8.5 SNP Imputation

The rapid progress of next-generation DNA sequencing (NGS) platforms, especially those based on re-sequencing a smaller genomic fraction, has bypassed the steps for marker assay development, and allowed direct comparison of large numbers of sequences for identifying DNA polymorphisms in a wide range of species, such as conifers (Chen et al. 2013). GBS, a low-genome coverage NGS platform that simultaneously collects millions of short-read sequences for a number of multiplexed individuals, increases genotyping capacity by leveraging the strength in SNP imputation to overcome the issue of low average read count on SNP sites (Elshire et al. 2011).

Statistically speaking, imputation is a procedure that substitutes missing data with reasonable values derived from known data points. This has been a popular method for handling incomplete data problems (Barnard and Meng 1999). Genotype imputation traditionally describes the process of predicting genotypic values that are not directly assayed in a sample of individuals (Marchini and Howie 2010); now it can also refer to the process of using a reference population genotyped at a higher density to increase power by predicting the study samples genotyped at a lower density. Several landmark cases that employed imputation techniques with human haplotype map Phase II HapMap (Frazer et al. 2007) for fine mapping, genome-wide and in meta-analysis, have been reviewed in Marchini and Howie (2010).

The simplest approach for SNP imputation is to replace missing genotypic values with the mean value of the locus (i.e., mean imputation). The mean of an SNP locus, as a discrete random variable, is a weighted average of all possible genotypic values, which makes imputing missing values as the mean reasonable for the stochastic nature of a complexity reduction SNP genotyping method like GBS. As shown in Table 8.2, the averaged imputation accuracy across five interior spruce half-sib families was comparable to other imputation algorithms. However, this strategy can severely distort the allelic distribution for the imputed locus, underestimate the correlation between SNP variables and disrupt the identical-by-descent when there was significant population structure underlying the analyzed samples. Although computationally efficient, the disadvantage of taking the global mean from the locus, as a result, has largely influenced genomic selection performance (Table 8.1). A more sophisticated method to recover local expected values of the missing SNP loci, like the expectation–maximization (EM) approach, could

Table 8.2 Comparison of imputation accuracy in interior spruce half-sib families

	MI[a]	SVD_2[b]	SVD_5[b]	SVD_10[b]	EM[c]	*k*NN-fam[d]
Family 11	0.799	0.806	0.800	0.780	0.771	0.795
Family 17	0.808	0.812	0.803	0.779	0.778	0.809
Family 21	0.805	0.805	0.795	0.772	0.769	0.800
Family 6	0.800	0.822	0.802	0.766	0.767	0.802
Family 47	0.803	0.802	0.798	0.774	0.770	0.809
Average	0.803	0.810	0.799	0.774	0.771	0.803

[a]mean imputation; [b]SVD, singular value decomposition, with 2, 5, and 10 iterations; [c]expectation–maximization imputation; [d]family-based *k*-nearest neighbor imputation with optimized $K1$ and $K2$

potentially yield improved genomic selection accuracy at the cost of computation.

For breeding populations with pedigree and family structure, the most intuitive scenarios to impute un-typed genotypic values lies in the families' information, as family members usually share a high degree of genetic similarity, bearing long stretches of DNA sequences that are identical between close relatives. This imputation procedure has proven successful in handling the missing data issue even for distantly related individuals, especially for targeting tag-SNPs that are responsible for trait variation (Huang et al. 2009). These algorithms, such as Beagle 4, leverage genome structure from multi-population studies by creating long, phased haplotypes and model the homozygosity-by-descent (HBD) and the identity-by-descent (IBD) based on family and pedigree information (Browning and Browning 2013). The strength of this kind of haplotype reconstruction method for missing data imputation comes from local LD, where a reliable estimate for the unobserved genotype can be provided using observed linked loci. Since high-quality genome assembly is so far unavailable for conifer species, genotype imputation turns into a process of training descriptive models using unsupervised machine learning techniques (Pantanowitz and Marwala 2009). To incorporate the open-pollinated, half-sib spruce family structure into genotype imputation, a family-based k-nearest neighbor (*k*NN) method has been developed, namely *k*NN-fam. The *k*NN-fam approach is a two-step process: first, SNP values were used to construct a standardized genomic similarity matrix for all samples using the method described in VanRaden (2008), and Euclidean distances between SNP markers were obtained following Rutkoski et al. (2013). In principle, SNP genotypes involved in this step can be a subset of the SNP table that has little missing values; an alternative method is to impute missing data with mean imputation. The *k*NN-fam then models the missing SNP values by proportionately weighting on resemblance among individuals within families as well as the correlation between SNP markers in the second step by:

$$\hat{y} = \text{mode}\left(\frac{1}{K1 + K2} y\right) \quad (8.1)$$

where $\hat{y}$ is the updated genotypic value, $K1$ is the number of neighbors within the family based on the genomic similarity and $K2$ is the number of neighbors from outside the family based on the Euclidean distance, and y is the original locus mean. The optimal values for $K1$ and $K2$ in (8.1) are obtained by a grid search.

The strength of *k*NN-fam was validated in Gamal El-Dien et al. (2015), where an exhaustive search for the optimal values was performed by permuting the within-family neighbor $K1$ through 1–30 (the nearest neighbor being 1, and then 2, 5, 10, 15, 20 to all family members, i.e., 30), and out-of-family neighbor $K2$ from 1 to 250, as the total sample size of the panel was 1,146. The accuracy of *k*NN-fam imputation was determined for each permutation by randomly masking 1 million known SNPs and calculating the percentages of markers being imputed back to the correct genotypic values. Using the optimal values for $K1$ and $K2$ (respectively, 5 and 20), imputation accuracy for *k*NN-fam could be as high as 84.9% (Gamal El-Dien et al. 2015). Comparing other common approaches for such SNP datasets (mean imputation, EM and singular value decomposition (SVD)), *k*NN-fam outperformed others by 4–10%. It might be worthwhile noting that SVD, in a few cases, generates comparable imputation accuracy. However, the numerical output resulting from SVD imputation was unfavorable, as in GS, SNP predictors are often discrete.

When a phenotype is strongly correlated with the genotyping platform, imputing the un-typed SNPs in cases from SNPs that are present in a dense set of genotype data from controls, did not lead to increased false-positive rates (Howie et al. 2009). However, if cases and controls are typed on different chips, then imputing SNPs that are un-typed in both cases and controls from a haplotype panel can lead to false-positive associations. SNPs that are imputed accurately from one chip but poorly from another chip may lead to differences in allele frequency that solely reflect allele frequency differences between the haplotype reference panel and the study population. This is similar to the way population structure can cause problems in genome-wide association studies. Ideally, situations like this are best avoided by a sensible study design, as well as by typing all individuals using a common platform (Sun et al. 2019).

8.6 Conclusions

The results of current published studies indicate that GS is an effective tool for improvement in the genetic evaluation of the relatively undomesticated spruce breeding programs in Canada. Using realized relationships (GBLUP) in place of pedigree relationships (ABLUP) has yielded improvements in the analysis of open-pollinated testing designs through the unraveling of hidden relationships in the population. Building on this concept, single-step evaluation (HBLUP) has allowed information from realized relationships between genotyped individuals to be transferred to non-genotyped individuals providing improvement in the analysis of the testing population at a reduced genotyping effort.

GS prediction models provide the platform to perform selection on untested trees at a very early age with greater accuracy than pedigree based prediction. The prediction of phenotypes offers the greatest possibility of increasing the rate of genetic gain in spruce breeding programs since it is inversely proportional to the length of the breeding cycle. The current published studies have shown intrageneration genomic prediction accuracies that meet or exceed those of pedigree prediction and provide evidence for operational implementation of GS technology.

With the continued development of spruce functional genomic resources (i.e., annotated reference sequences) and improvement in phenotyping platforms (i.e., remote sensing), it is convincing to assume that accurate prediction of phenotypes will be further improved. High-throughput phenotyping will allow the joint inclusion of many correlated traits in prediction models allowing greater prediction accuracy for traits that are difficult to measure. Additionally, the inclusion of functional rare causal variants that are relatively stable across environments and generations in predictive models will also help boost the accuracy of predictive models.

References

Andrews M (2002) Wood quality measurement-son et lumière. N Z J For Sci 47:19–21

Askew GR, El-Kassaby YA (1994) Estimation of relationship coefficients among progeny derived from wind-pollinated orchard seeds. Theor Appl Genet 88:267–272

Barnard J, Meng XL (1999) Applications of multiple imputation in medical studies: from AIDS to NHANES. Stat Methods Med Res 8:17–36

Beaulieu J, Doerksen T, Clément S, MacKay J, Bousquet J (2014a) Accuracy of genomic selection models in a large population of open-pollinated families in white spruce. Heredity 113:343–352

Beaulieu J, Doerksen TK, MacKay J, Rainville A, Bousquet J (2014b) Genomic selection accuracies within and between environments and small breeding groups in white spruce. BMC Genomics 15:1048

Brown GR, Bassoni DL, Gill GP, Fontana JR, Wheeler NC, Megraw RA, Davis MF, Sewell MM, Tuskan GA, Neale DB (2003) Identification of QTLs influencing wood property traits in loblolly pine (*Pinus taeda* L.). III. QTL verification and candidate gene mapping. Genetics 164:1537–1546

Browning BL, Browning SR (2013) Improving the accuracy and efficiency of identity-by-descent detection in population data. Genetics 194:459–471

Buerstmayr H, Ban T, Anderson JA (2009) QTL mapping and marker-assisted selection for Fusarium head blight resistance in wheat: a review. Plant Breed 128:1–26

Calus MP, Veerkamp RF (2011) Accuracy of multi-trait genomic selection using different methods. Genet Sel Evol 43:26

Cappa EP, El-Kassaby YA, Muñoz F, Garcia MN, Villalba PV, Klápště J, Poltri SNM (2017) Improving accuracy of breeding values by incorporating genomic information in spatial-competition mixed models. Mol Breed 37:125

Cappa EP, El-Kassaby YA, Muñoz F, Garcia MN, Villalba PV, Klápště J, Poltri SNM (2018) Genomic-based multiple-trait evaluation in Eucalyptus grandis using dominant DArT markers. Plant Sci 271:27–33

Chen C, Mitchell SE, Elshire RJ, Buckler ES, El-Kassaby YA (2013) Mining conifers' mega-genome using rapid and efficient multiplexed high-throughput genotyping-by-sequencing (GBS) SNP discovery platform. Tree Genet Genomes 9:1537–1544

Christensen OF, Lund MS (2010) Genomic prediction when some animals are not genotyped. Genet Sel Evol 42:2

De La Torre AR, Puiu D, Crepeau MW, Stevens K, Salzberg SL, Langley CH, Neale DB (2019) Genomic architecture of complex traits in loblolly pine. New Phytol 221:1789–1801

Doerksen TK, Herbinger CM (2010) Impact of reconstructed pedigrees on progeny-test breeding values in red spruce. Tree Genet Genomes 9:591–600

El-Kassaby YA (1982) Association between allozyme genotypes and quantitative traits in Douglas-fir [*Pseudotsuga menziesii* (Mirb.) Franco]. Genetics 101:103–115

El-Kassaby YA (1995) Evaluation of the tree-improvement delivery system: factors affecting genetic potential. Tree Physiol 15:545–550

El-Kassaby YA, Lstiburek M (2009) Breeding without breeding. Genet Res 91:111–120

El-Kassaby YA, Cappa EP, Liewlaksaneeyanawin C, Klapste J, Lstiburek M (2011) Breeding without breeding: Is a complete pedigree necessary for efficient breeding? PLoS ONE 6:e25737

El-Kassaby YA, Klápšte J, Guy RD (2012) Breeding without breeding: selection using the genomic best linear unbiased predictor method (GBLUP). New For 43:631–637

Elshire RJ, Glaubitz JC, Sun Q, Poland JA, Kawamoto K, Buckler ES, Mitchell SE (2011) A robust, simple Genotyping-by-Sequencing (GBS) approach for high diversity species. PLoS ONE 6:e19379

Evans R, Stringer S, Kibblewhite RP (2000) Variation of microfibril angle, density and fibre orientation in twenty-nine *Eucalyptus nitens* trees. Appita J 53:450–457

Falconer DS, MacKay TF (1969) Introduction to quantitative genetics. Longmans Green, Harlow

Fernandes SB, Dias KO, Ferreira DF, Brown PJ (2018) Efficiency of multi-trait, indirect, and trait-assisted genomic selection for improvement of biomass sorghum. Theor Appl Genet 131:747–755

Franceschini T, Longuetaud F, Bontemps J-D, Bouriaud O, Caritey B-D, Leban J-M (2013) Effect of ring width, cambial age, and climatic variables on the within-ring wood density profile of Norway spruce *Picea abies* (L.) Karst. Trees. https://doi.org/10.1007/s00468-013-0844-6

Frazer KA, Ballinger DG, Cox DR et al (2007) A second generation human haplotype map of over 3.1 million SNPs. Nature 449:851–861

Frentiu FD, Clegg SM, Chittock J, Burke T, Blows MW, Owens IPF (2008) Pedigree-free animal models: the relatedness matrix reloaded. Proc R Soc B-Biol Sci 275:639–647

Funda T, El-Kassaby YA (2012) Seed orchard genetics. CAB Rev 7:013

Gamal El-Dien O, Ratcliffe B, Klápště J, Chen C, Porth I, El-Kassaby YA (2015) Prediction accuracies for growth and wood attributes of interior spruce in space using genotyping-by-sequencing. BMC Genomics 16:370

Gamal El-Dien O, Ratcliffe B, Klápště J, Porth I, Chen C, El-Kassaby YA (2016) Implementation of the realized genomic relationship matrix to open-pollinated white spruce family testing for disentangling additive from non-additive genetic effects. G3 -Genes Genomes Genet 6:743–753

Gamal El-Dien O, Ratcliffe B, Klápště J, Porth I, Chen C, El-Kassaby YA (2018) Multi-environment genomic variance decomposition analysis of open-pollinated Interior spruce (*Picea glauca* x *engelmannii*). Mol Breed 38:26

Grattapaglia D, Resende M (2010) Genomic selection in forest tree breeding. Tree Genet Genomes 7: 241–255

Grattapaglia D, Silva-Junior OB, Resende RT, Cappa EP, Müller PSF, Tan B, Isik F, Ratcliffe R, El-Kassaby YA (2018) Quantitative genetics and genomics coverage to accelerate forest tree breeding. Front Plant Sci 9:1693

Henderson CR (1976) Simple method for computing inverse of a numerator relationship matrix used in prediction of breeding values. Biometrics 32:69–83

Heslot N, Yang H-P, Sorrells ME, Jannink J-L (2011) Genomic selection in plant breeding: a comparison of models. Crop Sci 52:146–160

Hill WG, Weir BS (2011) Variation in actual relationship as a consequence of Mendelian sampling and linkage. Genet Res 93:47–64

Howie BN, Donnelly P, Marchini J (2009) A flexible and accurate genotype imputation method for the next

generation of genome-wide association studies. PLoS Genet 5:e1000529
Huang L, Li Y, Singleton AB, Hardy JA, Abecasis G, Rosenberg NA, Scheet P (2009) Genotype-imputation accuracy across worldwide human populations. Am J Hum Genet 84:235–250
La Mantia J, Klapste J, El-Kassaby YA, Azam S, Guy RD, Douglas CD, Mansfield SD, Hamelin R (2013) Association analysis identifies *Melampsora* x *columbiana* poplar leaf rust resistance SNPs. PLoS ONE 8:e78423
Legarra A, Aguilar I, Misztal I (2009) A relationship matrix including full pedigree and genomic information. J Dairy Sci 92:4656–4663
Lenz PRN, Beaulieu J, Mansfield SD, Clément S, Desponts M, Bousquet J (2017) Factors affecting the accuracy of genomic selection for growth and wood quality traits in an advanced-breeding population of black spruce (*Picea mariana*). BMC Genomics 18:335
Lenz PRN, Nadeau S, Mottet M-J et al (2019) Multi-trait genomic selection for weevil resistance, growth, and wood quality in Norway spruce. Evol Appl 00:1–19. https://doi.org/10.1111/eva.12823
Lippert C, Quon G, Kang EY, Kadie CM, Listgarten J, Heckerman D (2013) The benefits of selecting phenotype-specific variants for applications of mixed models in genomics. Sci Rep 3:1815
Marchini J, Howie B (2010) Genotype imputation for genome-wide association studies. Nat Rev Genet 11:499–511
Meuwissen T (2007) Genomic selection: marker-assisted selection on a genome wide scale. J Anim Breed Genet 124:321–322
Meuwissen THE, Hayes BJ, Goddard ME (2001) Prediction of total genetic value using genome-wide dense marker maps. Genetics 157:1819–1829
McKown AD, Guy RD, Qumme L, Klapste J, La Mantia J, Constabel CP, El-Kassaby YA, Hamelin RC, Zifkin M, Azam MS (2014) Association genetics, geography and ecophysiology link stomatal patterning in Populus trichocarpa with carbon gain and disease resistance trade-offs. New Phytol 23:5771–5790
McKown AD, Klapste J, Guy RD, El-Kassaby YA, Mansfield SD (2018) Ecological genomics of variation in bud-break phenology and mechanisms of response to climate warming in *Populus trichocarpa*. New Phytol 20:300–316
Misztal I, Legarra A, Aguilar I (2009) Computing procedures for genetic evaluation including phenotypic, full pedigree, and genomic information. J Dairy Sci 92:4648–4655
Muñoz PR, Resende MFR Jr, Gezan S, Resende MD, de los Campos G, Kirst M, Huber D, Peter GF (2014) Unraveling additive from nonadditive effects using genomic relationship matrices. Genetics 198:1759–1768
Pantanowitz A, Marwala T (2009) Missing data imputation through the use of the random forest algorithm. Springer, Berlin, pp 53–62
Pavy N, Gagnon F, Deschênes A, Boyle B, Beaulieu J, Bousquet J (2016) Development of highly reliable in silico SNP resource and genotyping assay from exome capture and sequencing: an example from black spruce (*Picea mariana*). Mol Ecol Resour 16:588–598
Porth I, Klapste J, Skyba O et al (2013) Genome-wide association mapping for wood characteristics in Populus identifies an array of candidate single nucleotide polymorphisms. New Phytol 200:710–726
Pszczola M, Veerkamp RF, De Haas Y, Wall E, Strabel T, Calus MPL (2013) Effect of predictor traits on accuracy of genomic breeding values for feed intake based on a limited cow reference population. Animal 7:1759–1768
Ratcliffe B, Gamal El-Dien O, Klápště J, Porth I, Chen C, Jaquish B, El-Kassaby YA (2015) A comparison of genomic selection models across time in interior spruce (*Picea glauca* x *engelmannii*) using unordered SNP imputation methods. Heredity 115:547–555
Ratcliffe B, Gamal El-Dien O, Cappa EP, Porth I, Klápště J, Chen C, El-Kassaby YA (2017) Single-step BLUP with varying genotyping effort in open-pollinated Picea glauca. G3-Genes Genomes Genet 7:935
Resende Jr MFR, Munoz P, Resende MDV, Garrick DJ, Fernando RL, Davis JM, Jokela DEJ, Martin TA, Peter GF, Kirts M (2012) Accuracy of genomic selection methods in a standard data set of loblolly pine (*Pinus taeda* L.). Genetics 190:1503–1510
Rutkoski JE, Poland J, Jannink J-L, Sorrells ME (2013) Imputation of unordered markers and the impact on genomic selection accuracy. G3-Genes Genomes Genet 3:427–439
Schulthess AW, Wang Y, Miedaner T, Wilde P, Reif JC, Zhao Y (2016) Multiple-trait-and selection indices-genomic predictions for grain yield and protein content in rye for feeding purposes. Theor Appl Genet 129:273–287
Sewell MM, Bassoni DL, Megraw RA, Wheeler NC, Neale DB (2000) Identification of QTLs influencing wood property traits in loblolly pine (*Pinus taeda* L.). I. Physical wood properties. Theor Appl Genet 101:1273–1281
Sewell MM, Davis MF, Tuskan GA, Wheeler NC, Elam CC, Bassoni DL, Neale DB (2002) Identification of QTLs influencing wood property traits in loblolly pine (*Pinus taeda* L.). II. Chemical wood properties. Theor Appl Genet 104:214–222
Sun S, Miao Z, Ratcliffe B, Campell P, Pasch B, El-Kassaby YA, Balsundaram B, Chen C (2019) SNP variable selection by generalized graph domination. PLoS ONE 14:e0203242
Thistlethwaite FR, Ratcliffe B, Klápště J, Porth I, Chen C, Stoehr MU, El-Kassaby YA (2017) Genomic prediction accuracies in space and time for height and wood density of Douglas-fir using exome capture as the genotyping platform. BMC Genomics 18:930
Thistlethwaite FR, Ratcliffe B, Klápště J, Porth I, Chen C, Stoehr MU, El-Kassaby YA (2019) Genomic selection of juvenile height across a single generational gap in Douglas-fir. Heredity 122:848–863

VanRaden PM (2008) Efficient methods to compute genomic predictions. J Dairy Sci 91:4414–4423

VanRaden PM, Van Tassell CP, Wiggans GR, Sonstegard TS, Schnabel RD, Taylor JF, Schenkel FS (2009) Reliability of genomic predictions for North American Holstein bulls. J Dairy Sci 92:16–24

Veerkamp RF, Mulder HA, Thompson R, Calus MPL (2011) Genomic and pedigree-based genetic parameters for scarcely recorded traits when some animals are genotyped. J Dairy Sci 94:4189–4197

Visscher PM, Medland SE, Ferreira MAR, Morley KI, Zhu G, Cornes BK, Montgomery GW, Martin NG (2006) Assumption-free estimation of heritability from genome-wide identity-by-descent sharing between full siblings. PLoS Genet 2:e41

White TL, Adams WT, Neale DB (2007) Forest genetics. CABI Publishing, Cambridge

Wright S (1922) Coefficients of inbreeding and relationship. Am Nat 56:330–338

Xu S (2003) Advanced statistical methods for estimating genetic variances in plants. Plant Breed Rev 22:113–163

Ying CC, Yanchuk AD (2006) The development of British Columbia's tree seed transfer guidelines: purpose, concept, methodology, and implementation. For Ecol Manag 227:1–13

Zapata-Valenzuela J, Whetten RW, Neale D, McKeand S, Isik F (2013) Genomic estimated breeding values using genomic relationship matrices in a cloned population of loblolly pine. G3-Genes Genomes Genet 3:909–916

Zobel BJ, Talbert JT (1984) Applied forest tree improvement. Wiley, New York

Drought Stress Adaptation in Norway Spruce and Related Genomics Work

9

Jaroslav Klápště, Jonathan Lecoy, and María del Rosario García-Gil

Abstract

Ongoing climate change has resulted in more frequent occurrences of stress events including droughts. Global warming affects all living organisms, especially forest trees. Norway spruce has become the dominant forestation species in Central Europe thanks to historical directions in forest stands management. However, the species is more drought stress sensitive compared to others such as silver fir or European beech. Therefore, health and productivity of local forests may be jeopardized when prolonged drought periods become a common phenomenon. Nevertheless, moderate levels of genetic variability in response to drought stress have been identified, between provenances, as well as among individuals within provenances, and thus, in this respect, selection for genetically superior planting stock is worthwhile. Drought stress response is mainly driven by the flexibility in stomata closure to prevent further water loss as a direct reaction to intrinsic water potential change; stomatal closure is controlled through the phytohormone abscisic acid (ABA). The development of a more complex root system via root tips and root hairs elongation to improve soil water search, as well as presence of specific pit membrane structures (torus) between xylem tracheids, which limits the propagation of embolisms, serve as additional protective measures against drought stress in spruce. To better understand the genetic architecture of drought tolerance, several genome-wide approaches such as quantitative trait locus (QTL) mapping, gene expression profiling, or association studies have been implemented. QTL analyses suggest a positive relationship between stomatal conductance and water use efficiency and a favorable relationship with growth. Among many GWAS derived genetic associations, genes encoding proteins for phytohormones biosynthesis and signaling implicated in abiotic stress responses (such as ABA) were identified. These recent genomics studies suggest several key players in genetic resistance to drought in Norway spruce, intriguingly, also some candidates conserved across distantly related conifer species. We conclude by

J. Klápště (✉)
Scion (New Zealand Forest Research Institute Ltd.), 49 Sala Street, Whakarewarewa, Rotorua 3010, New Zealand
e-mail: Jaroslav.Klapste@scionresearch.com

J. Lecoy · M. del Rosario García-Gil
Department of Forest Genetics and Plant Physiology, Umeå Plant Science Centre, Swedish University of Agricultural Sciences, 901 87 Umeå, Sweden

I. M. Porth and A. R. De la Torre (eds.), *The Spruce Genome*, Compendium of Plant Genomes,
https://doi.org/10.1007/978-3-030-21001-4_9

providing directions for future research into drought resistance under the pressing need to identify the best locally adapted planting stock for future climates.

9.1 Introduction

Forest trees represent essential elements of any terrestrial ecosystem. Because ongoing climate change affects each ecosystem around the world, it imposes substantial pressure on forests to remain resilient under more frequent occurrences of climatic extremes such as prolonged droughts (Sheffield and Wood 2008). The forecasted climatic development for Central Europe indicates continuous trends in increased annual mean temperature, temporal changes in precipitation distribution, and more frequent occurrence of extreme conditions such as drought (Beniston et al. 2007; Ballester et al. 2010, Thiele et al. 2017). This, in turn, will affect future planted material's ability to maintain current productivity and survival rate (Boisvenue and Running 2006; Allen et al. 2010; Neuman et al. 2017). Such climatic conditions could also trigger forests to become carbon sources rather than carbon sinks, and therefore, precise monitoring of net ecosystem production has to be implemented to inform ecosystem managements (Soloway et al. 2017). In addition, an understanding of regional vulnerability and adaptive capacity of forest populations has to be developed (Lindner et al. 2010).

Several parameters such as growth resistance (i.e., the ability to maintain growth during a drought period), recovery (i.e., a growth increase after a drought event), and resilience (i.e., the potential to return to growth levels before the drought event happened) are usually investigated as attributes defining the level of tolerance to drought stress. Chen et al. (2017) investigated growth-climate relationships using the process-based Vaganov–Shashkin Lite model and found an impact of temperature-induced drought (lower precipitation in combination with higher temperature) on radial growth of white spruce in Alberta (Canada). However, there is still uncertainty regarding the attributes relevant for drought stress quantification. Speich (2019) reviewed indices related to atmospheric conditions, soil moisture, and physiological thresholds and found that each parameter considered individually provides incomplete information about a plant's physiological response to drought. The author found that considering evaporative demand together with precipitation improves drought indicators compared to looking only at precipitation. For example, Swann et al. (2016) also suggested that the increased level of atmospheric CO_2 (one of the climate change attributes) leads to increased photosynthetic rates and therefore compensates for plant water demand and water stress due to transpiration.

Additionally, it is recommended that information on functional and structural diversity of trees as well as tree sizes should be considered to increase the proportion of the explained variance in growth resilience to drought stress. Serra-Maluquer et al. (2018) analyzed tree rings of pine species in the northeast of Spain and remarked species' progressing vulnerability to drought stress with an increasing number of drought events throughout their lifetimes. A comparative analysis of tree rings from different forest tree species indicated that Norway spruce has a higher vulnerability to drought stress compared to European beech or silver fir (Zang et al. 2014). Gazol et al. (2016) suggested the presence of a natural trade-off between forest resistance of growth to drought and growth recovery after the drought period. Nevertheless, the conifers in this study showed higher intraspecific variability compared to angiosperms (oaks) in terms of their growth resistance and their recovery after drought stress events, and thus, exploring species' phenotypic plasticity is required to fully grasp the underlying mechanisms.

9.2 Natural Drought Tolerance Variability in the Pinaceae Family

Based on a meta-analysis of climate variables and signatures of natural selection performed across many organisms and ecosystems, precipitation and evapotranspiration patterns have been identified as the main drivers for global variation in natural selection (Siepielski et al. 2017). However, a deeper understanding of mechanistic relationships' complexity is still required (Myers-Smith and Myers 2018). Studies comparing the response to drought across species found that mechanisms involved in drought tolerance are rather conserved within species and show very little variability between families, especially after a drought event (Chmura et al. 2016). Norway spruce is generally more sensitive to drought stress compared to other species such as Scots pine (Kellomaki et al. 2008), silver fir or Douglas-fir (Zang et al. 2014; Vitali et al. 2017; Levesque et al. 2013), or European beech (Schafer et al. 2017). Since the natural distribution of Norway spruce is relatively large and covers heterogeneous environmental conditions, it can be assumed that moderate genetic variability with regard to drought tolerance among and within populations is present. While there is currently a lack of evidence for genetic variability in drought resistance among provenances (Burczyk and Giertych 1991; Sonesson and Eriksson 2010), moderate genetic variance was found among individuals within each tested provenance (Trujillo-Moya et al. 2018; Sonesson and Eriksson 2010). Yet, other conifer species studies reported higher tolerance to drought stress for high elevations populations. These studies on *Pinus tabuliformis* and *P. pinaster*, respectively, suggested a lack of phenotypic plasticity in response to drought in these populations (Aranda et al. 2010; Ma et al. 2014).

The proportion of *P. abies*' phenotypic variance in traits related to drought tolerance explained by genetics (their heritability) reached moderate ~20–40% (Sonesson and Eriksson 2010). Trujillo-Moya et al. (2018) tested several provenances originating from the species' southern part of its natural distribution (Fig. 9.1); they found up to 44% of the phenotypic variance in drought stress tolerance-related dendrochronological traits

Fig. 9.1 Central European Norway spruce provenances test for drought resistance. Photo credit: Ing. Lambert Weissenbacher (BFW Austria). Authors detected negative effects of high temperatures on Norway spruce's incremental growth at the trial site and significant among-provenance variation in climate-growth response (Trujillo-Moya et al. 2018)

explained by genetic variation within certain Norway spruce provenances. Similarly, Rosner et al. (2007) found moderate broad-sense heritability estimates ranging from 0.14 to 0.31 for traits related to hydraulic conductivity, important in drought resistance mechanisms. However, such genetic differences may disappear when tested under extreme environmental conditions at the edges of the species' natural distribution (Klisz et al. 2019). A strong relationship was found between high tolerance to drought with late budburst and early budset (Sonesson and Eriksson 2010), wood density (Rosner et al. 2013), and freezing tolerance (Blödner et al. 2005). Nevertheless, growth resistance to a drought event is also affected by the level of competition within the forest stands (Kohler et al. 2010), the timing of drought events (George et al. 2015), or by the presence of a beneficial ectomycorrhizal community (Gehring et al. 2017). Kapeller et al. (2012) acknowledged that provenances in warmer and drought-prone areas, which seem to be well adapted to drought stress, represent an optimal material for future reforestation activity to cope with climate change. Similarly, Chakraborty et al (2019) found that the influence of the climatic effect was predominant for early growth performance in provenances originating mainly from colder areas, while in provenances from warmer areas, soil had an importance similar to climate.

9.3 Physiological Principles (Mechanisms) of Drought Tolerance

9.3.1 Plants responses to water deficit stress

Before tackling the issue of how Norway spruce (*Picea abies*) and other higher plants can cope with water deficit stress, it is useful to briefly introduce the forces affecting water transport within plants, notably, how the structure of water molecules, and their interactions with each other, contribute to the water potential (ψ). Water, known by its chemical formula H_2O, is a polar molecule and capable of forming a maximum of four hydrogen bonds, involving its two intramolecular bonds as well as two intermolecular hydrogen bonds from neighboring water molecules (i.e., tetrahedral configuration); this specific bonding pattern is propagated throughout liquid phase water (Brini et al., 2017). Such important level of interactions between molecules provide for high cohesion due to the strong attraction between water molecules (Taiz et al., 2015) and the resulting cohesive force is necessary for the plant to pull water from deep within the ground via its root system and upward within its stem through capillary action (provided by both cohesion and adhesive forces properties of water).

The water potential is the free energy of water and is measured in $J.m^{-3}$. It translates to the habit of water to move passively from an environment of high water potential (less negative) to one of lower potential (more negative), thus following a gradient (Nonami and Boyer, 1993). In plants, it has been established that total water potential ψ_{total} corresponds to the combination of three distinct forces, these being the osmotic potential (ψ_s), the pressure potential (ψ_p), and the gravity potential (ψ_g), thus, $\psi_{total} = \psi_s + \psi_p + \psi_g$. The water potential can be estimated at the whole plant or at specific organs levels; this can provide information regarding which organs are absorbing water at a given time. The osmotic potential is defined by the following equation $\psi_s = - R.T.c$ with R being the ideal gas constant ($8.32\ J.mol^{-1}.K^{-1}$), T being the temperature in Kelvin, and c the concentration of a solute dissolved in water. A more concentrated solution will have a lower osmotic potential than a less concentrated one, thus, if these two solutions are separated by a semi-permeable membrane, water from the less concentrated solution will transfer by osmosis to the solution with a higher solute concentration until the concentrations on both sides of the membrane are equal, in which case there will no longer be a net movement of water. The pressure potential ψ_p corresponds either to (a) the turgor pressure which is a positive pressure, when a full vacuole presses against the cell membrane that is contained by the cell wall, and is overall important for plant stature maintenance or (b) the tension pressure which is a negative pressure occurring in the xylem, caused by water evaporation at the leaf-atmosphere interface from the leaf mesophyll, and is the main upward pulling force for water in the xylem. The gravity potential (ψ_g) is defined by the equation $\psi_g = \rho . g.h$, where ρ equals the density of the liquid measured in kg.m^{-3}, g corresponds to gravitational acceleration, and h represents the column height of the liquid. Therefore, ψ_g corresponds to the downward acting pressure that the liquid exerts; consequently, a water column of 100-meter height will display a higher gravity potential than a column of only one meter; the ψ_g contribution to ψ_{total} then becomes more important the taller the plant, and consequently, this increased gravitational downward pull requires the opposing upward pulling force to be stronger in tall trees, for soil water to reach the leaves via xylem transport.

The maintained decreasing water potential (i.e., gradient along the soil-root-stem-leaf-atmosphere system) enables plants to passively absorb water from the soil even under dry conditions; thus, the plant root cells need to have a lower water potential than the surrounding soil; this is usually established as the plant can control such water movement by manipulating its metabolism to decrease solute (osmotic) potential within its root cells, and thereby maintains a lower overall water potential than the surrounding soil allowing for the influx of water via osmosis into the root. Well-watered plants display a relatively high (less negative) water potential compared to plants exposed to drought stress. A very low water potential is thus indicative of a plant undergoing severe drought stress and may result in a plant's foreseeable death. This negative outcome is mainly due to the formation of embolisms which are caused by air pockets when air instead of water is absorbed into the root, due to very low water potential at the roots. Savi et al. (2019) found that xylem vulnerability to embolism is critical for *Pinus nigra*'s survival from severe drought. Embolisms are potentially deadly for the plant since these can block the continuous passage of water up the xylem. However, plants have developed different strategies for coping with such risk and which shall be discussed in the Chapter's next section. For further reading on the physiology subject, the reader is advised to consult Taiz and Zeiger's textbook on Plant Physiology and Development (6th edition).

9.3.2 The General Mechanisms by Which Plants Respond to Drought

Responses to drought have been studied in various plant model organisms, ranging from *Arabidopsis* and tobacco and also to loblolly pine (*Pinus taeda*) and Norway spruce (*Picea abies*). Plants have indeed developed many mechanisms for coping with stressors, certainly out of the necessity due to plants' sessile nature. For example, plants can redirect metabolites from defense against their natural enemies toward a drought response (MacAllister et al. 2019) or develop strategies based on primary resources allocation either to pathways implicated in defense mechanisms to cope with threats, or to those for reproductive development (Lauder et al. 2019). Such responses can vary on a spatio-temporal scale and also depend on the plant species. In this section, the general molecular and physiological responses of plants to drought stress will be discussed followed by a more in-depth presentation on the state of knowledge on Norway spruce response to drought.

As outlined previously, water moves within plant's vasculature from the roots up the stem to the top, where it is absorbed by the leaves' mesophilic tissue, and is subsequently transpired according to the water potential gradient (Tardieu et al. 2018). Many factors are involved in changing water demands of the plant, notably the time of the day, temperature and the plant's stage of development. Water deficit can lead to a change in solute concentration within the cell, protein denaturation, and damage to plant cell structures via oxidative stress and a loss of turgor. The loss of turgor is characterized by stem wilting in non-wood-producing plants (stature loss), or the wilting of leaves in wood-producing plants like trees. This loss of turgor is hypothesized to be detected by one or several unknown membrane receptor(s). The water deficit condition induces a reaction cascade mediated by Ca^{2+}, several mitogen-activated protein kinases (MAPKs) and protein tyrosine phosphatases that support the induction of genes involved in the biosynthesis of the stress hormone abscisic acid (ABA), mostly found in the roots where water deficit stress is thought to be first detected (Zhang et al. 2006; Tuberosa 2012). ABA is commonly associated with responses to a multitude of environmental stresses alongside water deficit stress (such as salt stress and cold stress) and is synthesized as a derivative from the terpenoid pathway, in what is called the ABA-dependent pathway. ABA further acts as a factor for transcription induction of drought response genes by binding to ABA response elements (ABREs). Such elements are present within promoter regions of genes such as *rd29A*, found to be induced under conditions of drought (Zhang et al. 2006; Tardieu et al. 2018).

ABA is delivered through specialized channels (ABCG transporters) into the guard cells that surround the leaf stomata, and in concert with other phytohormones such as cytokinins and auxin, ABA provokes an osmotic adjustment within these specialized cells (Daszkowska-Golec and Szarejko 2013). Within these, water potential now changes via ABA-mediated efflux of Ca^{2+} from the guard cell's vacuole into its cytosol. This, in turn, leads to the activation of anion channels as well as the depolarisation of the cell membrane (inactivated H + -ATPase) resulting in a net efflux of K^+ ions. The consequence of anion loss and K^+ ions efflux is the shrinking and flattening out of the guard cells due to turgor pressure decrease (water loss by osmosis), and thus, the stoma (pore) closes (Zhang et al. 2006; Tardieu et al. 2018; Brodribb and McAdam 2013). While this prevents the loss of water by transpiration albeit at the cost of not being able to absorb gases such as CO_2 (or O_2) from the atmosphere anymore, in the long-term, this may lead to carbon starvation in the plant. Prolonged stomatal closure may also cause oxidative stress damage to the cell structures due to an inability of the plant to replenish antioxidant molecules (Dulamsuren et al. 2018; Farquhar and Sharkey 1982).

The timing of stomatal closure is pivotal to the determination of water reserves maintained in the plant and is known to follow genetic trends related to sensitivity to ABA and the activity of ABA-induced transcription factors. Plants use two broad strategies for stomata closure in response to water stress, separating them into isohydric and anisohydric lines, respectively. Isohydric plants close their stomata within a narrow range of decreasing water potential, whereas anisohydric plants have been shown to close their stomata across a wider range of water potential decrease. Isohydric plants are thus already sensitive to a slight drop in their water potentials (Tardieu et al. 2018; Lu et al. 1995). Hammond and Adams (2019) hypothesize that species exhibiting a low hydraulic safety threshold and prolonged stomata closure reach turgor loss later during drought progression but could experience mortality more quickly compared to species employing the opposite strategy. Norway spruce belongs to the isohydric family of plants; as such, the species follows the former of these two mechanisms under water stress. The two different responses to water deficit stress show that there is a defined link between stomatal closure and water potential (Brodribb et al. 2014). Isohydric plants are also thought to be less able to recover from drought stress than anisohydric plants due to a lag time between the end of the drought stress and the (slow) degradation of residual ABA (Brodribb and McAdam 2013). But, due to their rapid stomatal closure, they have a higher water use efficiency (WUE), which represents the ratio of CO_2 uptake against evapotranspiration, compared to anisohydric plants. As such, it is hypothesized that isohydric plants outperform anisohydric plants in conditions of severe drought but do less well under milder drought conditions (Attia et al. 2015). Nevertheless, WUE does not seem to be a reliable indicator for greater survival or growth during a drought episode (Moran et al. 2017).

Plants also respond to drought stress via osmotic adjustment to maintain cell turgor. This can be accomplished by the release of ions from the vacuole into the cytosol, in turn maintaining the water potential gradient between the cell and its exterior. Increasing the ion concentration in the cell can however be problematic as it may interfere with protein structure and denature cell components, which is why plants also produce osmotically compatible compounds such as the amino acid proline and the late embryogenic abundant (LEA) group of proteins to help protect vital cell structures (Tardieu et al. 2018; Ditmarova et al. 2010; Morgan 1984). However, not only plant hydraulics is affected by severe drought; also, phloem functioning, i.e., the transport of non-structural carbohydrates, can be seriously affected causing a decreased phloem transport rate, an increased sugar concentration in the phloem sap and anatomical changes toward smaller sieve elements (Salmon et al. 2019).

9.3.3 Norway Spruce's Responses to Drought

Norway spruce has a wide geographic distribution, extending from the Alpine region in Central Europe to the entire Eurasian continent between 47°N and 68°N latitudes and passing through the Scandinavian peninsula (Chen et al. 2012); it is one of the most actively used trees in reforestation and silvicultural exploitations around Europe; as such, it is exposed to all kinds of abiotic stresses including very low temperatures and drought, either due to winter freezing or because of climate change. Future climate simulations predicted that Norway spruce stands will be affected by drought to a higher extent compared to their current status (Thiele et al. 2017). While there is an obvious decrease in stem size and stem width following periods of significant drought stress, more is happening with Norway spruce under water deficit stress than meets the eye. Although evidence suggests cross-talk in abiotic stress resistances, it is noted that frost tolerant spruce populations are also more drought resistant (Blödner et al. 2005). The focus of the following section will be on how Norway spruce responds to drought stress.

Norway spruce exhibited considerable drought stress susceptibility in several experiments, yet surprisingly, higher resistance than some tree species native to dry areas (Battipaglia et al. 2009). The initial response of spruce to low leaf water potential is the rapid closure of needles stomata leading to a reduction in both stomatal conductance (g_s) and CO_2 gas exchange. To prevent oxidative damage, spruce produces higher quantities of the superoxide dismutase enzyme and several osmoprotectant molecules such as proline and also lowers the activity of peroxidase enzymes. Production of LEA proteins also occurs to prevent the aggregation of macromolecules within spruce's cells and to protect the essential proteins from being denatured by trapping a thin layer of water around them. This allows the tree to maintain its photosynthetic capacity for up to 30 days after the onset of mild drought stress. However, beyond this period, there will be damage to the photosystems (mainly PS II) and other cellular structures due to ongoing oxidative damage if the drought stress persists (Ditmarova et al. 2010; Blödner et al. 2005; Wallin et al. 2002). To maintain hydration during drought, Norway spruce has also evolved physical adaptations to combat the effects of water deficit stress, notably by increasing its root/shoot ratio via the production and elongation of root tips and root hairs, thereby increasing its ability to search for and absorb water deep within the ground. Coupled with such adaptive changes, Norway spruce (and conifers more generally) also developed specific pit membrane structures called torus between the narrow (less than 30 μm wide) xylem tracheids, which limits the propagation of embolisms should they occur. Additionally, Montwé et al. (2014) found a plastic response of Norway spruce's hydraulic system to drought stress; individuals that were exposed to drought developed a more safety-oriented wood structure at the expense of their water-conducting system's efficiency. It should be noted however that plant's repeated exposure to freeze-thaw cycles, as it is often experienced by the perennials from boreal or alpine regions during winter, can lead to an increase in the frequency of embolisms within the plant's tracheids, which in turn may be an important limiting factor regarding the expansion of Norway spruce further north beyond 68°N or above a certain altitude (Mayr et al. 2003; Mottonen et al. 2001). On the other hand, Tomasella et al. (2017) suggested that Norway spruce can repair xylem embolism through the depletion of non-structural carbohydrates when water reappears. Similarly, Hartmann et al. (2013) found cyclic recovery of the hydraulic system under a repeated drought induction scenario. However, growth was sacrificed for maintenance of respiration and osmoprotection. The study also found differences in the dynamics of carbon allocation between roots and shoots; while carbon depletion resulting in carbon starvation is considered as the cause of roots death, this is not the case for above-ground plant parts.

Moreover, drought induces differences in the qualitative and quantitative chemical composition of spruce, notably with regard to an increasingly produced quantity of defensive molecules such as monoterpenes. While this comes at the cost of decreased tree growth overall, however, it may be considered a worthwhile trade-off as trees that produce more defensive secondary compounds are more likely to survive the onslaught of insect infestations that usually occur during or following drought (Turtola et al. 2003; Walling et al. 2002).

9.4 The Genetic Architecture of Drought Resistance Mechanisms

9.4.1 QTL Mapping

Quantitative trait loci (QTL) mapping represents a statistical procedure using phenotypes and genetic data to dissect the genetic basis of the phenotypic variation of interest. Genotypic data were historically derived from easy to generate and mostly anonymous simple sequence repeats (SSRs) or restriction fragment length polymorphism (RFLP) molecular markers, later, markers also included single nucleotide polymorphisms (SNPs) from candidate genes. Therefore, earlier QTL maps in forest trees suffered from very low genome coverage and consequently, large QTL interval regions and low resolution. The QTL mapping method itself requires two or more parental individuals that are genetically sufficiently diverse regarding their phenotypic expression of the investigated trait (segregation) in the F1 generation. The heterozygous individuals generated in an F1 generation are subsequently crossed with each other using different mating schemes to create an F2 generation; individuals at the F2 generation are then scored for phenotypes and genotypes (Miles and Wayne 2008). The statistical analysis of the data can then identify which genetic markers are associated with the trait's variability to indicate the most likely QTL position as well as the proportion of phenotypic variance explained by the uncovered QTLs.

However, this approach suffers from many caveats connected to conifer biology such as spruce's long time to reach sexual maturity in order to develop the F2 generation with sufficient sample size (González-Martínez et al. 2006) as well as conifer genomes' distinct nature with regard to their large sizes and the fast decay in linkage disequilibrium (LD) along their genomes (Neale and Kremer 2011). Therefore, dense marker arrays are required to achieve efficient capture of QTLs. Nevertheless, several QTL mapping studies using controlled environmental conditions were successfully performed for traits related to drought tolerance in conifers. Brendel et al. (2002) investigated the level of inheritance in wood cellulose carbon isotope composition $\delta^{13}C$ and mean ring width for the studied maritime pine population and found significant and moderate heritability estimates of 0.17 and 0.19, respectively. This was the first study of QTLs related to the $\delta^{13}C$ value as an indirect measure of WUE and the authors found four significant and four suggestive QTLs involved in the genetic architecture of this attribute. Additionally, the study found two significant and four suggestive QTLs for ring width but no co-location of QTLs with those involved in WUE. Marguerit et al.

(2014) found co-locations of QTLs involved in $\delta^{13}C$ with QTLs involved in intrinsic WUE and stomatal conductance. These co-localizations of QTLs provide important evidence of their functional relationship and the potential phenotypic integration between these traits (Plomion et al. 2016). In addition, the study did not find any unfavorable correlations between WUE and growth which would indicate possible selection for resistance to drought without compromising growth potential. In fact, the same study identified a positive relationship between WUE and growth for trees grown at the dry site, a finding attributed to trees' nutrient-use efficiency that positively affected their photosynthetic activity. De Miguel et al. (2014) identified photosynthesis and WUE-related QTLs in *Pinus pinaster* (the major QTL explained 67% of the phenotypic variance), genomic regions which contained several interesting candidate genes such as *MYXTA-LIKE TRANSCRIPTION FACTOR* involved in multiple physiological processes including photosynthesis-related signaling (Saibo et al. 2009) or *HISTONE CHAPERON*, a gene encoding a heat protection enzyme (Kotak et al. 2007). However, with regard to addressing the genetics of drought tolerance, the use of WUE and its indirect proxy $\delta^{13}C$ is discouraged because WUE is not necessarily related to greater survival or growth potential for conifers growing on a dry site, but it is rather these latter traits that are of interest to determine drought resistance in a species (Moran et al. 2017).

9.4.2 Gene Expression Studies

With the advent of next-generation sequencing platforms, great opportunities arose to study the entire transcriptome even in organisms lacking solid reference genomes (Margulies et al. 2005) as is the case for most conifers. One method of choice is based on whole-transcriptome shotgun sequencing where extracted RNA is sequenced to produce random short reads (due to current technologies' technical limitations), and which are used to reconstruct the original transcripts (de novo assembly). Alternatively, if the reference genome or the transcriptome has already been developed, the newly generated sequence reads are aligned directly to this reference.

The implementation of these methods in conifers can prove challenging for several reasons; the construction of the de novo assembly can be difficult due to the high heterozygosity levels, large genome sizes, the requirement for highly demanding computational resources, and a low transcriptional change to correct sequencing errors in poorly expressed genes. These challenges require specific bioinformatics skills and pipelines (López de Herdia and Váyquez-Poletti 2016). The counting of reads aligned to the reference then provides inference about each transcript's abundance (Wolf 2013). Du et al. (2018) used this approach to study gene expression differences in *Pinus massoniana* under different levels of drought stress and identified several transcription factors associated with circadian rhythms, defense responses, and signal transduction as being potentially involved in drought stress resistance mechanisms (Table 9.1). A similar approach implemented for *Pinus halepensis* found downregulated transcripts from genes involved in photosynthesis, cell wall biosynthesis, stomatal conductance, and biosynthesis of flavonoids and terpenoids. On the other hand, upregulated transcripts were identified for genes involved in chlorophyll degradation, ABA response, and the accumulation of heat shock proteins (Fox et al. 2018), see Table 9.1. Similarly, Dubos et al. (2003) found 28 over-represented and 20 under-represented transcript-derived gene fragments related to photosynthesis, carbohydrate, cell wall synthesis, and plant defense metabolisms in maritime pine (*Pinus pinaster*) seedlings grown under drought stress. Le Provost et al. (2013) investigated the effect of drought tolerance on cuticular wax content in maritime pine and found both higher level of cuticular wax content and overexpression of candidate genes involved in cuticle biosynthesis. Therefore, these results indicate cuticular wax modification as part of a mechanism that facilitates coping with drought stress in

Table 9.1 List of selected gene associations with drought resistance in several coniferous species of the Pinaceae (*Picea*; *Pinus*) from seven most recent genomics studies

Association	Gene annotation ID	Putative function in conifers	Method	Species	References
Phytohormone related					
NCED3	At3g14440 homolog	Upregulated in roots and needles by drought stress; ABA synthesis	1	*Picea abies*	Haas et al. (2019)
NCED5	At1g30100 homolog	Upregulated in roots and needles by drought stress; ABA synthesis	1	*P. abies*	Haas et al. (2019)
NCED—9-cis-epoxycarotenoid dioxygenase	c71048_g1	ABA biosynthesis; upregulated under light drought; drought tolerance development	2	*Pinus massoniana*	Du et al. (2018)
PP2CA	At3g11410 homolog	Upregulated in roots and needles by drought stress; ABA signaling	1	*Picea abies*	Haas et al. (2019)
PP2C, type 2 protein phosphatase	c68631_g1	Negative regulator of SnRK2 protein kinases; downregulated upon drought stress	2	*Pinus massoniana*	Du et al. (2018)
Serine/threonine-kinase SnRK2 (SNF1 related protein kinase)	At1g78290 homolog	Upregulated in roots and needles by drought stress; ABA signaling	1	*Picea abies*	Haas et al. (2019)
SnRK2	c85767_g1 and c85767_g2	ABA signaling enhancer; upregulated during light drought	2	*Pinus massoniana*	Du et al. (2018)
Serine threonine-kinase SRK2E	Contig3427	Activator of ABA signaling pathway; upregulation at stages of minimum transpiration and post irrigation	3	*Pinus halapensis*	Fox et al. (2018)
E3 ubiquitin-ligase RGLG1	TR46403|c0_g2_i1	Upregulated at stage of minimum transpiration; promotes ABA signaling by mediating PP2C degradation	3	*P. halapensis*	Fox et al. (2018)

(continued)

Table 9.1 (continued)

Association	Gene annotation ID	Putative function in conifers	Method	Species	References
ERF53	At2g20880 homolog	Upregulated in roots; drought stress related; involved in ethylene-regulated signaling pathway	1	*Picea abies*	Haas et al. (2019)
ARF auxin responsive factor	c83733_g1	DNA binding; constitutively downregulated expression with increasing drought stress; stimulates root–shoot ratio increase	2	*Pinus massoniana*	Du et al. (2018)
AUX/IAA repressor protein	c77087_g1 and c92989_g1	Auxin sensing; constitutively downregulated expression with increasing drought stress; stimulates root–shoot ratio increase	2	*P. massoniana*	Du et al. (2018)
PYL ABA receptor protein	c87460_g1	ABA response activation; upregulated under moderate drought	2	*P. massoniana*	Du et al. (2018)
PYL8 homolog	TR76718|c0_g1_i4	Upregulated at stage of minimum transpiration, active ABA receptor at this drought stage	3	*Pinus halapensis*	Fox et al. (2018)
Heat shock proteins HSP 70, 90	TR58521|c0_g1_i2 and TR759|c6_g1_i5	Upregulated at stage of pre-stomatal response, upstream of the ABA-induced stomatal closure	3	*P. halapensis*	Fox et al. (2018)
ABI5 ABA-responsive transcript	Phal_PtaS25591072	ABA response; upregulated at stage of minimum transpiration	3	*P. halapensis*	Fox et al. (2018)
CHY ZINC-FINGER AND RING FINGER PROTEIN1 (CHYR1)	TR40635|c2_g2_i5	ABA-mediated stomatal closure; gradual upregulation during drought and slow down-regulation during	3	*P. halapensis*	Fox et al. (2018)

(continued)

Table 9.1 (continued)

Association	Gene annotation ID	Putative function in conifers	Method	Species	References
		recovery; slow response of *P. halepensis* stomata (R-type)			
ARM repeat protein interacting with ABF2	MA_90007g0010	Associated with Recovery Rc00 and Resistance Rt03; ABA response	4	*Picea abies*	Trujillo-Moya et al. (2018)
Gibberellin 2-beta-dioxygenase 3 GA2ox3	MA_10289324g0010	Associated with CorJunT (June temperature and annual ring width); gibberellin biosynthesis	4	*P. abies*	Trujillo-Moya et al. (2018)
Hexose:hydrogen symporter	At5g26340 homolog	Response to ABA; associated with aridity gradient	5	*Pinus taeda*	Eckert et al. (2010)
Transcription (and stress)					
WRKY33 TF	c83644_g4	Systematic defense response; upregulated under light drought	2	*Pinus massoniana*	Du et al. (2018)
PLATZ TF family	TR2842|c0_g1_i6	Suppresses transcription via non-specific binding to A/T-rich sequences; upregulated at pre-stomatal response	3	*Pinus halapensis*	Fox et al. (2018)
ERF/AP2 family TF	c76570_g1	DNA binding; constitutively upregulated expression with increasing drought stress; drought tolerance	2	*Pinus massoniana*	Du et al. (2018)
ERF/AP2 DREB2B TF	At3g11020 homolog	Upregulated in roots; has well-defined function in drought stress (heat acclimation; response to water deprivation)	1	*Picea abies*	Haas et al. (2019)

(continued)

Table 9.1 (continued)

Association	Gene annotation ID	Putative function in conifers	Method	Species	References
ANAC072/RD26 TF, stress-specific NAC (SNAC)	At4g27410 homolog	Upregulated in needles and roots upon drought; response to desiccation; response to ABA	1	*P. abies*	Haas et al. (2019)
MYB2 TF	At2g47190 homolog	Root-specific TF; upregulation in early response to drought; regulates ABA-inducible genes; response to abiotic stress	1	*P. abies*	Haas et al. (2019)
MYB96 TF	At5g62470 homolog	Upregulated in roots, strongly induced by ABA; upon severe drought, arrest of lateral root growth	1	*P. abies*	Haas et al. (2019)
MYB15 TF	At3g23250 homolog	Upregulated in roots during recovery; improved drought tolerance; ABA sensitivity	1	*P. abies*	Haas et al. (2019)
Stress related					
Catalase isozyme 1	TR23182|c0_g3_i1	Strong expression of this ROS-scavenging enzyme at pre-stomatal response and strongly downregulated upon recovery	3	*Pinus halapensis*	Fox et al. (2018)
Ferredoxin-thioredoxin reductase catalytic chloroplastic	TR72875|c0_g1_i2	Strong expression at pre-stomatal response of this ROS-scavenging enzyme	3	*P. halapensis*	Fox et al. (2018)
Glutathione transferase (GST)	c92901_g1 and c93725_g3	Glutathione metabolism pathway and generic detoxification; constitutively upregulated under drought stress; positive regulator of drought tolerance	2	*Pinus massoniana*	Du et al. (2018)

(continued)

Table 9.1 (continued)

Association	Gene annotation ID	Putative function in conifers	Method	Species	References
GPX protein	c92413_g1, c92413_g2, and c88279_g1	Glutathione metabolism pathway and response to oxidative stress; constitutively upregulated with increasing drought stress; abiotic stress tolerance enhancement	2	*P. massoniana*	Du et al. (2018)
GSR protein	c83219_g1	Glutathione metabolism pathway and plant tolerance to stress conditions; upregulated under severe drought; defense against ROS	2	*P. massoniana*	Du et al. (2018)
Dehydrins PgDhn 10 N1AESK2, 16 K4, 33 N1K2, 35 N1K2	GQ0067_P11, GQ04103_G2, GQ03719_H02 and GQ03903_F04	Water stress response; strongly upregulated under drought	6	*Picea glauca*	Stival Sena et al. (2018)
L-ascorbate oxidase homolog	MA_161013g0010	Associated with Recovery Rc03; homozygous genotypes GG (of ss538944271) show better drought recovery	4	*Picea abies*	Trujillo-Moya et al. (2018)
Development (and stress)					
Photosystem II D1 (chloroplast)	TR37804|c4_g3_i3	At the early stage of pre-stomatal response strongly down-regulated; this impairs PSII function, leading to reduced CO_2 fixation and increased ROS production	3	*Pinus halapensis*	Fox et al. (2018)
Photosynthesis II protein	At2g06520 homolog	Photosynthesis; associated with aridity gradient	5	*Pinus taeda*	Eckert et al. (2010)
C3HC4-type RING finger protein	At4g26400 homolog	Protein ubiquitination; associated with aridity gradient	5	*P. taeda*	Eckert et al. (2010)

(continued)

Table 9.1 (continued)

Association	Gene annotation ID	Putative function in conifers	Method	Species	References
MATE efflux family protein	At425640 homolog	Multidrug and toxin efflux family transporter involved in flavonoid metabolism; associated with aridity gradient	5	*P. taeda*	Eckert et al. (2010)
UDP-galactose transporter	At3g59360 homolog	Carbohydrate transport; associated with aridity gradient	5	*P. taeda*	Eckert et al. (2010)
Pyrroline-5-carboxylate synthase	c93699_g2	Membrane stabilization; constitutively upregulated expression with increasing drought stress; proline accumulation	2	*Pinus massoniana*	Du et al. (2018)
HY5—BZIP TF	c84637_g1	Hormone and light signaling pathway; downregulated under light drought; circadian rhythm	2	*P. massoniana*	Du et al. (2018)
LHY protein	c85168_g1 and c91081_g2	Central oscillator component of the light input pathway; upregulated under light drought; circadian rhythm	2	*P. massoniana*	Du et al. (2018)
Gamma aminobutyrate transaminase chloroplastic	TR20760|c0_g1_i4	Upregulated at pre-stomatal response; catalyzes the breakdown of GABA; GABA functions as a long-term carbon and nitrogen reserve that facilitates recovery and confers a priming effect against future stress	3	*Pinus halapensis*	Fox et al. (2018)

(continued)

Table 9.1 (continued)

Association	Gene annotation ID	Putative function in conifers	Method	Species	References
Guard cell S-type anion channel SLAC1	clc_contig_4642 and Contig5129	Stomatal closure; downregulated during drought treatment (possibly related to R-type ABA response)	3	*P. halapensis*	Fox et al. (2018)
Histone-lysine N-methyltransferase SUVR5	TR4553|c0_g2_i1	At post irrigation, dramatic decrease in expression, suggesting reduction in H3K9me and hence transcriptional activation.	3	*P. halapensis*	Fox et al. (2018)
Lysine-specific histone demethylase 1 homolog 3	TR57204|c0_g2_i4	At post-irrigation, dramatic increase (~130-fold) in the expression, which specifically demethylates H3K4me, the positive regulator of gene expression.	3	*P. halapensis*	Fox et al. (2018)
Chalcone synthase	TR38771|c7_g4_i3	Strongly downregulated at pre-stomatal response and under partial recovery; strongly upregulated under full recovery; major flavonoid biosynthesis-related transcript	3	*P. halapensis*	Fox et al. (2018)
Homeobox-leucine zipper ANTHOCYANINLESS 2-like isoform X2	TR45014|c0_g3_i4	Strongly downregulated at pre-stomatal response	3	*P. halapensis*	Fox et al. (2018)
Homeobox-leucine zipper protein ANTHOCYANINLESS 2	MA_38472g0010	Associated with Recovery Rc03	4	*Picea abies*	Trujillo-Moya et al. (2018)
Aquaporin AQPs NIP; PIP; PIP	c85755_g1; c91691_g1 and c101640_g1	Radial water transport; constitutively downregulated expression with increasing	2	*Pinus massoniana*	Du et al. (2018)

(continued)

Table 9.1 (continued)

Association	Gene annotation ID	Putative function in conifers	Method	Species	References
		drought stress; water conservation			
Aquaporin AQP PIP 1, 2-plasma membrane intrinsic protein	GQ03401_M18, GQ03610_A06, GQ03111_E12, GQ02901_B20	Radial water transport; down-regulation during drought; change in hydraulic conductivity in bundle sheet cells and potential implication in embolism repair	7	*Picea glauca*	Laur and Hacke (2014)
Pentatricopeptide repeat-containing protein	MA_100637g0010	Associated with Recovery Rc00; highly expressed in new wood-forming tissue (cambium and xylem expansion layers)	4	*Picea abies*	Trujillo-Moya et al. (2018)
Short-chain dehydrogenase TIC 32, chloroplastic	MA_10432182g0010	Associated with Latewood Width LW	4	*P. abies*	Trujillo-Moya et al. (2018)
Translocon-associated protein subunit alpha	MA_109428g0010	Associated with Resistance Rt03	4	*P. abies*	Trujillo-Moya et al. (2018)
Calcium-binding protein CML49	MA_17539g0010	Associated with Latewood Width LW	4	*P. abies*	Trujillo-Moya et al. (2018)
Hydroxycinnamoyl-CoA shikimate hydroxycinnamoyltransferase	MA_39636g0010	Associated with Ring Width RW, RWt, Earlywood Width EW; cell wall lignification to minimize water loss and cell dehydration	4	*P. abies*	Trujillo-Moya et al. (2018)
Clathrin heavy chain 1	MA_47260g0010	Associated with Recovery Rc03	4	*P. abies*	Trujillo-Moya et al. (2018)
Predicted protein	MA_475589g0010	Associated with Resilience Rs03, Resistance Rt03	4	*P. abies*	Trujillo-Moya et al. (2018)

(continued)

Table 9.1 (continued)

Association	Gene annotation ID	Putative function in conifers	Method	Species	References
no similarity	MA_588952g0010	Associated with Earlywood Density EDt	4	*P. abies*	Trujillo-Moya et al. (2018)
Unknown protein	MA_59794g0010	Associated with "Resistance 1993" and "Resistance 2000" (Scen1, i.e., continuous drought resistance) and Latewood Width LW	4	*P. abies*	Trujillo-Moya et al. (2018)
no similarity	MA_10434266g0010	Associated with "Resistance 1993" and "Resistance 2000" (Scen2, i.e., continuous drought resistance)	4	*P. abies*	Trujillo-Moya et al. (2018)
Serine/threonine-protein kinase sepA	MA_6741659g0010	Associated with Latewood Width LW	4	*P. abies*	Trujillo-Moya et al. (2018)
Phytochrome, PHYN	MA_73153g0010	Associated with Latewood Width LW	4	*P. abies*	Trujillo-Moya et al. (2018)
Unknown protein	MA_8047450g0010	Associated with Recovery Rc00	4	*P. abies*	Trujillo-Moya et al. (2018)
ERBB-3 BINDING PROTEIN 1-like	MA_10426694g0020	Associated with mean Ring Density RD and RDt	4	*P. abies*	Trujillo-Moya et al. (2018)
EGF domain-specific O-linked N-acetylglucosamine transferase-like	MA_10434370g0040	Associated with Resistance Rt03	4	*P. abies*	Trujillo-Moya et al. (2018)
Unknown protein	MA_2267030g0010	Associated with CorJunT (June temperature and annual ring width)	4	*P. abies*	Trujillo-Moya et al. (2018)
Unknown protein	MA_6521216g0010	Associated with CorJunT (June temperature and annual ring width)	4	*P. abies*	Trujillo-Moya et al. (2018)

(continued)

Table 9.1 (continued)

Association	Gene annotation ID	Putative function in conifers	Method	Species	References
WD repeat-containing protein VIP3	MA_10432150p0010	Associated with Recovery Rc03	4	*P. abies*	Trujillo-Moya et al. (2018)

[1]RNA-seq; 3-year-old seedlings; mild drought (30% field capacity) and severe drought (with shoot water potential −2.1 MPa), re-irrigation (to 80% field capacity); needles, fine and small roots sampling (Haas et al. 2019)
[2]RNA-seq; drought-resistant genotype; stem apex needles sampling (Du et al. 2018)
[3]RNA-seq; rooted clones produced from 20-year-old tree growing in semi-arid environment; no irrigation for 34 d (stomatal conductance reduced to 94 mmol s^{-1} g^{-1}); mature needles sampling (Fox et al. 2018)
[4]GWAS; 1,700 variable SNPs; 147 trees from 11 provenances (complete Central European range of Norway spruce); needles sampling (Trujillo-Moya et al. 2018)
[5]GWAS; 3,059 SNPs; 622 trees; needles sampling (Eckert et al. 2010)
[6]Genome data mining; quantitative real-time PCR; young trees generated by somatic embryogenesis; drought stress foliage water potential ~ −35 bar; foliage sampling (Stival Sena et al. 2018)
[7]Genome data mining; quantitative real-time PCR; 3-year-old plants; drought stress stem water potential ~ −3 MPa; needle tissue analyses (Laur and Hacke 2014)

maritime pine. More recently, Haas et al. (2019) investigated gene expression patterns across several tissues in Norway spruce seedlings exposed to drought and found different trends between above- and below-ground tissues. Their analysis identified preferential gene expression of certain MYB-, ERF-, C2H2-, and NAC-transcription factor family members in needles and of other ERF-, MYB-, and SAP-TF family members in roots (Table 9.1). Severe drought influenced expression levels of NAC-, ERF-, WRKY-, and MYB-transcription factor gene family members in the roots (upregulated). Some ERF- and MYB-transcription factors were down-regulated. Conserved (across tissues) upregulated expression involved genes with suggested functions as dehydrins, malate, and ROS scavengers. Laur and Hacke (2014) retrieved thirty *Picea glauca* genes with aquaporin activity (from plasma membrane intrinsic protein (PIP) subfamilies PIP1 and PIP2, respectively) using BLASTn homology search and investigated their expression patterns; increased relative air humidity following a drought period triggered high aquaporin gene expression activity localized within white spruce's needle tissues. The study confirmed aquaporins' involvement in mediating radial water flow from foliage tissue toward the stem xylem and furthermore provided a functional link of aquaporins with embolism repair. Similarly, Stival Sena et al. (2018) used gene-specific amplification of RNA transcripts to investigate the roles of dehydrins in response to drought stress in white spruce. They found differential gene expression patterns between tissues and conditions, and identified *PgDhn* 10, 16, 33, and 35 upregulated during drought. A summary of all these results is provided in Table 9.1.

9.4.3 Genome-Wide Associations (GWAS)

The advancement of high-throughput genotyping platforms propelled the development of diverse genomic resources for non-model organisms (Neale and Kremer 2011). As a consequence, the implementation of genomic studies to dissect the genetic architecture of complex traits was also initiated for conifers. The main aim of the association mapping approach is to uncover genetic variants that significantly correlate with variation in the studied traits and support marker-assisted selection through breeding using the preferred alleles that originate from the phenotype associated genetic variants. Moreover, conifers' characteristics of relatively unstructured populations following random mating assumptions, their high levels of nucleotide diversity and the fast decay in species' genome-wide linkage disequilibrium seem to be ideal for such statistical analyses' implementation in the species (Neale and Savolainen 2004). Here, we summarize the most intriguing results from such recent GWAS in conifers (including *Picea abies*) related to drought stress adaptation (see also Table 9.1).

Trujillo-Moya et al. (2018) tested for differences in drought stress response among Norway spruce provenances that originated from Poland, The Czech Republic, Bulgaria, and Austria, respectively, and found a strong genetic signal in several provenances (Fig. 9.1). They performed GWAS using a customized SNP array and dendrochronological traits related to resistance to drought stress. In total 29 statistically significant SNP associations affiliated to 24 genes were identified when only provenances with strong genetic signal were included in the association mapping population. Among these, the identified spruce genes could be annotated among others as putative *hydroxycinnamoyl-CoA shikimate/quinate hydroxycinnamoyl transferase* (HCT) from lignin biosynthesis pathway, and *ARM repeat protein interacting with ABF2* which encodes a protein involved in ABA-mediated response (Table 9.1). The latter finding confirms the involvement of ABA in adaptation to a wide range of abiotic stresses (Vishwakarma et al. 2017). Genetic associations from this study were also localized within the intronic region of a putative *multicopper oxidase/L-ascorbate oxidase* gene implicated in ascorbate oxidation and a potential contributing factor to a plant's deleterious response to stress (Batth et al. 2017); furthermore, within an exon of the *gibberellin 2-beta-dioxygenase 3* gene whose product is implicated in gibberellin biosynthesis (Table 9.1). The phytohormone gibberellin is a proven contributor in the plant's physiological response to drought stress (Zawaski and Busov 2014), and therefore, the above result confirms gibberellin's implication in the drought stress response for Norway spruce. Expression profiling for all identified genes from the association work localized gene upregulation to cambium and xylem expansion cells (Trujillo-Moya et al. 2018).

A common approach to search for adaptive genetic variation is to undertake landscape genomics studies; Eckert et al. (2010) performed such GWAS in loblolly pine (*Pinus taeda* L.) by testing for significant correlations between genetic SNP markers within 3,059 genes and environmental variables. They found five statistically significant associations, among which a putative ortholog of an *Arabidopsis* gene implicated in ABA responsiveness was identified (Table 9.1); these results again confirmed this hormone's involvement in plant's response to environmental stress (Wasternack 2007), here, in a pine.

9.4.4 Major Genetic Players in Norway Spruce Drought Resistance

We compiled a comprehensive list of candidate genes for drought resistance originating from recent genomics studies querying (Norway) spruce's genome and transcriptomes, respectively, and that we put into context with results involving Pinaceae from around the world. Table 9.1 clearly indicates the consistency of some candidates involved in phytohormone-related drought stress signaling (foremost ABA, but also auxin, gibberellin, and ethylene) such as *9-cis-epoxycarotenoid dioxygenase (NCED)* in ABA biosynthesis and implicated in the development of drought tolerance (in spruce and pine); *Serine/threonine-kinase, SNF1 related protein kinase*, for enhancement of the ABA signal (in spruce and pine); *ERF/AP2 family transcription factor* implicated in acclimation and drought tolerance (spruce and pine); moreover the importance of aquaporins for water conservation and embolism repair (spruce and pine), and the implication of flavonoids in drought recovery (*ANTHOCYANINLESS 2*, e.g., found important in both spruce and pine species). Strikingly, we sometimes found overlaps among the same statistically supported hits of candidates from the different methods (GWAS vs. transcriptome analyses) and querying distantly related species of the Pinaceae in the drought experiments. Therefore, this provides convincing evidence that these candidate genes are involved in drought resistance across a wider range of conifer species. Including such strong candidates could also strengthen genomic predictions for drought resistance in Norway spruce and thus, could better inform practitioners on selecting the best planting material (Moran et al. 2017).

9.5 Future Directions to Integrate Findings

Climate change's increasing intensity puts more pressure on resilience management of forest genetic resources. More frequent drought event occurrences will require a better understanding regarding plant's physiological mechanisms and their genetic underpinnings with regard to drought stress resistance including more holistic approaches to the problem. Körner (2019) recommends the investigation of not only the mechanisms involved in the direct response to drought but also suggests further explorations into plant's water storage capacity and water accessibility within plants. In addition, the mechanisms underlying growth recovery after a drought period are still not well understood and will need more attention (Ruehr et al. 2019). Perhaps equally important, understanding species' phenotypic plasticity regarding how changes in individual environmental cues and their combinatory effects impact the expression of mechanisms underlying species' resistance to drought, is critical for defining the species and the populations within species' physiological optimums; and subsequently, species' or populations' limits for deployment strategies under changing climatic conditions (Woodall et al. 2010; Chakraborty et al. 2019). Altogether, with the implementation of current climate models projecting future climate, a reasonably developed adaptive management should be proposed to minimize the inevitable impacts of climate change (Albert and Schmidt 2010; Albert et al. 2017). Forest trees are not standalone units but represent parts of complex interacting ecosystems, and thus, their mechanisms underlying resistance to drought should be investigated within the context of such interactions (for example, through the analysis of environmental DNA (eDNA) to investigate the benefits of local ectomycorrhizal communities and other parts of the local soil microbiome). Additionally, attention should be paid to the consequences of drought-induced tree function failure on whole ecosystem's functioning (Salmon et al. 2019). Finally, Heinimann (2010) proposed an approach for adaptive forest ecosystem management focusing on three main aspects: keeping dynamics of the ecosystem within acceptable boundaries, looking at ecosystem-wide processes rather than more localized events, as well as considering all interactions between processes across the space-time scale.

References

Albert M, Schmidt M (2010) Climate-sensitive modelling of site-productivity relationships for Norway spruce (*Picea abies* (L.) Karst.) and common beech (*Fagus sylvatica* L.). For Ecol Manag 259:739–749

Albert M, Nagel R-V, Nuske RS, Sutmöller J, Spellmann H (2017) Tree species selection in the face of drought risk – uncertainty in forest planning. Forests 8:363

Allen CD, Macalady AK, Chenchouni H, Bachelet D, McDowell N, Vennetier M, Kitzberger T, Rigling A, Breshears DD, Hogg EH (Ted), Gonzalez P, Fensham R, Zhang Z, Castro J, Demidova N, Lim J-H, Allard G, Running SW, Semerci A, Cobb N (2010). A global overview of drougth and heat induced tree mortality reveals emerging climate change risks for forests. For Ecol Manag 259:660–684

Aranda I, Alia R, Ortega U, Dantas AK, Majada J (2010) Intra-specific variability in biomass partitioning and carbon isotopic discrimination under moderate drought stress in seedlings from four *Pinus pinaster* populations. Tree Genet Genomes 6:169–178

Attia Z, Domec JC, Oren R, Way DA, Moshelion M (2015) Growth and physiological responses of isohydric and anisohydric poplars to drought. J Exp Bot 66:4373–4381

Ballester J, Rodó X, Giorgi F (2010) Future changes in Central Europe heat waves expected to mostly follow summer mean warming. Clim Dyn 35:1191–1205

Batth R, Singh K, Kumari S, Mustafiz A (2017) Transcript profiling reveals the presence of abiotic stress and developmental stage specific ascorbate oxidase genes in plants. Front Plant Sci 8:198

Battipaglia G, Saurer M, Cherubini P, Siegwolf RTW, Cotrufo MF (2009) Tree rings indicate different drought resistance of a native (*Abies alba* Mill.) and a non-native (*Picea abies* (L.) Karst.) species co-occurring at a dry site in Southern Italy. For Ecol Manag 257:820–828

Beniston M, Stephenson DB, Christensen OB, Ferro CAT, Frei C, Goyette S, Halsnaes K, Holt T, Jylhä K, Koffi B, Palutikof J, Schöll R, Semmler T, Woth K (2007) Future extreme events in European climate: an exploration of regional climate model projections. Clim Chang 81:71–95

Blödner C, Skroppa T, Johnsen Ø, Polle A (2005). Freezing tolerance in two Norway spruce (*Picea abies* [L.] Karst.) progenies is physiologically correlated with drought tolerance. J Plant Physiol 162(5):549–558

Boisvenue C, Running SW (2006) Impacts of climate change on natural forest productivity – evidence since the middle of the 20th century. Glob Chang Biol 12:862–882

Brendel O, Pot D, Plomion C, Rozenberg P, Guehl J-M (2002) Genetic parameters and QTL analysis of $\delta^{13}C$ and ring width in maritime pine. Plant Cell Environ 25 (8):945–953

Brini E, Fennell CJ, Fernandez-Serra M, Hribar-Lee B, Lukšič M, Dill KA (2017) How water's properties are encoded in its molecular structure and energies. Chem Rev 117:12385–12414

Brodribb TJ, McAdam SAM (2013) Abscisic acid mediates a divergence in the drought response of two conifers. Plant Physiol 162:1370–1377

Brodribb TJ, McAdam SAM, Jordan GJ, Martins SCV (2014) Conifer species adapt to low-rainfall climates by following one of two divergent pathways. Proc Natl Acad Sci USA 111:14489–14493

Burczyk J, Giertych M (1991) Response of Norway spruce (*Picea abies* (L.) Karst.) annual increments to drought for various provenances and locations. Silvae Genet 40:146–152

Chakraborty D, Jandl R, Kapeller S, Schueler S (2019) Disentangling the role of climate and soil on tree growth and its interaction with seed origin. Sci Total Environ 654:393–401

Chen J, Källman T, Ma X, Gyllenstrand N, Zaina G, Morgante M, Bousquet J, Eckert A, Wegrzyn J, Neale D, Lagercrantz U, Lascoux M (2012) Disentangling the roles of history and local selection in shaping clinal variation of allele frequencies and gene expression in Norway spruce (*Picea abies*). Genetics 191:865–881

Chen L, Huang J-G, Stadt KJ, Comeau PG, Zhai L, Dawson A, Ashraful Alam S (2017) Drought explains variation in the radial growth of white spruce in western Canada. Agric For Meteorol 233:133–142

Chmura DJ, Guzicka M, McCulloh KA, Zytkowiak R (2016) Limited variation found among Norway spruce half-sib families in physical response to drought and resistance to embolism. Tree Physiol 36(2):252–266

Daszkowska-Golec A, Szarejko I (2013) Open or close the gate – stomata action under the control of phytohormones in drought stress conditions. Front Plant Sci 4:138

de Miguel M, Cabezas J-A, de María N, Sánchez-Gómez D, Guevara M-A, Vélez M-D, Sáez-Laguna E, Díaz L-M, Mancha J-A, Barbero M-C, Collada C, Díaz-Sala C, Aranda I, Cervera M-T (2014). Genetic control of functional traits related to photosynthesis and water use efficiency in *Pinus pinaster* Ait. drought response: integration of genome annotation, allele association and QTL detection for candidate gene identification. BMC Genomics 15:464

Ditmarova L, Kurjak D, Palmroth S, Kmet J, Strelcova K (2010) Physiological responses of Norway spruce (*Picea abies*) seedlings to drought stress. Tree Physiol 30:205–213

Du M, Ding G, Cai Q (2018) The transcriptomic response of *Pinus massoniana* to drought stress. Forests 9 (6):326

Dubos C, Le Provost G, Pot D, Salin F, Lalane C, Madur D et al (2003) Identification and characterization of water-stress-responsive genes in hydroponically grown maritime pine (*Pinus pinaster*) seedlings. Tree Physiol 23:169–179

Dulamsuren C, Abilova SB, Bektayeva M, Eldarov M, Schuldt B, Leuschner C, Hauck M (2018). Hydraulic architecture and vulnerability to drought-induced embolism in southern boreal tree species of Inner Asia. Tree Physiol tpy116, https://doi.org/10.1093/treephys/tpy116

Eckert AJ, van Heerwaarden J, Wegrzyn JL, Nelson CD, Ross-Ibarra J, Gonzalez-Martinez SC, Neale DB (2010). Patterns of population structure and environmental associations to aridity across the range of Loblolly pine (*Pinus taeda* L., Pinaceae). Genetics 185(3):969–982

Farquhar GD, Sharkey TD (1982) Stomatal conductance and photosynthesis. Ann Rev Plant Physiol 33:317–345

Fox H, Doron-Faigenboim A, Kelly G, Bourstein R, Attia Z, Zhou J, Moshe Y, Moshelion M, David-Schwartz R (2018) Transcriptome analysis of *Pinus halepensis* under drought stress and during recovery. Tree Physiol 38(3):423–441

Gazol A, Camarero JJ, Anderegg WRL, Vicente-Serrano SM (2016) Impacts of droughts on the growth resilience of Northern Hemisphere forests. Glob Ecol Biogeogr. https://doi.org/10.1111/geb.12526

Gehring CA, Sthultz CM, Flores-Renteria L, Whipple AV, Whitham TG (2017) Tree genetics defines fungal partner communities that may confer drought tolerance. Proc Natl Acad Sci USA 114:11169–11174

George J-P, Schueler S, Karanitsch-Ackerl S, Mayer K, Klumpp RT, Grabner M (2015) Inter- and intraspecific variation in drought sensitivity in Abies spec, and its relation to wood density and growth traits. Agric For Meteorol 214–215:430–443

González-Martínez SC, Krutovsky KV, Neale DB (2006) Forest-tree population genomics and adaptive evolution. New Phytol 170(2):227–238
Haas JC, Vergara A, Hurry V, Street NR (2019). Candidate regulators and target genes of drought stress in needles and roots of Norway spruce. https://doi.org/10.1101/517151
Hammond WM, Adams HD (2019). Dying on time: traits influencing the dynamics of tree mortality risk from drought. Tree Physiol 00:tzp050
Hartmann H, Ziegler W, Trumbore S (2013) Lethal drought leads to reduction in non-structural carbohydrates in Norway spruce tree roots but not in the canopy. Funct Ecol 27:413–427
Heinimann HR (2010) A concept in adaptive ecosystem management – an engineering perspective. For Ecol Manag 259:848–856
Kapeller S, Lexer MJ, Geburek T, Hiebl J, Schueler S (2012) Intraspecific variation in climate response of Norway spruce in the eastern Alpine range: selecting appropriate provenances for future climate. For Ecol Manag 271:46–57
Kellomaki S, Peltola H, Nuutinen T, Korhonen KT, Strandman H (2008) Sensitivity of managed boreal forests in Finland to climate change, with implications to adaptive management. Philos Trans R Soc B 363:2341–2351
Klisz M, Buras A, Sass-Klaassen U, Puchalka R, Koprowski M, Ukalska J (2019) Limitations at the limits? Diminishing of genetic effects in Norway spruce provenance trials. Front Plant Sci 10:306. https://doi.org/10.3389/fpls.2019.00306
Kohler M, Sohn J, Nagele G, Bauhus J (2010). Can drought tolerance of Norway spruce (*Picea abies* (L.) Karst.) be increased through thinning? Eur J For Res 129(6):1109–1118
Körner C (2019) No need for pipes when the well is dry – a comment on hydraulic failure in trees. Tree Physiol 39:695–700
Kotak S, Larkindale J, Lee U, von Koskull-Döring P, Vierling E, Scharf K-D (2007) Complexity of the heat stress response in plants. Curr Opin Plant Biol 10:310–316
Lauder JD, Moran EV, Hart SC (2019). Fight or flight? Potential tradeoffs between drought defense and reproduction in conifers. Tree Physiol 00:tpz031
Laur J, Hacke UG (2014) Exploring *Picea glauca* aquaporins in the context of needle water uptake and xylem refilling. New Phytol 203:388–400
Le Provost G, Domergue F, Lalanne C, Campos PR, Grosbois A, Bert D, Meredieu C, Danjon F, Plomion C, Gion J-M (2013) Soil water stress affects both cuticular wax content and cuticle-related gene expression in young saplings of maritime pine (*Pinus pinaster* Ait). BMC Plant Biol 13:95
Levesque M, Saurer M, Siegwolf R, Eilmann B, Brang P, Bugmann H, Rigling A (2013) Drought response of five conniver species under contrasting water availability suggests high vulnerability of Norway spruces and European larch. Glob Chang Biol 19:3184–3199
Lindner M, Maroschek M, Netherer S, Kremer A, Barbati A, Garcia-Gonzalo J, Seidl R, Delzon S, Corona P, Kolström M, Lexer MJ, Marchetti M (2010) Climate change impacts, adaptive capacity, and vulnerability of European forest ecosystems. For Ecol Manag 259:698–709
López de Herdia U, Váyquez-Poletti JL (2016) RNA-seq analysis in forest tree species: bioinformatic problems and solutions. Tree Genet Genomes 12:30
Lu P, Biron P, Bréda N, Granier A (1995) Water relations of adult Norway spruce (*Picea abies* (L) Karst) under soil drought in the Vosges mountains: water potential, stomatal conductance and transpiration. Ann For Sci 52:117–129
Ma F, Xu TT, Ji MF, Zhao CM (2014). Differential drought tolerance in tree populations from contrasting environments. AoB Plants 6:plu069. https://doi.org/10.1093/aobpla/plu069
MacAllister S, Mencuccini M, Sommer U, Engel J, Hudson A, Salmon Y, Dexter KG (2019). Drought-induced mortality in Scots pine: opening the metabolic black box. Tree Physiol 00:tpz049
Marguerit E, Bouffier L, Chancerel E, Costa P, Lagane F, Guehl J-M, Plomion C, Brendel O (2014) The genetics of water-use efficiency and its relation to growth in maritime pine. J Exp Bot 65(17):4757–4768
Margulies M, Egholm M, Altman WE, Attiya S, Bader JS, Bemben LA, Berka J, Braveman MS, Chen Y-J, Chen Z, Dewell SB, Du L, Fierro JM, Gomes XV, Godwin BC, He W, Helgesen S, He Ho C, Irzyk GP, Jando SC, Alenquer MLI, Jarvie TP, Jirage KB, Kim J-B, Knight JR, Lanza JR, Leamon JH, Lefkowitz SM, Lei M, Li J, Lohman KL, Lu H, Makhijani VB, McDade KE, McKenna MP, Myers EW, Nickerson E, Nobile JR, Plant R, Puc BP, Ronan MT, Roth GT, Sarkis GJ, Simons JF, Simpson JW, Srinivasan M, Tartaro KR, Tomasz A, Vogt KA, Volkmer GA, Wang SH, Wang Y, Weiner MP, Yu P, Begley RF, Rothberg JM (2005) Genome sequencing in microfabricated high-density picolitre reactors. Nature 437:376–380
Miles C, Wayne M (2008) Quantitative trait locus (QTL) analysis. Nat Educ 1(1):208
Mayr S, Gruber A, Bauer H (2003) Repeated freeze-thaw cycles induce embolism in drought stressed conifers (Norway spruce, stone pine). Planta 217:436–441
Montwé D, Spiecker H, Hamann A (2014) An experimentally controlled extreme drought in a Norway spruce forest reveals fast hydraulic response and subsequent recovery of growth rates. Trees 28:891
Moran E, Lauder J, Musser C, Stathos A, Shu M (2017) The genetics of drought tolerance in conifers. New Phytol 216(4):1034–1048

Morgan JM (1984) Osmoregulation and water stress in higher plants. Ann Rev Plant Physiol 35:299–319
Mottonen M, Aphalo PJ, Lehto T (2001) Role of boron in drought resistance in Norway spruce (*Picea abies*) seedlings. Tree Physiol 21:673–681
Myers-Smith IH, Myers JH (2018). Comment on "Precipitation drives global variation in natural selection". Science 359(6375):eaan5028. https://doi.org/10.1126/science.aan5028
Neale DB, Savolainen O (2004) Association genetics of complex traits in conifers. Trends Plant Sci 9(7):325–330
Neale DB, Kremer A (2011) Forest tree genomics: Growing resources and applications. Nat Rev Genet 12(2):111–122
Neuman M, Mues V, Moreno A, Hasenauer H, Seidl R (2017) Climate variability drives recent tree mortality in Europe. Glob Chang Biol 23(11):4788–4797
Nonami H, Boyer JS (1993) Direct demonstration of a growth-induced water potential gradients. Plant Physiol 102:13–19
Plomion C, Bartholomé J, Bouffier L, Brendel O, Cochard H, de Miguel M, Delzon S, Gion J-M, Gonzalez-Martinez SC, Guehl J-M, Lagraulet H, Le Provost G, Marguerit E, Porté A (2016) Understanding the genetic bases of adaptation to soil water deficit in trees through the examination of water use efficiency and cavitation resistance: maritime pine as case study. J Plant Hydraul 3:e008
Rosner S, Klein A, Müller U, Karlsson B (2007) Hydraulic and mechanical properties of young Norway spruce clones related to growth and wood structure. Tree Physiol 27(8):1179–1188
Rosner S, Svetlik J, Andreassen K, Borja I, Dalsgaard L, Evans R, Karlsson B, Tollefsrud MM, Solberg S (2013) Wood density as a screening trait for drought sensitivity in Norway spruce. Can J For Res 44 (2):154–161
Ruehr NK, Grote R, Mayr S, Arneth A (2019). Beyond the extreme: recovery of carbon and water relations in woody plants following heat and drought stress. Tree Physiol 00:tpz032
Saibo NJM, Lourenço T, Oliveira MM (2009) Transcription factors and regulation of photosynthetic and related metabolism under environmental stresses. Ann Bot 103:609–623
Salmon Y, Dietrich L, Sevanto S, Hölttä T, Dannoura M, Epron D (2019) Drought impacts on tree phloem: from cell-level responses to ecological significance. Tree Physiol 39:173–191
Savi T, Casolo V, Dal Borgo A, Rosner S, Torboli V, Stenni B, Bertoncin P, Martllos S, Pallavicini A, Nardini A (2019). Drought-induced dieback of *Pinus nigra*: a tale of hydraulic failure and carbon starvation. Conserv Physiol 7(1):coz012
Schaefer C, Grams TEE, Rotzer T, Feldermann A, Pretzsch H (2017) Drought stress reaction of growth and Δ^{13} C in tree rings of European beech and Norway spruce in monospecific versus mixed stands along precipitation gradient. Forests 8(6):177. https://doi.org/10.3390/f8060177
Serra-Maluquer X, Mencuccini M, Martinez-Vilalta J (2018) Changes in tree resistance, recovery and resilience across three successive extreme droughts in the northeast Iberian Peninsula. Oecologia 187:343–354
Sheffield J, Wood EF (2008) Projected changes in drought occurrence under future global warming from multi-model, multi-scenario, IPCC AR4 simulations. Clim Dyn 31:79–105
Siepielski AM, Morrissey MB, Buoro M, Carlson SM, Caruso CM, Clegg SM, Coulson T, DiBattista J, Gotanda KM, Francis CD, Hereford J, Kingsolver JG, Augustine KE, Kruuk LE, Martin RA, Sheldon BC, Sletvold N, Svensson EI, Wade MJ, MacColl AD (2017) Precipitation drives global variation in natural selection. Science 355(6328):959–962
Soloway AD, Amiro BD, Dunn AL, Wofsy SC (2017) Carbon neutral or a sink? Uncertainty caused by gap - filling long-term flux measurements for an old-growth boreal black spruce forest. Agr For Meteorol 233: 110–121
Sonesson J, Eriksson G (2010) Genetic variation in drought tolerance in *Picea abies* seedlings and its relationship to growth in controlled and filed experiments. Scand J For Res 18:7–18
Speich MJR (2019). Quantifying and modeling water availability in temperate forests: a review of drought and aridity indices. iForest 12:1–16
Stival Sena J, Giguère I, Rigault P, Bousquet J, Mackay J (2018) Expansion of the gene family in the Pinaceae is associated with considerable structural diversity and drought-responsive expression. Tree Physiol 38 (3):442–456
Swann ALS, Hoffman FM, Koven CD, Randerson JT (2016) Plant responses to increasing CO_2 reduce estimates of climate impacts on drought severity. Proc Natl Acad Sci USA 113(36):10019–10024
Taiz L, Zeiger E, Møller IM, Murphy A (2015). Plant physiology and development, 6th edn. Sinauer Associates, Inc.
Tardieu F, Simonneau T, Muller B (2018) The physiological basis of drought tolerance in crop plants: a scenario-dependent probabilistic approach. Ann Rev Plant Biol 69:733–759
Thiele JC, Nuske RS, Ahrends B, Panferov O, Albert M, Staupendahl K, Junghans U, Jansen M, Saborowski J (2017) Climate change impact assessment – a simulation experiment with Norway spruce for a forest district in Central Europe. Ecol Model 346:30–47
Tomasella M, Häberle K-H, Nardini A, Hesse B, Machlet A, Matyssek R (2017) Post-drought hydraulic recovery is accompanied by non-structural carbohydrate depletion in the stem wood of Norway spruce samplings. Sci Rep 7:14308
Trujillo-Moya C, George J-P, Fluch S, Geburek T, Grabner M, Karanitsch-Ackerl S, Konrad H, Mayer K, Sehr EM, Wischnitzki E, Schueler S (2018). Drought sensitivity of Norway spruce at the species' warmest fringe: quantitative and molecular analysis reveals high genetic variation among and

within provenances. G3-Genes Genomes Genet 8:1225–1245

Tuberosa R (2012) Phenotyping for drought tolerance of crops in the genomics era. Front Physiol 3:347

Turtola S, Manninen A-M, Rikala R, Kainulainen P (2003) Drought stress alters the concentration of wood terpenoids in Scots pine and Norway spruce seedlings. J Chem Ecol 29:1981–1995

Vishwakarma K, Upadhyay N, Kumar N, Yadav G, Singh J, Mishra RK, Kumar V, Verma R, Upadhyay RG, Pandey M, Sharma S (2017) Abscisic acid signaling and abiotic stress tolerance in plants: a review on current knowledge and future prospects. Front Plant Sci 8:161

Vitali V, Buntgen U, Bauhus J (2017) Silver fir and Douglas fir are more tolerant to extreme droughts than Norway spruce in south-west Germany. Glob Chang Biol 23:5108–5119

Wallin G, Karlsson PE, Selldén G, Ottosson S, Medin E-L, Pleijel H, Skärby L (2002) Impact of four years exposure to different levels of ozone, phosphorus and drought on chlorophyll, mineral nutrients, and stem volume of Norway spruce, *Picea abies*. Physiol Plant 114:192–206

Wasternack C (2007) Jasmonates: an update on biosynthesis, signal transduction and action in plant stress response, growth and development. Ann Bot 100: 681–697

Wolf JBW (2013) Principles of transcriptome analysis and gene expression quantification: an RNA-seq tutorial. Mol Ecol Res 13:559–572

Woodall CW, Oswalt CM, Westfall JA, Perry CH, Nelson MD, Finley AO (2010) Selecting tree species for testing climate change migration hypotheses using forest inventory data. For Ecol Manag 259:778–785

Zang C, Hartl-Meier C, Dittmar C, Rothe A, Menzel A (2014) Patterns of drought tolerance in major European temperate forest trees: climatic drivers and levels of variability. Glob Chang Biol 20(12):3767–3779

Zawaski C, Busov VB (2014) Roles of gibberellin catabolism and signaling in growth and physiological response to drought and short-day photoperiods in *Populus* trees. PLoS ONE 9(1):e86217

Zhang J, Jia W, Yang J, Ismail AM (2006) Role of ABA in integrating plant responses to drought and salt stresses. Field Crop Res 97:111–119

10 Local Adaptation in the Interior Spruce Hybrid Complex

Jonathan Degner

Abstract

Interior spruce is a hybrid species complex composed of white spruce, Engelmann spruce, and their hybrids, distributed across a wide area in western Canada. In both interior spruce and allopatric white spruce, half a century of study on local adaptation to climate has been complemented by a large number of genomic studies within the past decade. This chapter reviews the genomic evidence for local adaptation in interior spruce and allopatric white spruce, comparing the results gleaned from these studies with phenotypic studies when possible. In interior spruce, genomic and phenotypic studies have consistently identified local adaptation to cold temperatures, particularly through autumn cold hardiness, with mixed results for other traits and climatic variables. Studies in allopatric white spruce have found variable evidence for local adaptation, with genomic studies largely finding clinal variation with mean annual temperature. Several genomic approaches have been used to infer local adaptation in this species complex, and these have often found contrasting patterns of adaptation between allopatric white spruce and interior spruce, and weak correlation between genetic markers identified across multiple studies. While genomic and phenotypic analyses are often complementary, novel genomic approaches have also been applied to understand aspects of local adaptation that are difficult or impossible to measure exclusively phenotypically, including hybrid fitness, the genetic architectures of adaptive traits, and phenotypic plasticity.

J. Degner (✉)
Department of Forest and Conservation Science, Faculty of Forestry, University of British Columbia, Vancouver, BC, Canada
e-mail: jonathan.degner@ubc.ca

10.1 Introduction

In Darwin's *On the Origin of Species*, he noted that pine seeds acquired from different elevations in the Himalayas differed in their resistances to cold when grown in England, concluding that species can sometimes diverge over the course of generations to become acclimated to different temperatures (Darwin 1859). Other prominent botanists also noted this phenomenon, casting doubt on the prevailing idea that all individuals of a species were more or less capable of the same growth under all conditions in which that species is placed; although this important finding was initially largely ignored. Less than two decades later, several botanists reciprocally transplanted seedlings of herbaceous species that were sampled at drastically differing elevations or geographies, noting that local seed tended to produce healthier plants (Naudin and Radlkofer 1876; Kerner von Marilaun 1895). These early

I. M. Porth and A. R. De la Torre (eds.), *The Spruce Genome*, Compendium of Plant Genomes,
https://doi.org/10.1007/978-3-030-21001-4_10

observational experiments all hinted at an unexplored aspect of natural selection, while it took several decades before their results were first effectively synthesized (Turesson 1922; Clausen et al. 1940). Through these early experiments on reciprocal transplants across multiple environments, we began to understand that many plant species are composed of natural populations that are not only genetically adapted to their habitats, but are uniquely adapted to that habitat. This concept has been codified into the present use of the term "local adaptation" (Kawecki and Ebert 2004).

Current evolutionary theory suggests that local adaptation is, in most cases, produced by divergent selection acting across one or more selective gradients throughout a species' range (Linhart and Grant 1996; but see Blanquart et al. 2012). If selection pressures do not change substantially over short temporal scales but rather differ spatially, populations should become increasingly well-adapted to the selective environment in which they thrive, and local adaptation may thus come at the cost of their ability to adapt to other environments (Felsenstein 1976). This is generally thought to occur through the selection of alleles that are favorable in one environment but neutral or detrimental in another (Colautti et al. 2012). As allele frequencies gradually shift in response to variable strengths of selection across these environmental gradients, clinal relationships in adaptive phenotypes and alleles are expected to form (Huxley 1939; Barton 1999), and are commonly cited as evidence of local adaptation (Rellstab et al. 2015).

As many species experience differing selective pressures across their distribution (McKay and Latta 2002), local adaptation is often viewed as a default path for natural selection, one which is bound to occur if it is not constrained by other pressures. Theoretical and empirical studies have identified many factors that can prevent species from becoming locally adapted, including a lack of standing genetic variation, high gene flow, small population size, or high temporal heterogeneity in selection pressures (Brown and Pavlovic 1992; Kassen 2002; Willi et al. 2006; Leimu and Fischer 2008). Additionally, some theoretical models predict that even if a species is capable of becoming locally adapted, phenotypic plasticity or generalism may be favorable under some evolutionary scenarios (Sultan and Spencer 2002; Spichtig and Kawecki 2004). Despite these caveats, local adaptation is common in nature in general (Hereford 2009) and in plants in particular (Leimu and Fischer 2008).

Local adaptation was first identified in plants, where it may be viewed as an obvious solution to some of the challenges of being a plant: a sessile growth form that must tolerate its environment, a general correlation in habitat between parents and offspring due to limited dispersal capabilities, flexibility in mating systems, and a propensity for intra and interspecific hybridization (Bradshaw 1972; Linhart and Grant 1996). It is unsurprising, then, that local adaptation has been repeatedly identified in plants (Savolainen et al. 2007; Leimu and Fischer 2008), and less commonly in animals (Hereford 2009; Sanford and Kelly 2011).

Many plants are locally adapted to climate, particularly widespread forest trees (Savolainen et al. 2007). As a consequence of their long life spans, trees must endure their local climate for decades or centuries, including all of the extreme climatic events that may occur over this substantial timescale. As injury and mortality from climatic events can often be mitigated through phenological timing (Richardson et al. 2013), local adaptation related to tree's phenology and seasonal characteristics is common (Howe et al. 2003; Alberto et al. 2013). Given climate's importance to local adaptation and recent tools allowing many climatic variables to be quickly estimated for precise geographic locations (e.g., Wang et al. 2016; Fick and Hijmans 2017), local adaptation to climate has been extensively studied also in interior spruce, and is therefore the primary driver of local adaptation that will be discussed here.

Local adaptation plays a central role in our understanding of ecological speciation, and more broadly in our understanding of the role natural selection plays in a species' evolutionary history (Nosil 2012). It is also of critical concern to conservation biologists, as species that are locally adapted to climate may suffer substantially due to climate change without further assistance (Aitken et al. 2008), and translocations of individuals for restoration and genetic rescue should be designed with knowledge of local adaptation. In particular, assisted migration of forest trees to mitigate climate change has received considerable attention in recent years, given the large impact of forests on global climate (Aitken and Whitlock 2013; Sáenz-Romero et al. 2016; Aitken and Bemmels 2016). Having a firm understanding of local adaptation is critical to effectively achieving these species protection aims.

10.2 The Interior Spruce Species Complex

From the vast expanse of Canada and Alaska's boreal forest to the ragged tree line of the US and Canadian Rocky Mountains, spruces thrive where few other trees can survive. White spruce (*Picea glauca* (Moench) Voss) and Engelmann spruce (*Picea engelmannii* Parry ex Engelm.) are North American trees that occupy distinct ecological niches, where they represent typically dominant or co-dominant canopy species. White spruce is a boreal species, distributed widely across North America, covering 26° of latitudinal and 111° of longitudinal range (Little 1971). Engelmann spruce is a subalpine species, growing from mid-elevation to treeline throughout many of the mountainous regions of western North America, a broad and highly fragmented area spanning 26° in latitude and 23° in longitude (Little 1971; Coupé et al. 1991). The climate of these regions is generally cool and humid, with short growing seasons and high snowfall, and populations are often geographically isolated on mountaintops (Alexander and Shepperd 1984). Relative to white spruce, the habitat of Engelmann spruce tends to be warmer and wetter (Liepe et al. 2016). Molecular phylogenetic analyses have consistently recognized white and Engelmann spruce as sister taxa (Sigurgeirsson and Szmidt 1993; Ran et al. 2006; Bouillé et al. 2011; Lockwood et al. 2013), with the most recent estimates of divergence around 2.5—3 Mya based on chloroplast DNA (Lockwood et al. 2013, Ran et al. 2015).

Much of the present range of these species was covered by ice during the last glaciation (Shafer et al. 2010), and both species were forced south of their present distribution, where they may have come into secondary contact (Ledig et al. 2006; De La Torre et al. 2014a). Since the last glacial maximum, both species recolonized large areas made available by glacial retreat and have formed several stable hybrid zones, covering an extensive area in the central plateaus of British Columbia, as well as the Rocky Mountains, extending eastward to Alberta and southward to Wyoming (Roche 1969; Daubenmire 1974).

Hybrids between white and Engelmann spruce have been identified through morphological markers such as cone and branch morphology (e.g., Garman 1957; Daubenmire 1974) and confirmed through genetic markers (e.g., Rajora and Dancik 2000; De La Torre et al. 2014a; Conte et al. 2017; MacLachlan et al. 2018). The spatial and climatic extent of the hybrid zone is extensive, spanning at least 13° of latitude and 20° of longitude (Haselhorst and Buerkle 2013; Conte et al. 2017), with mean annual temperatures ranging from −3 to +6 °C, and mean annual precipitation regimes from 350 to 1500 mm in Western Canada (Yeaman et al. 2016). The hybrid zone spans all degrees of admixture, with most of the central zone containing hybrids with more white spruce ancestry, and hybrids in southern montane regions with higher Engelmann ancestry (Fig. 10.1). Both white spruce and interior spruce further hybridize with Sitka spruce (*Picea sitchensis* (Bong.) Carr) in coastal and submaritime river drainages in British Columbia and Alaska (Little 1953; Daubenmire 1968), producing three-way hybrids in some western parts of central British Columbia (Roche 1969; Hamilton et al. 2015; Fig. 10.1). Local adaptation has been less studied in this system and will not be covered here (but see O'Neill et al. 2002; Hamilton et al. 2013; Hamilton and Aitken 2013).

Given the relatively recent range expansions of white and Engelmann spruce into western

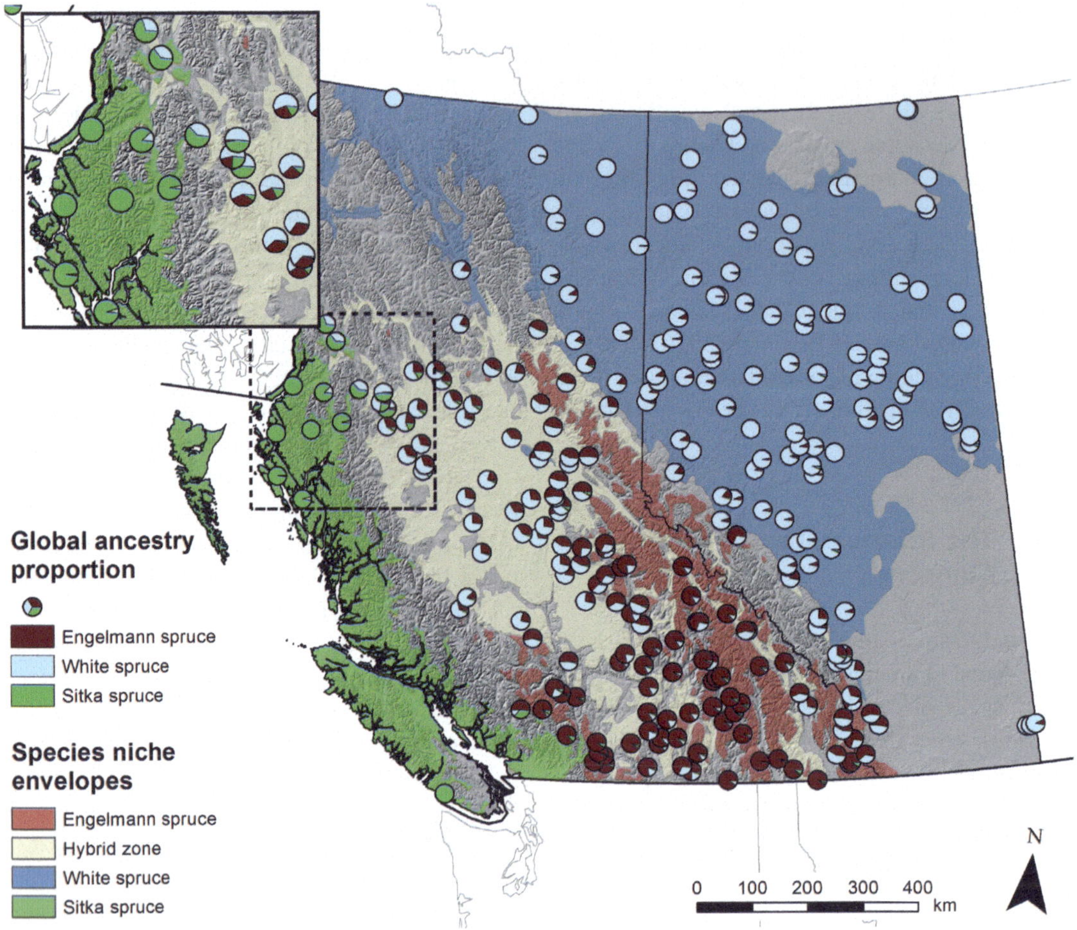

Fig. 10.1 Admixture between white, Engelmann, and Sitka spruce across western Canada. Ancestry proportions calculated with ADMIXTURE using 20,000 unlinked SNPs (n = 7–12 per provenance) (Degner et al., manuscript in prep.)

Canada, and subsequent secondary contact, it is possible that populations still bear strong signatures of these events and may not have inhabited their present geographic ranges long enough to locally adapt to those environments. However, analyses of interspecific heterozygosity throughout the hybrid zone suggest that these zones are at least 5,000 years old, with populations mostly composed of multigenerational hybrids and backcrosses (De La Torre et al. 2015). Paleoclimatic modeling suggests that these species may have been in secondary contact even during the last glacial maximum (De La Torre et al. 2014b). Although some traces of demographic processes and recent introgression may exist and obscure genomic patterns of local adaptation, there is still potential for local adaptation in this system.

Local adaptation has been of interest to foresters for centuries to ensure maximum tree health and timber value (Langlet 1971), and has received more attention over recent decades due to a reliance on replanting seed that is sourced from relatively broad geographic areas and the need to develop geographic seed zones or seed transfer guidelines (e.g., Ying and Yanchuk 2006). As interior spruce is an important timber tree in western Canada, and white spruce is economically important across much of eastern Canada, local adaptation in this species complex has received considerable study attention and produced a number of valuable insights into climatic adaptation (discussed below). Given pressing concerns about local adaptation in these species in the face of climate change (O'Neill et al. 2014), genomic research has been pursued in this system to rapidly gain insights into the underlying adaptive drivers present in this species complex, and to inform land managers and conservationists about the potential need for assistance to maintain healthy and productive forests for the future.

10.3 Phenotypic Evidence for Local Adaptation

10.3.1 Methods for Detection

In a strict sense (sensu strictu, Kawecki and Ebert 2004), local adaptation is a form of genotype-by-environment interaction that can only be *directly* inferred from the reciprocal transplantation of genotypes across two or more environments, comparing fitness or some of its components. As Darwinian fitness is notoriously difficult to estimate for any but the simplest systems (Benton and Grant 2000), phenotypic proxies for fitness are often used, especially for long-lived species like trees. In plants, these include parameters related to survival, growth, reproductive output, or stress from biotic and abiotic factors (e.g., Clausen et al. 1940; Bradshaw 1960; Waser and Price 1985; Grøndahl and Ehlers 2008; Kim and Donohue 2013).

Several statistical methods for characterizing local adaptation in quantitative traits have been developed and compared across species in various biological systems (Hoeksema and Forde 2008; Leimu and Fischer 2008; Fraser et al. 2011). These fall into two general categories (Kawecki and Ebert 2004; Blanquart et al. 2013). "Home versus away" analyses measure local adaptation by comparing the fitness of genotypes grown in their local environment to those same genotypes grown in one or more foreign environments. "local versus foreign" analyses instead compare the fitness of genotypes grown in their local environment to foreign genotypes grown in the same environment (Fig. 10.2). Depending on the fitness measure used and the species system under study, the choice of analysis may result in differing interpretations of identical data (Blanquart et al. 2013), although either is typically acceptable to confirm the presence of local adaptation.

In the absence of a reciprocal transplant experiment, there are other methods to infer local adaptation, although without the strong evidence provided by reciprocal transplant. As populations of locally adapted species are expected to show divergence in fitness or its components under any environment, a single common garden, where individuals from multiple sampled locations (often referred to in the literature as provenances) are raised and evaluated in a common environment, can still yield evidence of local adaptation (Mátyás 1996). Locally adapted species should show strong clines in fitness corresponding to selective gradients across the tested populations (Huxley 1939), and there is expected to be a relationship between an individual's fitness and the distance of transplanting i.e., populations closer to the common garden (either geographically or in relationship to selective gradients) should perform better than more distant populations (Williams 1966). The presence and strength of these clinal relationships have been used to infer local adaptation across many taxa, particularly for plants given the ease of coaxing them into a common garden relative to animals (many plant taxa as reviewed in Linhart and Grant 1996; Alberto et al. 2013).

10.3.2 Phenotypic Evidence for Local Adaptation in Allopatric White and Engelmann Spruce

As white spruce is an ecologically and economically important tree species in eastern Canada, local adaptation has been assessed extensively in its allopatric range. The results of these studies are mixed, generally suggesting weak but extant local adaptation. In eastern Canada, height and diameter of mature trees are best for provenances

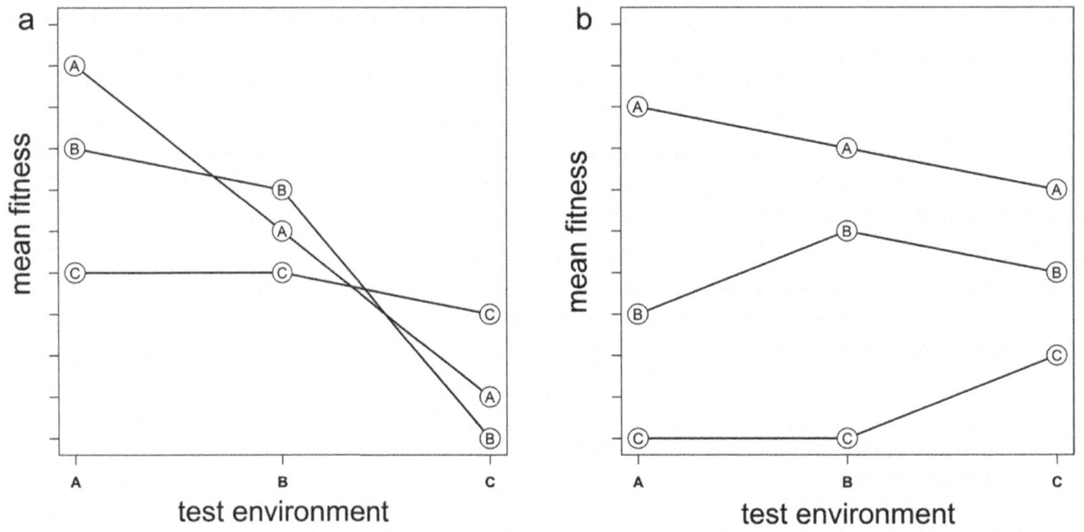

Fig. 10.2 Theoretical patterns of local adaptation as shown by the fitness of three populations (A, B, and C) reciprocally transplanted and tested in each population's home environment. The pattern shown in **a** demonstrates local adaptation in the "local versus foreign" contrast. Although each population may not be at its optimal fitness in its home environment, it is fitter than any other population measured there. The pattern shown in **b** demonstrates local adaptation in the "home versus away" contrast. Despite population A having superior fitness across all test environments, each population is fitter in its home environment than other test environments

with climates similar to the test sites, although climatically distant provenances still performed relatively well (Andalo et al. 2005). Rweyongeza et al. (2011) found similar patterns for a series of provenance trials in western Canada, although with generally steeper clines in performance. Clinal relationships have also been observed for traits related to photosynthetic capacity, water use efficiency, and cold hardiness in relation to provenance climates (Benomar et al. 2015; Sebastian-Azcona et al. 2018).

The most direct evidence for local adaptation in white spruce comes from a large series of provenance trials across eastern Canada, analyzed explicitly for local adaptation by Lu et al. (2014, 2016). They found in general that local provenances did not outperform non-local ones i.e., in a local-foreign comparison. However, when provenances were combined into southern, central, and northern groups, divergent clines emerged, showing that provenances tended to grow best nearer their provenance mean annual temperature, indicative of local adaptation in the home-away contrast (Fig. 10.3). The results of Lu et al. echo those of Li et al. (1997) and Lesser and Parker (2004), finding that northern provenances tend to perform poorly across all test environments while southern provenances excel, which the authors attributed to reduced genetic variation in northern populations.

There is substantially less data available for local adaptation in allopatric Engelmann spruce, but it has generally suggested weak or no local adaptation. In a trial established in Colorado, near the center of the species range, provenances from the northern and westernmost portions of the range outperformed local provenances for growth after ten years (Shepperd et al. 1981). However, a series of provenance trials farther north found that southwestern provenances grew exceptionally well at four years of age, and that clines in growth were present for elevational and climatic gradients relating to frost-free period, suggesting adaptation along these axes (Rehfeldt 1994, 2004). This early age of measurement leaves open the possibility that well-performing provenances may be susceptible to infrequent episodes of cold injury that would cause

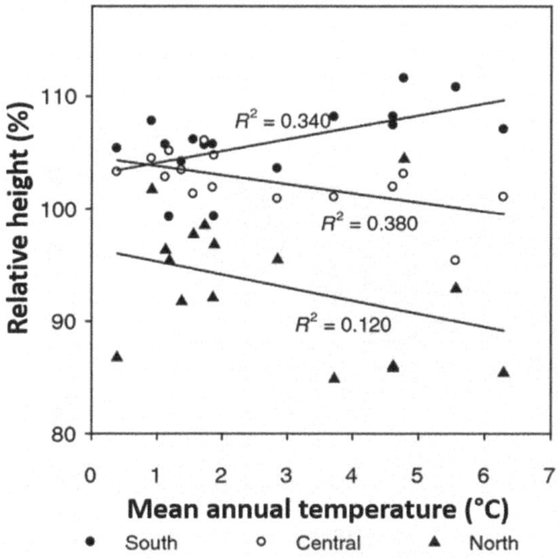

Fig. 10.3 Relative height at age 16 of white spruce provenances compared to mean annual temperature of test sites. Provenances are grouped based on latitude (modified from Lu et al. 2014)

mortality or hinder long-term growth. A recent study also found no local adaptation for recruitment success in a reciprocal transplant across an elevational gradient, which could be limiting the development of local adaptation in these highly heterogeneous environments (Kueppers et al. 2017).

10.3.3 Phenotypic Evidence for Local Adaptation in Interior Spruce

There is strong evidence for local adaptation in interior spruce via multi-environment experiments. Liepe et al. (2016) partitioned the variance of growth, phenology, and cold injury traits in seedlings from across Western Canada grown in growth chambers under various climatic regimes. They identified 11 adaptive syndromes—groups of provenances with similar patterns across the measured traits—that correspond to provenance climates and ecosystems (Fig. 10.4). They found the strongest clines for cold injury, suggesting strong local adaptation for this trait in particular. MacLachlan et al. (2018) likewise identified strong clinal variation in cold hardiness in a single outdoor seedling common garden trial, also noting strong clines in this and other phenotypic variables associated with provenance mean annual temperature.

Ukrainetz et al. (2011) and O'Neill et al. (2014) assessed local adaptation via seed transfer distance, a metric reflecting the climatic distance at which provenances become maladapted, across 4 and 15 provenance trials, respectively. They found that transfer distances depended heavily on the climate of the test site, where sites with moderate climates showed little provenance differentiation, and sites with more extreme climates showed strong local adaptation for height and survival (O'Neill et al. 2014) and for a composite fitness measure (Ukrainetz et al. 2011). This suggests that local adaptation in this species complex may express strongly within the allopatric ranges of white and Engelmann spruce, but weaker within the actual hybrid zones. Interestingly, Xie (2003) found little evidence of local growth superiority for ten-year-old trees across a large series of provenance trials in interior spruce, despite significant genotype-by-environment interactions.

In summary, these experiments suggest relatively weak local adaptation in allopatric white and Engelmann spruce, but stronger local adaptation among interior spruce populations. Several explanations for this trend have been put forth, including selective pressures being higher in the highly heterogenous montane environments faced by interior spruce, differential selection acting upon interspecific hybrids, or an increase in standing genetic variation generated by the multi-species complex (Ledig et al. 2006; De La Torre et al. 2015; Liepe et al. 2016). In general, local adaptation appears stronger for growth, autumn phenology-related traits, and cold hardiness than for survival or spring phenology, a pattern observed across many other tree species (Alberto et al 2013).

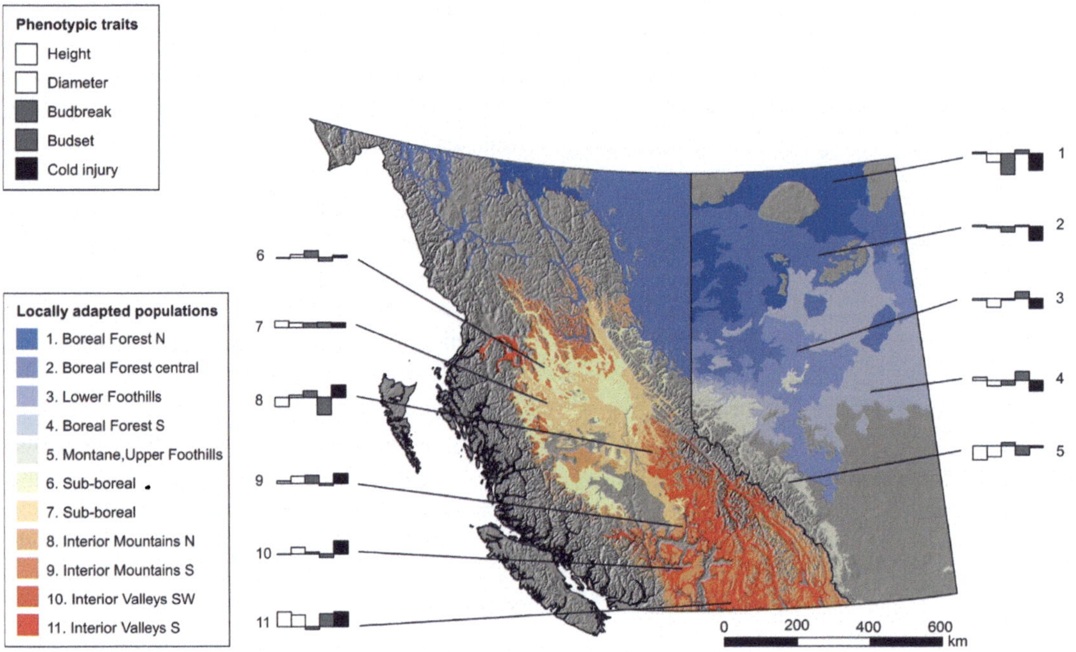

Fig. 10.4 Geographic patterns of adaptation in white, Engelmann, and interior spruce, partitioned by ecosystem variants in western Canada. Bars represent relative means in five seedling phenotypes measured in a growth chamber experiment (modified from Liepe et al. 2016)

10.4 Genomic Evidence for Local Adaptation

10.4.1 Methods for Detection

While reciprocal transplant experiments constitute the "gold standard" for measuring local adaptation, this may be difficult or impossible to implement for many species due to their life histories or ecological constraints (Savolainen et al. 2013). Also, the age at which local adaptation is measured can have a profound impact on results, with studies measuring the same individuals frequently finding weak or even inverse relationships across years (e.g., Conkle 1973; Hannerz et al. 1999; Weng et al. 2007; Stackpole et al. 2010; De La Torre et al. 2014b). Fortunately, the processes that generate local adaptation leave signatures throughout the genome, providing an alternate means to detection that should be comparable regardless of age (Fournier-Level et al. 2011).

Although genomic data alone cannot conclusively confirm local adaptation, an understanding of the genetic underpinnings of local adaptation can broaden our understanding of adaptation beyond what can be derived from reciprocal transplant experiments (Sork 2018). For example, genetic information may be required to fully describe local adaptation when the selective drivers are uncertain, or when identifying parallel evolution of local adaptation between taxa, or when determining the genomic architecture of local adaptation is the goal (Savolainen et al. 2013; Rausher and Delph 2015; Hoban et al. 2016). Additionally, many selective pressures are expected to act over long temporal windows, especially for long-lived species including trees (Mátyás 1996), and rare, extreme events may be important. Local adaptation to these pressures may be undetectable in the relatively short time span over which transplant experiments are conducted, but should still be detectable in genomic data.

As divergent selection occurs across a species' range, markers in association with such selection should shift in allele frequency, while neutral markers should follow prevailing patterns of drift or gene flow and show little divergence between populations (Wright 1931). This is often measured by Wright's fixation index, F_{ST} (Wright 1949), or one of its analogues (Meirmans and Hedrick 2011). Markers showing unusually high F_{ST} may be local adaptation candidates for causal or linked variation, and the proportion and divergence of outliers may be evidence of the strength of selection occurring there (Lotterhos and Whitlock 2014). The detection of local adaptation through these analyses may be tenuous, as they are easily confounded by demography (Hoban et al. 2016), and may not be generalizable or reproducible due to sampling and linkage effects (Ćalić et al. 2016).

In much the same way that polygenic traits tend to show strong clines along environmental gradients under local adaptation, the alleles responsible for those phenotypes should show similar patterns (Huxley 1939). Associations between allele frequencies and environmental variables have been used to infer the presence of local adaptation across a variety of taxa (e.g., Evans et al. 2014; Schweizer et al. 2016; Lind et al. 2017). Genotype-environment clines suffer similar drawbacks to outlier analysis in terms of confounding by demography and issues of reproducibility, but have the added benefit of highlighting environmental gradients that may be responsible for local adaptation. As such, the presence of strong clinal variation should suggest candidate local adaptation or support other lines of evidence. It should be noted that, while genomic divergence and clinal associations are not in and of themselves strong evidence for local adaptation, a lack of these patterns is generally seen as strong evidence against local adaptation (Barton 1999).

10.4.2 Genomic Divergence Approaches

Among the first successful attempts at identifying genomic markers associated with local adaptation in this spruce species study system came from candidate SNP panels. Namroud et al. (2008) successfully genotyped 534 SNPs distributed across 345 genes previously identified as candidates for regulatory functions, growth, phenology, and wood formation (Pavy et al. 2005). Across six populations in as many distinct ecosystems in southern Quebec, there was extremely low divergence across all SNPs ($F_{ST} = 0.006$), but the authors identified 20 SNPs across 19 genes with exceptionally high F_{ST} (range = 0.04–0.13). Forty-nine SNPs were considered putative candidates for local adaptation based on their abnormally high or low allele frequency in a particular population, as measured by the interaction between locus and population effects on F_{ST} for a given locus (Beaumont and Balding 2004). None of these interactions were significant at 95 or 99% confidence levels, which the authors attributed to the limited size of the study area relative to the species range of white spruce and the highly polygenic nature of the traits associated with the candidate SNP panel used.

De La Torre et al. (2015) developed a SNP panel based on candidate genes for cold hardiness and herbivory resistance in Sitka spruce (Holliday et al. 2008). They were able to successfully genotype a panel of 86 candidate SNPs across interior spruce and allopatric populations. Approximately 6.5% of these candidate SNPs showed signals of selection based on F_{ST} outlier analysis or genotype-environment association, although there was little overlap in individual candidates between analyses across these species (De La Torre et al. 2014a). Most SNPs showed signals of divergent selection, although a small number showed signatures of balancing selection. This detection rate is similar in scale to allopatric white spruce and many other plant species (Namroud et al. 2008; Strasburg et al. 2012). Unsurprisingly, F_{ST} values for outliers within the hybrid zone (range = 0.08–0.18) and across the hybrid zone and both parental species (range = 0.04–0.29) are substantially higher than those found for white spruce alone. While this is certainly in part due to the multi-species nature of interior spruce, it may also suggest stronger divergent selection among interior spruce populations than for white spruce.

A separate approach of genomic divergence has been employed in white spruce, comparing population divergence in phenotypes (Q_{ST}) to overall genomic divergence, as measured by G_{ST}, a multi-allelic F_{ST} analogue (Jaramillo-Correa et al. 2001). They found high phenotypic divergence, particularly in bud set, compared to low genetic divergence measured by 11 expressed sequence tags. Although this pattern is suggestive of local adaptation (McKay and Latta 2002; Sæther et al. 2007; but see Whitlock 2008), the limited genetic resolution provided by these few molecular markers did not allow for direct inference of local adaptation.

10.4.3 Genotype-Environment Clines

Early studies of molecular variation identified a number of clinal relationships for interior spruce and its parent species. Latitudinal clines in allele frequency were identified for 12 of the 24 studied isozyme markers in Engelmann spruce, as well as a general latitudinal cline in allelic diversity across all loci (Ledig et al. 2006). This supports the earlier conclusions by Shepperd et al. (1981) that genetic diversity in this species increases northward, with both authors suggesting hybridization as a putative cause. The described relationships were generally attributed to patterns of population structure, as isozymes are expected to behave neutrally under most circumstances (Weeden and Wendel 1989, but see Falkenhagen 1985).

The first of genome-wide studies of clinal variation in this species complex came from Hornoy et al. (2015) in allopatric white spruce. Using a custom SNP array, they genotyped 11,085 markers across 7,819 genes, likely representing between one quarter and one half of the transcriptome (Yeaman et al. 2014; Warren et al. 2015; Suren et al. 2016). Using a linear and quadratic regression approach and population structure as a covariate, the authors identified 31 genes showing strong clinal relationships with mean annual temperature, two of which were also associated with precipitation. Although the proportion of genes showing a clinal relationship with temperature is very low (0.004%) compared to the F_{ST}-based approaches discussed above, this study was the first to use a genome-wide panel, rather than a candidate gene approach which would presumably enrich the panel for targets under selection. Additionally, the weak local adaptation signals in phenotypic studies for white spruce may be due to smaller shifts in allele frequencies than detectable within the study (Le Corre and Kremer 2012).

The most comprehensive study of genome-wide adaptation to the environment in interior spruce was performed by Yeaman et al. (2016). With 888,000 SNPs across nearly 23,000 genes and 254 provenances, Yeaman et al. truly characterized genomic signatures of climate adaptation in this species complex. Using a simple univariate marker-environment raw correlation method, Yeaman et al. identified 854 candidate genes for local adaptation (based on enrichment for SNP-environment association outliers compared to a null distribution). As their approach did not account for population structure, it is expected to have a high false-positive rate, providing an overestimate of the number of genes with a true effect on local adaptation to climate. However, the authors expected adaptive variation to follow the same environmental gradients as population structure due to the hybrid nature of interior spruce (Fig. 10.1), and statistically removing that population structure was expected to reduce their power to detect local adaptation (Lotterhos and Whitlock 2015). Like De La Torre et al. (2014b), Yeaman et al. found strong clinal relationships with precipitation as snow and mean annual precipitation, as well as strong relationships with winter temperatures, supporting phenotypic results showing strong local adaptation for cold hardiness in this species complex (Liepe et al. 2016; Yeaman et al. 2016;

MacLachlan et al. 2018) and in other temperate tree species (Savolainen et al. 2007). After these initial correlations were corrected for population structure, the number of detected candidate genes dropped by an order of magnitude, as expected. Associations with winter temperatures remained strong, while associations with precipitation variables did not. This may suggest that local adaptation is stronger for temperature-associated genes, or that temperature is simply more orthogonal to population structure in this species complex.

Yeaman et al. also directly compared their results to those of Hornoy et al., identifying shared genes in their dataset. They were able to compare 38 of the 43 genes identified by Hornoy et al., although none of these appeared as top candidates in the analyses by Yeaman et al. However, 17 genes had weak-to-moderate associations with climatic variables. This may indicate that allopatric white spruce faces differing selective pressures than interior spruce, or that the different analyses employed by the two studies are not sensitive to the same adaptive signatures.

The correlation between individual markers and environmental variables may be ineffective without sampling a large proportion of the genome and can be prone to false-positives, especially in systems with strong population structure or that recently underwent range expansion (de Villemereuil et al. 2014; Lotterhos and Whitlock 2015; Hoban et al. 2016). To circumvent these concerns, a popular approach in interior spruce has been to correlate the hybrid index, a measure of the overall proportion of ancestry an individual shares with one parental species or the other, with provenance environments. Hybrid index is, in this sense, a composite of population structure produced by migration and gene flow, as well as adaptation through natural selection (Barton and Gale 1993). While this index has the benefit of averaging the effects of many markers, it also confounds the effects of local adaptation and population structure.

De La Torre et al. (2015) identified a cline in hybrid index with elevation in their candidate SNP data set (86 markers), in agreement with relationships based on morphological features (Garman 1957) and allozymes (Rajora and Dancik 2000). This cline was not detected by De La Torre et al. (2015) with a separate estimate of hybrid index based on 10 microsatellite loci, emphasizing the importance of marker density in accurately describing genomic ancestry and therefore environmental clines in this species complex. This SNP-based estimate of hybrid index was also regressed against a suite of twenty measured and derived climate variables (Wang et al. 2006), finding that variation in hybrid index was best described by a model containing two climate variables: the amount of precipitation falling as snow, and a measure of summer aridity (De La Torre et al. 2014b). In this model, ancestry from Engelmann spruce increases as snowfall and summer aridity both increase, although snowfall was found to be the most important predictor in this system. A second study, using a 6,500 SNP panel and a broader range of provenances, again found a strong relationship between snowfall and hybrid index, but did not identify a relationship with summer aridity (MacLachlan et al. 2018). Instead, the authors found strong correlations between hybrid index and winter temperatures, likely as a result of more extensive sampling in the range of allopatric white spruce than De La Torre et al. The results of MacLachlan et al. closely mirror those of Yeaman et al. (2016) in finding strong axes of genetic adaptation associated with winter temperatures and weaker relationships with precipitation.

10.4.4 Genotype-Phenotype Clines

Due to generally small allelic effect sizes in polygenic traits, studies linking genomic variants and these traits typically require large sample sizes, in terms of both loci and sampled individuals (Hong and Park 2012). Due to these limitations, only one study has comprehensively assessed these associations in interior spruce (Yeaman et al. 2016). In line with the environmental associations discussed above, the strongest relationships were found for traits related to cold adaptation. Only a handful of genes were identified as candidates for growth and phenology, but hundreds were identified in strong association with cold hardiness (Yeaman et al. 2016). Correcting for population structure identified candidate genes in association with more traits, but results had almost no overlap with candidates identified without correction for population structure. As discussed below, cold hardiness is tightly correlated with hybrid index, and as such many of the loci found to be associated with these traits are likely indicative of broader adaptive divergence between white and Engelmann spruce. This again highlights the coincident clines of adaptive variation and population structure in this species complex.

As with the environmental association analyses discussed above, some studies have used hybrid index as a composite genomic measure to identify clinal relationships in phenotypes. De La Torre et al. (2014b) split provenances from eleven field common gardens into parental and hybrid classes based on hybrid index, finding higher survival for provenances matching the environment of the common garden, e.g., pure white spruce provenances performed best in common gardens planted in the allopatric range of white spruce. This provides good evidence of an adaptive advantage of hybridization in this species complex, as provenances with hybrid ancestry generally outperformed pure parental species within the present hybrid zone. Additionally, De La Torre et al. (2014b) identified clinal relationships between hybrid index and growth, with a higher proportion of ancestry from Engelmann spruce generally translating to less growth across most of the common gardens.

MacLachlan et al. (2018) utilized a similar approach, finding a very strong cline in cold hardiness with hybrid index, with a higher proportion of Engelmann ancestry corresponding to lower cold tolerance (Fig. 10.5). However, these authors found only a weak trade-off with growth ($p = 0.07$), and with regard to the clinal relationships giving opposing results to those by De La Torre et al. (2014b). This is interesting, as Engelmann ancestry has also been associated with depressed growth as well as lower cold tolerance (Liepe et al. 2016). These discrepancies may be, in part, due to differences in age of testing and growing conditions between the studies. MacLachlan et al. and Liepe et al. tested younger individuals than De La Torre et al., and all studies differed with respect to their growing environments. The common garden used by MacLachlan et al. was in Vancouver, BC, well outside the range of white or Engelmann spruce with warm, mild temperatures.

Bud phenology has been implicated in local adaptation for many temperate tree species (Alberto et al. 2013), including within the genus *Picea* (Mimura and Aitken 2007; Chen et al. 2012), and bud break and bud set are critical components of cold hardiness (Aitken and Hannerz 2001). Interestingly, despite strong evidence for local adaptation to cold, little genomic evidence of phenological adaptation has been observed in interior spruce. Although clinal relationships have been observed between autumn bud phenology and climate in interior spruce (De La Torre et al. 2014b; Liepe et al. 2016; MacLachlan et al. 2018) and allopatric white spruce (Lesser and Parker 2004), no genomic studies have identified relationships between bud phenology and hybrid index, or strong associations with genomic markers (De La Torre et al. 2014b; Yeaman et al. 2016; MacLachlan et al. 2018).

10.4.5 Genomic Architecture of Local Adaptation

The potentially locally adaptive phenotypes that have received attention so far, include survival, growth, phenology, and cold hardiness. Survival and growth are both to such large extent polygenic that most genes in the genome may have some near-immeasurably small effect, causing these traits to potentially be "omnigenic" (Boyle et al. 2017). As such, it is unsurprising that Yeaman et al. (2016) identified only a few genomic loci in relation to height. Despite the adaptive significance of faster growth within the limitations of a given environment, the ability to detect associated loci declines with increasing

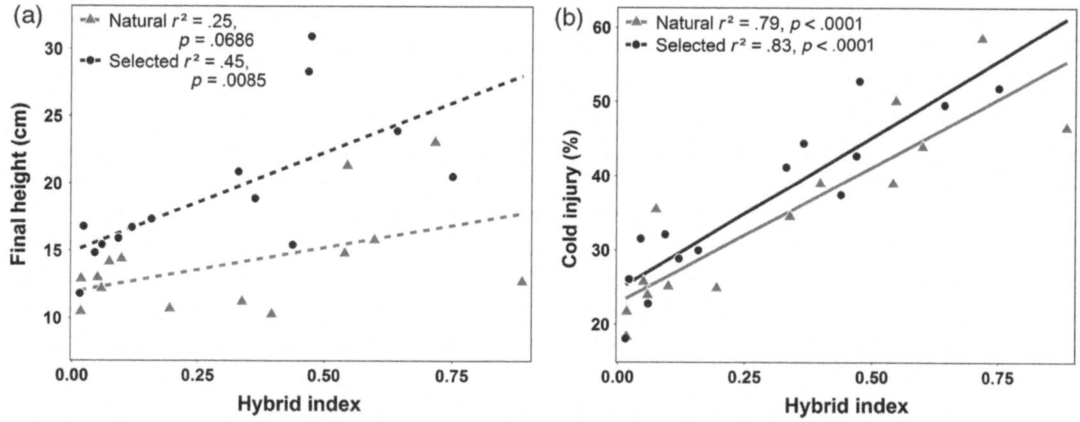

Fig. 10.5 Height and autumn cold injury at age three for wild and selected seedlots of interior spruce compared to hybrid index. Points represent breeding zone means. Dashed regression lines are not significant at $\alpha = 0.05$ following Bonferroni correction (MacLachlan et al. 2018)

polygenic control of traits (Yeaman 2015). In allopatric white spruce, quantitative trait loci for height are numerous, with small effect sizes and little consistency between crosses (Pelgas et al. 2011), an expected outcome for highly polygenic traits (McKay and Latta 2002).

Across most of the genome, linkage disequilibrium (LD) decays rapidly in allopatric white spruce and other conifers (Pavy et al. 2012). However, interior spruce does have detectable amounts of LD across many loci due to hybridization (Yeaman et al. 2016). As such, it is difficult to disentangle linkage caused by introgression from linkage caused by selective sweeps or epistasis. However, approximately one-sixth of the genes associated with cold hardiness show LD above the 95th percentile of background (Yeaman et al. 2016), suggesting that while the trait is highly polygenic and composed of largely independent genes, there may be more linked or strongly selected gene complexes present for this trait than for growth. As cold hardiness is a composite phenotype and produced by the confluence of many phenological traits coordinating growth initiation and cessation (Bigras et al. 2001), and thus is highly polygenic (Howe et al. 2003), some of its component phenotypes have presumably simpler genomic architectures (Cooke et al. 2012; Singh et al. 2017) that could be responsible for the pattern seen here. However, these architectures have not been directly and independently identified for interior spruce.

A paradox of widespread, outcrossing tree species is that they frequently exhibit local adaptation in the face of high gene flow through wind-dispersed pollen (Savolainen et al. 2007). It has long been expected that high levels of gene flow should homogenize genetic variation, limiting local adaptation (Lenormand 2002). However, local adaptation appears nearly ubiquitous in these species (Alberto et al. 2013; Lind et al. 2018) and there is strong evidence for its presence in interior spruce. Recent theoretical work has suggested models that allow for the maintenance of local adaptation in the presence of high gene flow (Yeaman and Otto 2011; Yeaman 2015; Tigano and Friesen 2016). Factors that can act to produce or maintain local adaptation in the face of high gene flow include adaptive introgression, selection of linked loci, low recombination, and highly polygenic architectures (Yeaman 2015; Tigano and Friesen 2016).

Adaptive introgression seems likely for this study system. Out of 86 SNP markers tested, 7 showed narrow geographic clines in allele frequency across the hybrid zone (De La Torre et al. 2015), a pattern expected under local adaptation (Bridle et al. 2010). Of these seven SNPs, four also exhibited signs of adaptive introgression, generally in the form of directional selection for one parental allele over the other. While these only constitute a few observations, detection of any adaptive introgression, combined with other evidence for hybrid superiority and complementation (De La Torre et al. 2014b; Conte et al. 2017), suggests that novel combinations of white and Engelmann spruce alleles or parental alleles in novel environments, are contributing to increased hybrid fitness and providing adaptive fodder for local adaptation.

The large number of loci associated with cold hardiness and high background LD in interior spruce suggest that, while selection may be acting on some groups of linked loci, adaptation is primarily occurring through selection on independent genes (Yeaman et al. 2016). Additionally, conifers tend to have low rates of recombination (Jaramillo-Correa et al. 2010), with white spruce having an average recombination rate of 0.08–0.11 cM/Mb (Pavy et al. 2017), an order of magnitude below most angiosperm trees (Jaramillo-Correa et al. 2010). Together, these lines of evidence suggest that, despite high levels of gene flow, interior spruce has the requisite genomic capacity to adapt to local conditions.

10.4.6 Hybrid Fitness

As many hybrid zones are geographically narrow and mobile (Hewitt 1988; Buggs 2007), and many interspecific hybrids are expected to show reduced fitness compared to their parents (Arnold and Hodges 1995), the apparent persistence and scale of the interior spruce zone begs for an explanation. As discussed above, De La Torre et al. (2014b) found some phenotypic evidence for bounded hybrid superiority, i.e., hybrids having higher fitness than either parent within the hybrid zone but lower fitness outside of it, a model under which stable hybrid zones can form (Moore 1977).

Another potential explanation for hybrid superiority is complementation, wherein the effects of recessive deleterious alleles present in a parental species are mitigated in hybrids by the addition of non-deleterious alleles from the other parent species (Crow 1948; Birchler et al. 2006). As the effects of any single deleterious allele are expected to be small (Agrawal and Whitlock 2012), their detection and impact on fitness have been challenging to measure. However, the effects of nonsynonymous amino acid changes on protein function are reasonably predictable across taxa (Choi et al. 2012). Using the dataset of Yeaman et al. discussed above, Conte et al. (2017) identified 165,576 potential nonsynonymous SNP variants, of which about 13% contained a putatively deleterious allele. They found a relationship between the proportion of loci predicted to be homozygous for deleterious alleles and hybrid index, with hybrids having generally lower values than the parental species, as predicted under complementation. Two-year seedling biomass was used as a fitness proxy, showing a weak but significant parabolic relationship with hybrid index ($R^2 = 0.04$, $p < 0.001$), lending some evidence to a fitness effect of hybridization in this system.

10.4.7 Plasticity

Phenotypic plasticity provides an alternative strategy to local adaptation, allowing the expression of traits within a single individual to vary depending on its environment. The degree of plasticity differs between traits (Sonia 2000), species (Funk 2008), and often between populations (Mägi et al. 2011; Molina-Montenegro and Naya 2012), and is difficult to predict based on species characteristics or ecology (Nicotra et al. 2010). Despite the importance of plasticity in understanding species' historical adaptation and potential responses to climate change (Aitken et al. 2008; Valladares et al. 2014), it has received limited direct study in interior spruce. However, the effects of phenotypic plasticity may be inferred from other experiments measuring phenotypes across environmental gradients. For example, while Liepe et al. (2016) partitioned phenotypic variance in several seedling traits across multiple environments with the interest of determining genetic effects, plasticity can also be inferred here from the environmental effects on these phenotypes. The results suggest a wide variation in plasticity among traits. Bud break, for example, was found to be almost entirely plastic with environmental variation explaining 93.3% of the observed phenotypic variation, and only 0.5% attributable to population variation. Fall cold hardiness showed almost no plasticity, with 36.6% of variation attributable to genetic variation but only 0.3% to environmental variation.

Another approach to studying plasticity has been through plasticity in gene expression. Yeaman et al. (2014) analyzed whole transcriptomic data from 39 interior spruce seedlings from a single population grown under seven climatic regimes, identifying 6,413 genes (27%) differing in expression across treatments. These genes clustered into 15 co-expression networks, suggesting a wide variety of traits may have experienced plastic responses across treatments. The largest differences in gene expression appeared in response to combined heat plus drought stress, suggesting a large plastic response to drought, with more moderate levels of plasticity for temperature treatments alone. To determine the relationship between plasticity and local adaptation in this species complex, Yeaman et al. (2016) compared the expression patterns of top candidate genes for environmental and phenotypic associations (discussed above) with non-candidate genes, finding no difference between these groups ($p = 0.34$). This suggests that genes involved in local adaptation may also play roles in adaptive plasticity and vice versa, and that the genomic architectures underlying these evolutionary processes are not necessarily at odds in interior spruce.

10.5 Conclusions

10.5.1 Toward a Synthesis in Understanding Local Adaptation in Interior Spruce

To date, the strongest evidence for local adaptation in this species complex comes from separate analyses identifying similar environmental drivers associated with adaptive genomic variation. The research reviewed here suggests that, for interior spruce, these tend to largely be adaptations to cold temperature. Analyses of cold hardiness phenotypes in common garden experiments have revealed very strong clines in association with provenance winter temperatures (Ukrainetz et al. 2011; Liepe et al. 2016; Yeaman et al. 2016; MacLachlan et al. 2018; Sebastian-Azcona et al. 2018), while genomic signatures of adaptation are also strong for these variables (Yeaman et al. 2016; MacLachlan et al. 2018). Additionally, cold hardiness appears to be under strong genetic control, with almost no plasticity (Liepe et al. 2016).

Adaptation to precipitation and aridity is less clear. Hybrid index appears to be correlated with precipitation regimes and snowfall (De La Torre et al. 2014b; MacLachlan et al. 2018), and aridity was found to be correlated with cold tolerance in one study (Ukrainetz et al. 2011), but only weakly in others (Yeaman et al. 2016; MacLachlan et al. 2018; Sebastian-Azcona et al. 2018). Precipitation may be a stronger driver of adaptation in allopatric white spruce, showing clines in bud set, cold hardiness, and survival (Jaramillo-Correa et al. 2001; Sebastian-Azcona et al. 2018).

Trade-offs between growth and cold acclimation have been identified in many conifer species, as trees that grow for longer generally grow larger, but have delayed phenology and cold acclimation (Howe et al. 2003). This provides a strong fitness driver for local adaptation, as individuals must sacrifice growth for the ability to withstand colder temperatures when necessary, or neglect these adaptations to increase growth potential and competitive ability. However, this trade-off is not well characterized in interior spruce. In seedlings, most studies have found weak trade-offs (Ukrainetz et al. 2011; De La Torre et al. 2014b; Yeaman et al. 2016), while MacLachlan et al. (2018) found strong trade-offs. However, there is a clear trade-off in mature allopatric white spruce (Sebastian-Azcona et al. 2018). Unfortunately, no comparable data exists for mature interior spruce, so the true extent of trade-offs between these traits in natural systems remains unknown. If trends identified in seedlings persist to adulthood, a lack of trade-offs between these ecologically relevant traits may explain some of the adaptive advantages of hybridization in this system. Additionally, spring and autumn cold hardiness appear to have differing environmental drivers and levels of genetic control (Bigras et al. 2001; Alberto et al. 2013), and may show opposing clinal variation (Simpson 1994), complicating the relationship of these trade-offs.

10.5.2 Future Directions for Research

While our understanding of local adaptation in interior spruce is much more complete than for many other non-model wild species, there are still large gaps that remain unfilled. While many studies have attempted to infer the presence of local adaptation through clinal variation, only Lu et al. (2014, 2016) have *directly* tested for the presence of local adaptation within this species system, and no studies in interior spruce have fully synthesized genomic data with reciprocal transplant data. Additional studies combining the inferential power of reciprocal transplants with genomic data could reveal genes directly associated with genotype-by-environment interactions and therefore likely candidates for local adaptation (Lasky et al. 2018). Furthermore, few studies have assessed local adaptation in allopatric Engelmann spruce. While the evidence gathered so far suggests that interior spruce is more strongly locally adapted than its parent species, this may simply be due to a lack of comparable studies in the allopatric range of white and Engelmann spruce.

Plasticity also remains under-explored in this species complex (but see Benomar et al. 2016). In general, plasticity is the assumed response to environmental variation unless evidence exists for genetic control, instead of being directly quantified (Merilä and Hendry 2014). Studies coherently synthesizing analyses of adaptive plasticity and genetic variation will produce a more holistic understanding of adaptation to climate in interior spruce, and may have profound impacts on our ability to predict future population responses to climate (Valladares et al. 2014). Although we have learned that plasticity and adaptation are not mutually exclusive at the gene level (Yeaman et al. 2016), we still do not understand the interplay of these forces at the landscape level.

As adaptation in the interior spruce hybrid zone appears to have historically been in relation to cold temperatures, climate change has the potential to massively disrupt the adaptive strategies of natural populations. A consensus of 15 global circulation models predicts that the mean annual temperature of western Canada is expected to increase by 1.6–1.8 °C over the next 70 years, and that the temperature of the coldest month will increase by 1.8–2.1 °C, based on the RCP 4.5 and 8.5 emissions scenarios (Wang et al. 2016). Climatic modeling for hybrid index suggests that, assuming populations are currently locally adapted, most areas of the interior spruce hybrid zone will experience large shifts in adaptive optima within the next century, resulting in widespread maladaptation (Degner et al., manuscript in prep.). Assisted migration trials are necessary to track the effects of this ongoing climate change, and to inform reforestation and conservation practices.

The genomic resources available for interior spruce are developing rapidly (reviewed elsewhere in this book), but there are still critical areas limiting our ability to understand the genomic basis of local adaptation. The lack of a complete reference genome is a major hurdle, as analyses of the genomic architecture of local adaptation to date have relied on statistical linkage, and generally lack information on physical linkage beyond a few kilobases (Yeaman et al. 2016). Understanding how candidate local adaptation genes cluster across the genome may help narrow our focus to particular genomic regions enriched for adaptive genes and help better understand the genomic architectures of selection for traits and their interactions across the genome. Additionally, functional characterization for conifer genes is still rare, and gene annotations tend to come from distantly related species. This has limited the practical identification of genes related to local adaptation, with many studies instead of characterizing genes by shared ontologies (e.g., Namroud et al. 2008; Yeaman et al. 2014). While this has the benefit of characterizing large numbers of genes and being more generalizable than annotations from distant relatives (Ashburner et al. 2000), they are difficult to interpret in an adaptive context.

10.5.3 Concluding Remarks

Genomic data has provided unprecedented insight into the patterns and processes underpinning local adaptation in interior spruce. Through an increased understanding of the dynamics of hybridization and a better grasp on the climatic drivers that have shaped populations for millennia, genomic data for interior spruce has proved to be an invaluable companion to phenotypic experiments measuring local adaptation in this complex. Specifically, the many adaptations to cold temperature that are only at times apparent in phenotypic data become abundantly clear in the genomic signal. While there is much yet to learn about the nature and extent of local adaptation in the interior spruce species complex, the countless hours of effort contributed to understanding this species complex at both the phenotypic and genomic levels have made it a strong non-model study system. The knowledge gained from the studies discussed above contributes not just to forest management and conservation practices, but also to our understanding of climate adaptation in trees and hybrid evolution more broadly.

References

Agrawal AF, Whitlock MC (2012) Mutation load: the fitness of individuals in populations where deleterious alleles are abundant. Annu Rev Ecol Evol Syst 43:115–135

Aitken SN, Whitlock MC (2013) Assisted gene flow to facilitate local adaptation to climate change. Annu Rev Ecol Evol Syst 44:367–388

Aitken SN, Hannerz M (2001) Genecology and gene resource management strategies for conifer cold hardiness. In: Bigras FJ, Colombo SJ (eds) Conifer cold hardiness. Springer, Dordrecht, pp 23–53

Aitken SN, Bemmels JB (2016) Time to get moving: assisted gene flow of forest trees. Evol Appl 9:271–290

Aitken SN, Yeaman S, Holliday JA, Wang T, Curtis-McLane S (2008) Adaptation, migration or extirpation: climate change outcomes for tree populations. Evol Appl 1:95–111

Alberto FJ, Aitken SN, Alía R, González-Martínez SC, Hänninen H, Kremer A, Lefèvre F, Lenormand T, Yeaman S, Whetten R, Savolainen O (2013) Potential for evolutionary responses to climate change - evidence from tree populations. Glob Chang Biol 19:1645–1661

Alexander RR, Shepperd WD (1984) Silvical characteristics of Engelmann spruce. RM-114, USDA Forest Service, Fort Collins

Andalo C, Beaulieu J, Bousquet J (2005) The impact of climate change on growth of local white spruce populations in Québec, Canada. For Ecol Manag 205:169–182

Arnold ML, Hodges SA (1995) Are natural hybrids fit or unfit relative to their parents? Trends Ecol Evol 10:67–71

Ashburner M, Ball CA, Blake JA, Botstein D, Butler H, Cherry JM, Davis AP, Dolinski K, Dwight SS, Eppig JT, Harris MA, Hill DP, Issel-Tarver L, Kasarskis A, Lewis S, Matese JC, Richardson JE, Ringwald M, Rubin GM, Sherlock G (2000) Gene ontology: tool for the unification of biology. Nat Genet 25:25–29

Barton NH (1999) Clines in polygenic traits. Genet Res 74:223–236

Barton NH, Gale KS (1993) Genetic analysis of hybrid zones. In: Harrison R (ed) Hybrid zones and the evolutionary process. Oxford University Press, Oxford

Beaumont MA, Balding DJ (2004) Identifying adaptive genetic divergence among populations from genome scans. Mol Ecol 13:969–980

Benomar L, Lamhamedi MS, Villeneuve I, Rainville A, Beaulieu J, Bousquet J, Margolis HA (2015) Fine-scale geographic variation in photosynthetic-related traits of *Picea glauca* seedlings indicates local adaptation to climate. Tree Physiol 35:864–878

Benomar L, Lamhamedi MS, Rainville A, Beaulieu J, Bousquet J, Margolis HA (2016) Genetic adaptation vs. ecophysiological plasticity of photosynthetic-related traits in young *Picea glauca* trees along a regional climatic gradient. Front Plant Sci 7:1–15

Benton TG, Grant A (2000) Evolutionary fitness in ecology: comparing measures of fitness in stochastic, density-dependent environments. Evol Ecol Res 2:769–789

Bigras FJ, Ryyppö A, Lindström A, Stattin E (2001) Cold acclimation and deacclimation of shoots and roots of conifer seedlings. In: Bigras FJ, Colombo SJ (eds) Conifer cold hardiness. Springer, Dordrecht, pp 57–88

Birchler J, Yao H, Chudalayandi S (2006) Unraveling the genetic basis of hybrid vigor. Proc Natl Acad Sci 103:12957–12958

Blanquart F, Gandon S, Nuismer SL (2012) The effects of migration and drift on local adaptation to a heterogeneous environment. J Evol Biol 25:1351–1363

Blanquart F, Kaltz O, Nuismer SL, Gandon S (2013) A practical guide to measuring local adaptation. Ecol Lett 16:1195–1205

Bouillé M, Senneville S, Bousquet J (2011) Discordant mtDNA and cpDNA phylogenies indicate geographic speciation and reticulation as driving factors for the diversification of the genus *Picea*. Tree Genet Genomes 7:469–484

Boyle EA, Li YI, Pritchard JK (2017) An expanded view of complex traits: from polygenic to omnigenic. Cell 169:1177–1186

Bradshaw AD (1972) Some evolutionary consequences of being a plant. In: Dobzhansky T, Hecht MK, Steere WC (eds) Evolutionary biology. Appleton-Century-Crofts, New York, pp 25–47

Bradshaw AD (1960) Population differentiation in *Agrostis tenuis* Sibth. III. Populations in varied environments. New Phytol 59:92–103

Bridle JR, Polechová J, Kawata M, Butlin RR (2010) Why is adaptation prevented at ecological margins? New insights from individual-based simulations. Ecol Lett 13:485–494

Brown JS, Pavlovic NB (1992) Evolution in heterogeneous environments: effects of migration on habitat specialization. Evol Ecol 6:360–382

Buggs RJA (2007) Empirical study of hybrid zone movement. Heredity 99:301–312

Ćalić I, Bussotti F, Martínez-García PJ, Neale DB (2016) Recent landscape genomics studies in forest trees—what can we believe? Tree Genet Genomes 12:1–7

Chen J, Källman T, Ma X, Gyllenstrang N, Zaina G, Morgante M, Bousquet J, Eckert A, Wegrzyn J, Neale D, Lagercrantz U, Lascoux M (2012) Disentangling the roles of history and local selection in shaping clinal variation of allele frequencies and gene expression in Norway spruce (*Picea abies*). Genetics 191:865–881

Choi Y, Sims GE, Murphy S, Miller JR, Chan AP (2012) Predicting the functional effect of amino acid substitutions and indels. PLoS ONE 7:e46688

Clausen J, Keck DD, Hiesey WM (1940) Experimental studies on the nature of species I. Effect of varied environments on western North American plants. Carnegie Institution of Washington, Washington

Colautti RI, Lee CR, Mitchell-Olds T (2012) Origin, fate, and architecture of ecologically relevant genetic variation. Curr Opin Plant Biol 15:199–204

Conte GL, Hodgins KA, Yeaman S, Degner JC, Aitken SN, Rieseberg LH, Whitlock MC (2017) Bioinformatically predicted deleterious mutations reveal complementation in the interior spruce hybrid complex. BMC Genomics 18:1–12

Conkle MT (1973) Growth data for 29 years from the California elevational transect study of Ponderosa pine. For Sci 19:31–39

Cooke JEK, Eriksson ME, Junttila O (2012) The dynamic nature of bud dormancy in trees: environmental control and molecular mechanisms. Plant Cell Environ 35:1707–1728

Correa JJ, Verdu M, Martinez SG (2010) The contribution of recombination to heterozygosity differs among plant evolutionary lineages and life-forms. BMC Evol Biol 10

Coupé R, Stewart AC, Wikeem BM (1991) Engelmann spruce - subalpine fir zone. In: Meidinger D, Pojar J (eds) Ecosystems of British Columbia. BC Ministry of Forests, Victoria BC

Crow JF (1948) Alternative hypotheses of hybrid vigor. Genetics 33:447–487

Darwin C (1859) On the origin of species by means of natural selection, or the preservation of favoured races in the struggle for life. Murray, London

Daubenmire RF (1968) Some geographic variations in *Picea sitchensis* and their ecological interpretation. Can J Bot 46:787–798

Daubenmire RF (1974) Taxonomic and ecologic relationships between *Picea glauca* and *Picea engelmannii*. Can J Bot 52:1545–1560

De La Torre AR, Roberts DR, Aitken SN (2014a) Genome-wide admixture and ecological niche modelling reveal the maintenance of species boundaries despite long history of interspecific gene flow. Mol Ecol 23:2046–2059

De La Torre AR, Wang T, Jaquish B, Aitken SN (2014b) Adaptation and exogenous selection in a *Picea glauca* x *Picea engelmannii* hybrid zone: implications for forest management under climate change. New Phytol 201:687–699

De La Torre A, Ingvarsson PK, Aitken SN (2015) Genetic architecture and genomic patterns of gene flow between hybridizing species of *Picea*. Heredity 115:153–164

de Villemereuil P, Frichot É, Bazin É, François O, Gaggiotti OE (2014) Genome scan methods against more complex models: when and how much should we trust them? Mol Ecol 23:2006–2019

Evans LM, Slavov GT, Rodgers-Melnick E, Martin J, Ranjan P, Muchero W, Brunner AM, Schackwitz W, Gunter L, Chen JG, Tuskan GA, Difazio SP (2014) Population genomics of *Populus trichocarpa* identifies signatures of selection and adaptive trait associations. Nat Genet 46:1089–1096

Falkenhagen ER (1985) Isozyme studies in provenance research of forest trees. Theor Appl Genet 69:335–347

Felsenstein J (1976) The theoretical population genetics of variable selection and migration. Annu Rev Genet 10:253–280

Fick SE, Hijmans RJ (2017) WorldClim 2: new 1-km spatial resolution climate surfaces for global land areas. Int J Climatol 37:4302–4315

Fournier-Level A, Korte A, Cooper MD, Nordborg M, Schmitt J, Wilczek AM (2011) A map of local

adaptation in *Arabidopsis thaliana*. Science 334:86–90
Fraser DJ, Weir LK, Bernatchez L, Hansen MM, Taylor EB (2011) Extent and scale of local adaptation in salmonid fishes: review and meta-analysis. Heredity 106:404–420
Funk JL (2008) Differences in plasticity between invasive and native plants from a low resource environment. J Ecol 96:1162–1173
Garman EH (1957) The occurrence of spruce in the interior of British Columbia. British Columbia Forest Service, Victoria
Grøndahl E, Ehlers BK (2008) Local adaptation to biotic factors: reciprocal transplants of four species associated with aromatic *Thymus pulegioides* and *T. serpyllum*. J Ecol 96:981–992
Hamilton JA, Aitken SN (2013) Genetic and morphological structure of a spruce hybrid (*Picea sitchensis* x *P. glauca*) zone along a climatic gradient. Am J Bot 100:1651–1662
Hamilton JA, Lexer C, Aitken SN (2013) Differential introgression reveals candidate genes for selection across a spruce (*Picea sitchensis* × *P. glauca*) hybrid zone. New Phytol 197:927–938
Hamilton JA, De la Torre AR, Aitken SN (2015) Fine-scale environmental variation contributes to introgression in a three-species spruce hybrid complex. Tree Genet Genomes 11:817
Hannerz M, Sonesson J, Ekberg I (1999) Genetic correlations between growth and growth rhythm observed in a short-term test and performance in long-term field trials of Norway spruce. Can J For Res 29:768–778
Haselhorst MSH, Buerkle CA (2013) Population genetic structure of *Picea engelmannii*, *P. glauca* and their previously unrecognized hybrids in the central Rocky Mountains. Tree Genet Genomes 9:669–681
Hereford J (2009) A quantitative survey of local adaptation and fitness trade-offs. Am Nat 173:579–588
Hewitt GM (1988) Hybrid zones - natural laboratories for evolutionary studies. Trends Ecol Evol 3:158–167
Hoban S, Kelley JL, Lotterhos KE, Antolin MF, Bradburd G, Lowry DB, Poss ML, Reed LK, Storfer A, Whitlock MC (2016) Finding the genomic basis of local adaptation: Pitfalls, practical solutions, and future directions. Am Nat 188:379–397
Hoeksema JD, Forde SE (2008) A meta-analysis of factors affecting local adaptation between interacting species. Am Nat 171:275–290
Holliday JA, Ralph SG, White R, Bohlmann J, Aitken SN (2008) Global monitoring of autumn gene expression within and among phenotypically divergent populations of Sitka spruce (*Picea sitchensis*). New Phytol 178:103–122
Hong EP, Park JW (2012) Sample size and statistical power calculation in genetic association studies. Genomics Inform 10:117
Hornoy B, Pavy N, Gérardi S, Beaulieu J, Bousquet J (2015) Genetic adaptation to climate in white spruce involves small to moderate allele frequency shifts in functionally diverse genes. Genome Biol Evol 7:3269–3285
Howe GT, Aitken SN, Neale DB, Jermstad KD, Wheeler NC, Chen TH (2003) From genotype to phenotype: unraveling the complexities of cold adaptation in forest trees. Can J Bot 81:1247–1266
Huxley JS (1939) Clines: an auxiliary method in taxonomy. Bijdr tot Dierkd 27:491–520
Jaramillo-Correa JP, Beaulieu J, Bousquet J (2001) Contrasting evolutionary forces driving population structure at expressed sequence tag polymorphisms, allozymes and quantitative traits in white spruce. Mol Ecol 10:2729–2740
Kassen R (2002) The experimental evolution of specialists, generalists, and the maintenance of diversity. J Evol Biol 15:173–190
Kawecki TJ, Ebert D (2004) Conceptual issues in local adaptation. Ecol Lett 7:1225–1241
Kerner von Marilaun A (1895) Dependence of plant form on soil and climate. In: Oliver FW (ed) The natural history of plants. Blackie, London, pp 495–514
Kim E, Donohue K (2013) Local adaptation and plasticity of *Erysimum capitatum* to altitude: its implications for responses to climate change. J Ecol 101:796–805
Kueppers LM, Conlisk E, Castanha C, Moyes AB, Germino MJ, de Valpine P, Torn MS, Mitton JB (2017) Warming and provenance limit tree recruitment across and beyond the elevation range of subalpine forest. Glob Chang Biol 23:2383–2395
Langlet O (1971) Two hundred years genecology. Taxon 20:653–721
Lasky JR, Forester BR, Reimherr M (2018) Coherent synthesis of genomic associations with phenotypes and home environments. Mol Ecol Resour 18:91–106
Le Corre V, Kremer A (2012) The genetic differentiation at quantitative trait loci under local adaptation. Mol Ecol 21:1548–1566
Ledig FT, Hodgskiss PD, Johnson DR (2006) The structure of genetic diversity in Engelmann spruce and a comparison with blue spruce. Can J Bot 84:1806–1828
Leimu R, Fischer M (2008) A meta-analysis of local adaptation in plants. PLoS ONE 3:e4010
Lenormand T (2002) Gene flow and the limits to natural selection. Trends Ecol Evol 17:183–189
Lesser MR, Parker WH (2004) Genetic variation in *Picea glauca* for growth and phenological traits from provenance tests in Ontario. Silvae Genet 53:141–148
Li P, Beaulieu J, Bousquet J (1997) Genetic structure and patterns of genetic variation among populations in eastern white spruce (*Picea glauca*). Can J For Res 27:189–198
Liepe KJ, Hamann A, Smets P, Fitzpatrick CR, Aitken SN (2016) Adaptation of lodgepole pine and interior spruce to climate: implications for reforestation in a warming world. Evol Appl 9:409–419
Lind BM, Friedline CJ, Wegrzyn JL, Maloney PE, Vogler DR, Neale DB, Eckert AJ (2017) Water availability drives signatures of local adaptation in whitebark pine (*Pinus albicaulis* Engelm.) across fine

spatial scales of the Lake Tahoe Basin, USA. Mol Ecol 26:3168–3185

Lind BM, Menon M, Bolte CE, Faske TM, Eckert AJ (2018) The genomics of local adaptation in trees: are we out of the woods yet? Tree Genet Genomes 14:29

Linhart YB, Grant MC (1996) Evolutionary significance of local genetic differentiation in plants. Annu Rev Ecol Syst 27:237–277

Little EL (1953) A natural hybrid species in Alaska. J For 41:745–746

Little E (1971) Atlas of United States trees volume 1: conifers and important hardwoods. USDA Forest Service, Washington

Lockwood JD, Aleksić JM, Zou J, Wang J, Liu J, Renner SS (2013) A new phylogeny for the genus *Picea* from plastid, mitochondrial, and nuclear sequences. Mol Phylogenet Evol 69:717–727

Lotterhos KE, Whitlock MC (2014) Evaluation of demographic history and neutral parameterization on the performance of FST outlier tests. Mol Ecol 23:2178–2192

Lotterhos KE, Whitlock MC (2015) The relative power of genome scans to detect local adaptation depends on sampling design and statistical method. Mol Ecol 24:1031–1046

Lu P, Parker WH, Cherry M, Colombo S, Parker WC, Man R, Roubal N (2014) Survival and growth patterns of white spruce (*Picea glauca* [Moench] Voss) rangewide provenances and their implications for climate change adaptation. Ecol Evol 4:2360–2374

Lu P, Parker WC, Colombo SJ, Man R (2016) Restructuring tree provenance test data to conform to reciprocal transplant experiments for detecting local adaptation. J Appl Ecol 53:1088–1097

MacLachlan IR, Yeaman S, Aitken SN (2018) Growth gains from selective breeding in a spruce hybrid zone do not compromise local adaptation to climate. Evol Appl 11:166–181

Mägi M, Semchenko M, Kalamees R, Zobel K (2011) Limited phenotypic plasticity in range-edge populations: a comparison of co-occurring populations of two *Agrimonia* species with different geographical distributions. Plant Biol 13:177–184

Mátyás C (1996) Climatic adaptation of trees: rediscovering provenance tests. Euphytica 92:45–54

Mckay JK, Latta RG (2002) Adaptive population divergence: markers, QTL and traits. Trends Ecol Evol 17:285–291

Meirmans PG, Hedrick PW (2011) Assessing population structure: FST and related measures. Mol Ecol Resour 11:5–18

Merilä J, Hendry AP (2014) Climate change, adaptation, and phenotypic plasticity: the problem and the evidence. Evol Appl 7:1–14

Mimura M, Aitken SN (2007) Adaptive gradients and isolation-by-distance with postglacial migration in *Picea sitchensis*. Heredity 99:224–232

Molina-Montenegro MA, Naya DE (2012) Latitudinal patterns in phenotypic plasticity and fitness-related traits: assessing the climatic variability hypothesis (CVH) with an invasive plant species. PLoS One 7: e47620

Moore WS (1977) An evaluation of narrow hybrid zones in vertebrates. Q Rev Biol 52:263–277

Namroud MC, Beaulieu J, Juge N, Laroche J, Bousquet J (2008) Scanning the genome for gene single nucleotide polymorphisms involved in adaptive population differentiation in white spruce. Mol Ecol 17:3599–3613

Naudin C, Radlkofer L (1876) Recherches au sujet des influences que les changement de climats exercent sur les plantes. Libraire de L'Académie de Médecine, Paris

Nicotra AB, Atkin OK, Bonser SP, Davidson AM, Finnegan EJ, Mathesius U, Poo P, Purugganan MD, Richards CL, Valladares F, van Kleunen M (2010) Plant phenotypic plasticity in a changing climate. Trends Plant Sci 12:684–692

Nosil P (2012) Ecological speciation. Oxford University Press, Oxford

O'Neill GA, Aitken SN, King JN, Alfaro RI (2002) Geographic variation in resin canal defenses in seedlings from the Sitka spruce × white spruce introgression zone. Can J For Res 32:390–400

O'Neill GA, Stoehr M, Jaquish B (2014) Quantifying safe seed transfer distance and impacts of tree breeding on adaptation. For Ecol Manag 328:122–130

Pavy N, Paule C, Parsons L, Crow JA, Morency MJ, Cooke J, Johnson JE, Noumen E, Guillet-Claude C, Butterfield Y, Barber S, Yang G, Liu J, Stott J, Kirkpatrick R, Siddiqui A, Holt R, Marra M, Seguin A et al (2005) Generation, annotation, analysis and database integration of 16,500 white spruce EST clusters. BMC Genom 6:1–19

Pavy N, Namroud MC, Gagnon F, Isabel N, Bousquet J (2012) The heterogeneous levels of linkage disequilibrium in white spruce genes and comparative analysis with other conifers. Heredity 108:273–284

Pavy N, Lamothe M, Pelgas B, Gagnon F, Birol I, Bohlmann J, Mackay J, Isabel N, Bousquet J (2017) A high-resolution reference genetic map positioning 8.8 K genes for the conifer white spruce: Structural genomics implications and correspondence with physical distance. Plant J 90:189-203

Pelgas B, Bousquet J, Meirmans PG, Ritland K, Isabel N (2011) QTL mapping in white spruce: gene maps and genomic regions underlying adaptive traits across pedigrees, years and environments. BMC Genom 12:145

Rajora OP, Dancik BP (2000) Population genetic variation, structure, and evolution in Engelmann spruce, white spruce, and their natural hybrid complex in Alberta. Can J Bot 78:768–780

Ran JH, Wei XX, Wang XQ (2006) Molecular phylogeny and biogeography of *Picea* (Pinaceae): implications for phylogeographical studies using cytoplasmic haplotypes. Mol Phylogenet Evol 41:405–419

Ran JH, Shen TT, Liu WJ, Wang PP, Wang XQ (2015) Mitochondrial introgression and complex biogeographic history of the genus *Picea*. Mol Phylogenet Evol 93:63–76

Rausher MD, Delph LF (2015) Commentary: when does understanding phenotypic evolution require identification of the underlying genes? Evolution 69:1655–1664

Rehfeldt GE (1994) Adaptation of *Picea engelmannii* populations to the heterogeneous environments of the intermountain west. Can J Bot 72:1197–1208

Rehfeldt GE (2004) Interspecific and intraspecific variation in *Picea engelmannii* and its congeneric cohorts: biosystematics, genecology, and climate change. RMRS-GTR-134, USDA Forest Service, Fort Collins

Rellstab C, Gugerli F, Eckert AJ, Hancock AM, Holderegger R (2015) A practical guide to environmental association analysis in landscape genomics. Mol Ecol 24:4348–4370

Richardson AD, Keenan TF, Migliavacca M, Ryu Y, Sonnentag O, Toomey M (2013) Climate change, phenology, and phenological control of vegetation feedbacks to the climate system. Agric For Meteorol 169:156–173

Roche L (1969) A genecological study of the genus *Picea* in British Columbia. New Phytol 68:505–554

Rweyongeza DM, Barnhardt LK, Hansen CR (2011) Patterns of optimal growth for white spruce provenances in Alberta. Ref. T/255, Alberta Tree Improvement and Seed Centre, Alberta Sustainable Resource Development, Smoky Lake

Sáenz-Romero C, Lindig-Cisneros RA, Joyce DG, Beaulieu J, Bradley JSC, Jaquish BC (2016) Assisted migration of forest populations for adapting trees to climate change. Rev Chapingo, Ser Ciencias For y del Ambient 22:303–323

Sæther SA, Fiske P, Kålås JA, Kuresoo A, Luigujoe L, Piertney SB, Sahlman T, Höglund J (2007) Inferring local adaptation from QST–FST comparisons: neutral genetic and quantitative trait variation in European populations of great snipe. J Evol Biol 20:1563–1576

Sanford E, Kelly MW (2011) Local adaptation in marine invertebrates. Ann Rev Mar Sci 3:509–535

Savolainen O, Pyhäjärvi T, Knürr T (2007) Gene flow and local adaptation in trees. Annu Rev Ecol Evol Syst 38:595–619

Savolainen O, Lascoux M, Merilä J (2013) Ecological genomics of local adaptation. Nat Rev Genet 14:807–820

Schweizer RM, Vonholdt BM, Harrigan R, Knowles JC, Musiana M, Coltman D, Novembre J, Wayne RK (2016) Genetic subdivision and candidate genes under selection in North American grey wolves. Mol Ecol 25:380–402

Sebastian-Azcona J, Hacke UG, Hamann A (2018) Adaptations of white spruce to climate: strong intraspecific differences in cold hardiness linked to survival. Ecol Evol 8:1758–1768

Shafer ABA, Cullingham CI, Côté SD, Coltman DW (2010) Of glaciers and refugia: a decade of study sheds new light on the phylogeography of northwestern North America. Mol Ecol 19:4589–4621

Shepperd WD, Jeffers RM, Ronco F (1981) An Engelmann spruce seed source study in the Central Rockies. RM-231, USDA Forest Service, Fort Collins

Sigurgeirsson A, Szmidt AE (1993) Phylogenetic and biogeographic implications of chloroplast DNA variation in *Picea*. Nord J Bot 13:233–246

Simpson DG (1994) Seasonal and geographic origin effects on cold hardiness of white spruce buds, foliage, and stems. Can J For Res 24:1066–1070

Singh RK, Svystun T, AlDahmash B, Jönsson AM, Bhalerao RP (2017) Photoperiod- and temperature-mediated control of phenology in trees – a molecular perspective. New Phytol 213:511–524

Sork VL (2018) Genomic studies of local adaptation in natural plant populations. J Hered 109:3–15

Spichtig M, Kawecki TJ (2004) The maintenance (or not) of polygenic variation by soft selection in heterogeneous environments. Am Nat 164:70–84

Stackpole DJ, Vaillancourt RE, de Aguigar M, Potts BM (2010) Age trends in genetic parameters for growth and wood density in *Eucalyptus globulus*. Tree Genet Genomes 6:179–193

Strasburg JL, Sherman NA, Wright KM, Moyle LC, Willis JH, Rieseberg LH (2012) What can patterns of differentiation across plant genomes tell us about adaptation and speciation? Philos Trans R Soc Lond B Biol Sci 367:364–373

Sonia SE (2000) Phenotypic plasticity for plant development, function and life history. Trends Plant Sci 5:537–542

Sultan SE, Spencer HG (2002) Metapopulation structure favors plasticity over local adaptation. Am Nat 160:271–283

Suren H, Hodgins KA, Yeaman S, Nurkowski KA, Smets P, Rieseberg LH (2016) Exome capture from the spruce and pine giga-genomes. Mol Ecol Resour 16:1136–1146

Tigano A, Friesen VL (2016) Genomics of local adaptation with gene flow. Mol Ecol 25:2144–2164

Turesson G (1922) The genotypical response of the plant species to the habitat. Hereditas 3:211–350

Ukrainetz NK, O'Neill GA, Jaquish B (2011) Comparison of fixed and focal point seed transfer systems for reforestation and assisted migration: a case study for interior spruce in British Columbia. Can J For Res 41:1452–1464

Valladares F, Matesanz S, Guilhaumon F, Araújo MB, Balaguer L, Benito-Garzón M, Cornwell W, Gianoli E, van Kleunen M, Naya DE, Nicotra AB, Poorter H, Zavala MA (2014) The effects of phenotypic plasticity and local adaptation on forecasts of species range shifts under climate change. Ecol Lett 17:1351–1364

Wang T, Hamann A, Spittlehouse DL, Aitken SN (2006) Development of scale-free climate data for western Canada for use in resource management. Int J Climatol 26:383–397

Wang T, Hamann A, Spittlehouse D, Carroll C (2016) Locally downscaled and spatially customizable climate data for historical and future periods in North America. PLoS ONE 11:e0156720

Warren RL, Keeling CI, Saint YM, Raymond A, Taylor GA, Vandervalk BP, Mohamadi H, Paulino D,

Chiu R, Jackman SD, Robertson G, Yang C, Boyle B, Hoffmann M, Weigel D, Nelson DR, Ritland C, Isabel N, Jaquish B et al (2015) Improved white spruce (*Picea glauca*) genome assemblies and annotation of large gene families of conifer terpenoid and phenolic defense metabolism. Plant J 83:189–212

Waser NM, Price MV (1985) Reciprocal transplant experiments with *Delphinium nelsonii* (Ranunculaceae): evidence for local adaptation. Am J Bot 72:1726–1732

Weeden NF, Wendel JF (1989) Genetics of plant isozymes. In: Soltis D, Soltis P (eds) Isozymes in plant biology. Chapman and Hall, London, pp 46–72

Weng YH, Tosh KJ, Park YS, Fullarton MS (2007) Age-related trends in genetic parameters for jack pine and their implications for early selection. Silvae Genet 56:242–252

Whitlock MC (2008) Evolutionary inference from QST. Mol Ecol 17:1885–1896

Willi Y, Van Buskirk J, Hoffmann AA (2006) Limits to the adaptive potential of small populations. Annu Rev Ecol Evol Syst 37:433–458

Williams GC (1966) Adaptation and natural selection. Princeton University Press, Princeton

Wright S (1931) Evolution in Mendelian populations. Genetics 16:97–158

Wright S (1949) The genetical structure of populations. Ann Eugen 15:323–354

Xie C-Y (2003) Genotype by environment interaction and its implications for genetic improvement of interior spruce in British Columbia. Can J For Res 33:1635–1643

Yeaman S (2015) Local adaptation by alleles of small effect. Am Nat 186:S74–S89

Yeaman S, Otto SP (2011) Establishment and maintenance of adaptive genetic divergence under migration, selection, and drift. Evolution 65:2123–2129

Yeaman S, Hodgins KA, Suren H, Nurkowski KA, Rieseberg LH, Holliday JA, Aitken SN (2014) Conservation and divergence of gene expression plasticity following c. 140 million years of evolution in lodgepole pine (*Pinus contorta*) and interior spruce (*Picea glauca* x *Picea engelmannii*). New Phytol 203:578–591

Yeaman S, Hodgins KA, Lotterhos KE, Suren H, Nadeau S, Degner JC, Nurkowski KA, Smets P, Wang T, Gray LK, Liepe KJ, Hamann A, Holliday JA, Whitlock MC, Rieseberg LH, Aitken SN (2016) Convergent local adaptation to climate in distantly related conifers. Science 353:23–26

Ying CC, Yanchuk AD (2006) The development of British Columbia's tree seed transfer guidelines: purpose, concept, methodology, and implementation. For Ecol Manag 227:1–13

11 The Terpene Synthase Gene Family in Norway Spruce

Xue-Mei Yan, Shan-Shan Zhou, Ilga M. Porth, and Jian-Feng Mao

Abstract

Terpenes (isoprenes) represent a diverse group of chemical compounds that plants produce during normal development (e.g., phytohormones), and mostly as secondary metabolites with important roles in defense responses against diverse environmental stressors, either abiotic or biotic in nature. Studies with an aspect on conifer-derived terpenes have highlighted some of the underlying metabolic and molecular mechanisms in the implicated defense processes. Terpene synthases (TPSs), being the core enzymes for terpenes' functional diversity, therefore, gained attention as the key elements for molecular terpene studies. Norway spruce (*Picea abies*) is arguably Europe's native conifer with the most substantial economic and ecological value given its vast geographic distribution throughout the continent as an indigenous species and, introduced, as an important plantation species outside its natural range. In this study, we aimed at retrieving TPS genes from the genomes of Norway spruce and, in addition, of representative land plant lineages in order to resolve their phylogenetic relationship. We show that the majority of TPS genes from gymnosperms are distributed within the TPS-d subfamily. As expected, the "DDXXD" and the "RXR" motifs are highly conserved for TPS in general, and the structural characteristics of closely related TPS genes are highly similar. Concrete subfamily membership along with shared structural molecular properties was the main driver of gene expression variation among Norway spruce TPSs, indicating important functional divergence. In this study, the two key factors within TPS gene structure that were related to differential TPS gene expression were found to be motif

X.-M. Yan · S.-S. Zhou · J.-F. Mao (✉)
National Engineering Laboratory for Tree Breeding, School of Nature Conservation, Beijing Advanced Innovation Center for Tree Breeding by Molecular Design, College of Biological Sciences and Technology, Beijing Forestry University, Beijing 100083, China
e-mail: jianfeng.mao@bjfu.edu.cn

X.-M. Yan
e-mail: xuemeiyan@bjfu.edu.cn

S.-S. Zhou
e-mail: shanshan9@bjfu.edu.cn

I. M. Porth
Département des sciences du bois et de la forêt, Université Laval, 1030, Avenue de la Médecine, Quebec City, QC G1V 0A6, Canada
e-mail: ilga.porth@sbf.ulaval.ca

I. M. Porth
Institute for System and Integrated Biology, Pavillon Charles-Eugène-Marchand, Université Laval, 1030, Avenue de la Médecine, Quebec City, QC G1V 0A6, Canada

I. M. Porth
Centre d'Étude de la Forêt, Université Laval, 1030, Avenue de la Médecine, Quebec City, QC G1V 0A6, Canada

I. M. Porth and A. R. De la Torre (eds.), *The Spruce Genome*, Compendium of Plant Genomes,
https://doi.org/10.1007/978-3-030-21001-4_11

composition and intron size. Our study is valuable for further in-depth functional evaluation of these additionally uncovered spruce TPSs and will support future efforts in metabolic engineering involving terpenes.

11.1 Introduction

Compounds of the isoprenoid class of metabolites (terpenes, terpenoids) play numerous important roles in the interactions of plants with their environment, and constitute a large family of natural products with an estimated number of 50,000 distinct substrates. Additionally, they also serve as basic sources for the provision of many biomaterials and biofuels (Bohlmann and Keeling 2008). The large chemical space provided by the plethora of terpenes includes many molecules that can also be biologically active in humans such as the medical terpenes from cannabis (Booth and Bohlmann 2019). While some terpene-related compounds have well-characterized physiological functions that are common to most plant species, such as phytohormones as regulators of plant growth (gibberellic acid, abscisic acid), photosynthetic pigments (precursors of chlorophyll (phytol), carotenoids), and membrane structural components (phytosterols), in addition to these universal physiological, metabolic, and structural functions considered part of plants' primary metabolism (Chen et al. 2011), specialized volatile or semi-volatile terpenes (and secondary metabolism-related) provide important biological signals in the plant's interactions with its environment, such as attractants for pollinators and seed dispersers, deterrents of insect herbivory, and allelopathic agents (Külheim et al. 2015; McGarvey and Croteau 1995; Chaw et al. 2019).

Conifers have evolved elaborate, constitutive as well as inducible, terpene-based defense mechanisms to deter or kill insect pests and their symbiotic fungal pathogens (Phillips and Croteau 1999). In response to wounding by invading insects and pathogens, conifer bark can extrude substantial amounts of oleoresin from blends of monoterpenes, sesquiterpenes, and diterpene resin acids (Trapp and Croteau 2001a, b; McKay et al. 2003). With respect to the biosynthesis of such terpenes, terpene synthases (TPSs) are the core enzymes that usually catalyze the final step in the generation of the terpenic end product (Smit et al. 2019). Studies have shown that the differential transcript expression among the multiple TPS genes detected in the conifer bark during an induced defense response, is responsible for the domination of one terpenic compound over others within the oleoresin blend of 30–50 different terpenes as typically found for a given conifer species (McKay et al. 2003). Moreover, the significant diversity of terpenes present in conifer oleoresin can be attributed to the remarkable functional diversity of conifer TPSs including their potential for important functional plasticity (Keeling et al. 2008).

In Europe, as an important conifer species, *Picea abies* (Norway spruce) composes, together with *Pinus sylvestris* (Scots pine), the majority of the continuous boreal forests of the Northern hemisphere (Gardner 1993). Norway spruce has the most substantial economic value and ecological function among European forest tree species, both, within and outside of its native range forests by serving also as the dominant softwood plantation forest tree in Europe. Norway spruce provides a useful system for terpene research as it produces substantial quantities of diversified terpene-rich oleoresin and, moreover, the related research on terpenic compounds has been performed. A previous study aiming at the functional characterization of individual TPS genes has identified nine gene members of the TPS-d subfamily from Norway spruce and which were further grouped into three distinct categories (Martin et al. 2004). However, regarding *P. abies* TPSs, a systematic identification and phylogenetic reconstruction, gene structure analysis and expression features of all TPSs is still lacking. Therefore, a comprehensive analysis of the Norway spruce TPS gene family would be helpful for exploring terpene metabolism further and providing useful tools for understanding the evolution and formation of spruce defense mechanisms (Mewalal et al. 2017).

11.2 Phylogenetic Analysis of the Norway Spruce TPS Gene Family

A gene family is a group of homologous genes, providing generally the same or very similar biochemical functions. Therefore, genes are categorized into families based on shared nucleotide or protein sequences (sequence similarity and conservation). Gene duplication and gene conversion are sources of gene family evolution and are the major forces behind the expansion of gene families; gene duplication followed by sequence differentiation provides an important origin of new gene functions formation (Ohta 2000). With mutations accumulating among members within the same family, the differentiation among new functions will be further deepened. Therefore, a gene family can be further subdivided into several categories or subfamilies according to the variation in gene sequence characteristics and gene functions. With regard to the TPS gene family, it is mid-sized but, at the same time, highly diversified throughout plant lineages (Chen et al. 2011).

11.2.1 General Classification of the TPS Gene Family

All plant TPS genes share a common evolutionary origin; however, extant TPS functions evolved independently in gymnosperms and angiosperms (Bohlmann et al. 1998; Trapp and Croteau 2001a, b). Some plant lineages have the majority of their TPS genes in one or two clades, indicating lineage-specific expansion of specific types of genes (Chen et al. 2011).

The recent classification of the monophyletic plant TPSs recognizes seven subfamilies (TPS-a through TPS-g, with TPS-e and TPS-f designated as TPS-e/f) based on sequence relatedness as well as functional assessments (Dudareva et al. 2003; Martin et al. 2004; Chen et al. 2011; Hu et al. 2016). Among these subfamilies, TPS-a, TPS-b, and TPS-g are angiosperm-specific; the TPS-a subfamily is further divided into the two clades TPS-a1 and TPS-a2 (Shalev et al. 2018), and the TPS-b subfamily is further divided into TPS-b and TPS-b2 clades (Hu et al. 2016). TPS-d is gymnosperm-specific. Based on functional assessments, it was proposed to label clades within the TPS-d subfamily according to the three biochemical classes for TPSs, producing mono- (TPS-d1), sesqui- (TPS-d2), or di- (TPS-d3) terpenes, respectively (Martin et al. 2004). TPS-h thus far contains only eight genes from the lycophyte plant *Selaginella moellendorffii*, and almost all of the TPS genes in this subfamily contain both, “DXDD” or “DDXXD” motifs (Chen et al. 2011). The TPS-c subfamily is believed to be ancestral; it contains copalyl diphosphate synthases (CPS), bifunctional CPS/kaurene synthases (CPS/KS), and other diterpene synthases (Falara et al. 2011; Shalev et al. 2018). Closely related to the TPS-c subfamily are the TPS-e and TPS-f subfamilies. TPS-f is derived from TPS-e, therefore, these two subfamilies have always been combined into one, designated as the TPS-e/f subfamily. TPS-e/f contains gymnosperm as well as angiosperm kaurene synthases (KS) and other apparently ancestral mono-, sesqui-, and diterpene synthase genes (Chen et al. 2011; Falara et al. 2011).

11.2.2 TPS Sequence Data Retrieval, Filtering and Phylogenetic Tree Construction

To accurately classify the Norway spruce TPS gene family members in the broader context of the evolutionary relationships involving TPSs from different plant species, firstly, we collected TPS protein sequences from previous studies (Martin et al. 2004; Chen et al. 2011; Külheim et al. 2015; Hu et al. 2016; Shalev et al. 2018; Zhou et al. 2019) for 11 representative land plants, including five gymnosperm species (*Thuja plicata, Platycladus orientalis, Cupressus gigantea, Taxus brevifolia, Taxus baccata*), four angiosperms (*Populus trichocarpa, Arabidopsis thaliana, Eucalyptus grandis, Oryza sativa*), one lycophyte species (*Selaginella moellendorffii*), and one bryophyta (*Physcomitrella patens*). Additional TPS protein sequences from ten gymnosperm species were retrieved from the PLAZA database (Van Bel et al. 2017) (https://bioinformatics.psb.ugent.be/plaza/versions/gymno-plaza/). The details of these data sources are provided in Table 11.1.

In order to obtain reliable TPS genes, we used Hmmer-3.1b2 (Finn et al. 2011) to ensure each sequence contains two TPS domains (i.e., the N-terminal domain IPR001906 and the metal-binding domain IPR005630). The TPS HMMs (Hidden Markov Models) were obtained from the Pfam database (Bateman et al. 2004) (http://pfam.xfam.org/). MAFFT version 7.335 (Nakamura et al. 2018) and trimAL v1.3 (Capella-Gutiérrez et al. 2009) (trimal-gt 0.8-st 0.001-cons 60) were used to align the TPS sequences and undertake further filtering. Finally, we obtained 678 eligible TPS genes (containing the two TPS domains and having sequence lengths greater than 200 amino acids; see Table 11.1).

A recent study by Chaw et al. (2019), who performed a transcriptome screening, detected 62 TPS gene family members for Norway spruce, which is much higher than what we detected

Table 11.1 The number of TPS genes and data sources for each plant species. "IN", the number of TPS genes present in the database before in-depth screening; "F-1 (HMM)", the number of TPS genes containing two protein domains retained after the first screening using HMM analysis; "F-2 (L > 200 AAs)", the number of remaining genes after the second screening based on length > 200 AAs. (PLAZA: (https://bioinformatics.psb.ugent.be/plaza/versions/gymno-plaza/, Van Bel et al. 2017)). In bold are the three spruce species we used in the analyses

Species	IN	F-1 (HMM)	F-2 (L > 200 AAs)	Sources
Cycas micholitzii	10	0	0	PLAZA
Ginkgo biloba	8	1	1	PLAZA
Gnetum montanum	5	2	2	PLAZA
Picea abies	**133**	**31**	**26**	**PLAZA**
Picea glauca	**47**	**16**	**16**	**PLAZA**
Pseudotsuga menziesii	140	47	46	PLAZA
Picea sitchensis	**36**	**16**	**16**	**PLAZA**
Pinus taeda	264	77	70	PLAZA
Taxus baccata	24	3	3	PLAZA
Pinus lambertiana	146	64	64	PLAZA
Thuja plicata	33	33	33	Shalev et al. (2018)
Platycladus orientalis	43	43	43	Hu et al. (2016)
Cupressus gigantea	121	121	121	Zhou et al. (2019)
Selaginella moellendorffii	14	14	14	Chen et al. (2011)
Abies grandis	11	11	11	Martin et al. (2004)
Taxus brevifolia	1	1	1	Martin et al. (2004)
Populus trichocarpa	32	32	32	Chen et al. (2011)
Oryza sativa	32	32	32	Chen et al. (2011)
Arabidopsis thaliana	33	33	33	Chen et al. (2011)
Physcomitrella patens	1	1	1	Chen et al. (2011)
Eucalyptus grandis	113	113	113	Külheim et al. (2015)
Total	1247	691	678	

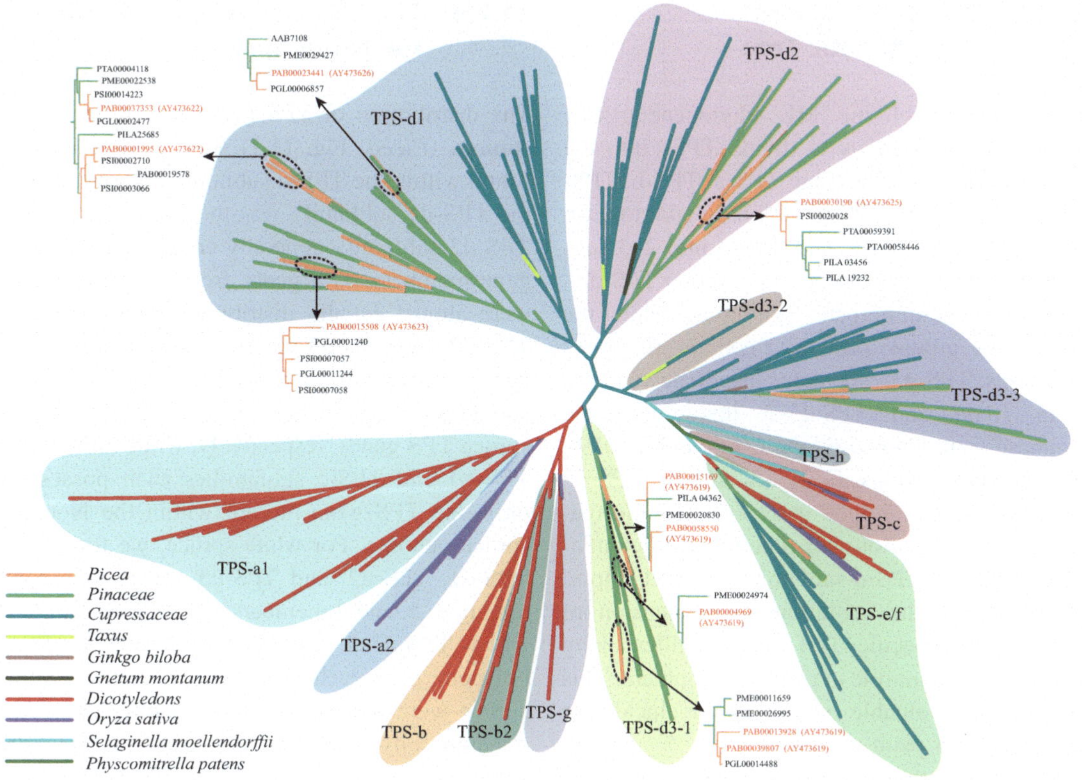

Fig. 11.1 Phylogenetic tree displaying 678 TPS genes from 20 land plants [three spruces (*Picea abies, Picea glauca, Picea sitchensis*); four other members of Pinaceae (*Pseudotsuga menziesii, Pinus taeda, Pinus lambertiana, Abies grandis*), three Cupressaceae members (*Thuja plicata, Platycladus orientalis, Cupressus gigantea*), two members from Taxus (*Taxus brevifolia, Taxus baccata*), *Ginkgo biloba*, *Gnetum montanum*, three dicotyledons (*Populus trichocarpa, Arabidopsis thaliana, Eucalyptus grandis*), one monocotyledon (*Oryza sativa*), one lycophyte (*Selaginella moellendorffii*), and one bryophyta (*Physcomitrella patens*)]. Colors on individual branches correspond to the different land plant groups from which mapped genes originated (see legend in the lower left corner), and the resolved clades' background colors correspond to the different subfamilies. Previously characterized Norway spruce TPSs (Martin et al. 2004) were also mapped to our phylogenetic tree with an imposed threshold of >80% amino acid sequence similarity; genes marked in red highlight these TPSs with matches to our genomic database. The NCBI gene IDs in parentheses also indicate those TPSs previously characterized (Martin et al. 2004)

here, yet similar to the number of TPSs (69) found previously in white spruce based on in silico assessment using a white spruce EST database (Keeling et al. 2011) and prior to the availability of a spruce reference genome assembly (Nystedt et al. 2013). The differences in the number of TPSs detected might be caused by the use of different data sources. We retrieved the TPS genes from the PLAZA database, and each gene was annotated against the genome assembly. Gene annotation against a transcriptome assembly is prone to deviations due to incidences of alternative splicing which may increase the number of sequences identified. Therefore, the total number of TPS members may be more reliable in this study. Finally, JTT+F+R8 was selected as the optimal sequence evolution model, and a maximum likelihood (ML) phylogenetic tree was constructed (Fig. 11.1) by running IQ-TREE version1.6.7 (Nguyen et al. 2014) and with ultrafast bootstrapping (-bb 1000).

11.2.3 The TPSs-Related Phylogenetic Tree

The topology of the ML phylogenetic tree (Fig. 11.1) allowed us to divide TPSs into the seven subfamilies following TPS-a, TPS-b, TPS-c, TPS-d, TPS-e/f, TPS-g, TPS-h. In addition, we divided TPS-a, TPS-b, and TPS-d into 2–3 subgroups. Among these subgroups, group TPS-d3–1 contains sesquiterpene synthases, while the other two groups primarily contain diterpene synthases: TPS-d3–2 contains taxadiene synthase from *Taxus* and TPS-d3–3 levopimaradiene synthase from *Ginkgo biloba* (Zhou et al. 2019). This distinct sub-classification within the TPS-d3 subfamily helps provide more specific information on the evolution and functional differentiation within the TPS-d subfamily and will help to further reveal the evolution and expansion of gymnosperm-specific TPS gene families along with its important role in plant development. This will provide valuable references for elucidating the process of plant evolution.

Regarding the number of TPSs, we found important differences among the three considered spruces, with the number of TPS genes consistently higher for *Picea abies* (Norway spruce); thus, in this phylogenetic tree, there are 26, 16, and again 16 TPS genes, respectively, represented from Norway spruce, white spruce (*P. glauca*), and Sitka spruce (*P. sitchensis*), respectively (see also Tables 11.1 and 11.2). However, the basic topology of this tree is consistent with what was postulated in previous studies (Dudareva et al. 2003; Martin et al. 2004; Chen et al. 2011; Külheim et al. 2015; Hu et al. 2016). TPS-a/b/g is a specific branch of angiosperms, while TPS-h only contains eight TPS genes from *Selaginalla moellendorffii*. The TPS-d subfamily, of which all members derive from gymnosperm gene sequences, is the largest branch of this tree, accounting for 58.4% of all TPS genes (Table 11.2).

11.2.4 The TPS Genes' Classification for Three Spruces

The distribution of TPS genes among the three spruces (*Picea*, Fig. 11.1) was similar, mainly found within the TPS-d subfamily, among which the TPS-d1 subfamily was the largest, and with TPS-d3 subfamily generally containing the least gene members. TPS genes from Norway spruce were almost equally distributed across TPS-d1, TPS-d2, and TPS-d3 subfamilies with 9, 8, and 7 genes, respectively. It was shown that TPS-d3–1 and TPS-d3–3 branches housed 2 and 5 Norway spruce TPS genes, respectively, while among the TPS-e/f and TPS-c subfamilies each possessed only one TPS gene present within the Norway spruce genome. For white spruce, we found the TPS-d1, TPS-d2, and TPS-d3 subfamilies containing 9, 5, and 2 members, respectively, among which TPS-d3–1 and TPS-d3–3 branches have only one gene each. The TPS genes from Sitka spruce were mostly distributed within the TPS-d1 subfamily, and the total numbers of TPS genes within the three branches TPS-d1, TPS-d2, and TPS-d3–3 were 11, 2, and 2, respectively. In addition, Sitka spruce was also found to contain one TPS-e/f gene (Table 11.2).

11.2.5 The Significance of TPSs Classification

Phylogenetic analyses of nucleotide sequences indicated that the earliest divergence of extant seed plants is likely represented by a split between conifer-cycad and angiosperm lineages (Savard et al. 1994). The TPS gene family is widely distributed in plants, and relevant studies have pointed at functional specialization of many TPSs that occurred after the separation of angiosperms and gymnosperms (Martin et al. 2004). More detailed functional classification and understanding of the evolutionary origin and

Table 11.2 TPS genes identified from different plant species. IspS…isoprene synthase; CPS… copalyl synthase; KS… kaurene synthase; isoprene units…mono (C10)-, sesqui (C15)-, di (C20)-terpenes

TPS subfamilies	a		b		c	d			e/f	g	h	Total
Function/species	C15		IspS, C10		CPS, C20	C10, C15, C20			KS, C20	C10		
	a1	a2	b	b2		d1	d2	d3				
Ginkgo biloba	–	–	–	–	–	–	–	1	–	–	–	1
Gnetum montanum	–	–	–	–	–	–	2	–	–	–	–	2
Picea abies	–	–	–	–	**1**	**9**	**8**	**7**	**1**	–	–	**26**
Picea glauca	–	–	–	–	–	**9**	**5**	**2**	–	–	–	**16**
Picea sitchensis	–	–	–	–	–	**11**	**2**	**2**	**1**	–	–	**16**
Pseudotsuga menziesii	–	–	–	–	2	24	5	7	8	–	–	46
Pinus taeda	–	–	–	–	1	19	23	26	1	–	–	70
Abies grandis	–	–	–	–	–	7	2	2	–	–	–	11
Pinus lambertiana	–	–	–	–	1	22	16	24	1	–	–	64
Thuja plicata	–	–	–	–	–	7	13	8	5	–	–	33
Platycladus orientalis	–	–	–	–	–	25	4	9	5	–	–	43
Cupressus gigantea	–	–	–	–	1	30	31	30	29	–	–	121
Taxus baccata	–	–	–	–	–	1	1	1	–	–	–	3
Taxus brevifolia	–	–	–	–	–	–	–	1	–	–	–	1
Selaginella moellendorffii	–	–	–	–	3	–	–	–	3	–	8	14
Physcomitrella patens	–	–	–	–	1	–	–	–	–	–	–	1
Eucalyptus grandis	52		17	19	2	–	–	–	10	13	–	113
Oryza sativa		19	–	–	3	–	–	–	9	1	–	32
Arabidopsis thaliana	23		6		1	–	–	–	2	1	–	33
Populus trichocarpa	13		8	4	2	–	–	–	3	2	–	32
Total	88	19	31	23	18	164	112	120	78	17	8	678
Proportion (%)	130	2.8	4.6	3.4	2.65	24.2	16.5	17.7	11.5	2.5	1.2	
	15.8		8			58.4						

expansion of the TPS-d subfamily will help to reveal its importance for plant development and will prove valuable for further elucidating the process of land plant evolution. In a previous study, Keeling et al. (2011) functionally characterized, using recombinant TPS enzyme assays, a total of 21 spruce TPSs (genes from Sitka, white and hybrid white spruces) covering mono-, sesqui-, and diterpene synthases. Because of important TPSs sequence variants even between closely related species and known within-species plasticity of TPSs, without proper functional characterization, we cannot infer gene function of TPSs by phylogeny alone (Keeling et al. 2008, 2011). Therefore, the here newly identified Norway spruce TPSs (Table 11.3) still await in-depth characterization to properly elucidate their specific mechanistic implications related to terpene synthesis and functioning in vivo (Martin et al. 2004).

11.3 Analysis of Gene Structure Within Norway Spruce TPSs

Several research groups have proposed that plant TPSs share evolutionary origins based on enzymes' similar reaction mechanisms, their

Table 11.3 Conserved domains identified with MEME (genes with ≥ 15 motifs accounted for 58% and genes with <10 motifs accounted for 27% of the total number of the identified TPSs, respectively)

Gene ID	motif3 (DDXXD)		motif11 (RXR)		motif10 (RX8W)		Motif number	Exon number	Accession No.	Function
	Start	End	Start	End	Start	End				
PAB00005566	–	–	–	–	–	–	7	5		Unknown
PAB00047162	508	536	472	492	–	–	15	13		Unknown
PAB00013348	515	543	479	499	–	–	15	15		Unknown
PAB00014674	553	581	–	–	–	–	12	14		Unknown
*PAB00058550**	551	579	515	535	24	67	17	12	AY473619	E-α-bisabolene synthase
*PAB00015169**	317	145	281	301	–	–	17	9	AY473619	
*PAB00004969**	–	–	–	–	–	–	5	5	AY473619	
*PAB00013928**	307	335	271	291	–	–	15	10	AY473619	
*PAB00039807**	164	192	128	148	–	–	8	6	AY473619	
PAB00006215	343	371	307	327	28	71	17	10		Unknown
PAB00013963	334	362	298	318	19	62	17	10		Unknown
PAB00006225	280	308	–	–	14	57	14	9		Unknown
PAB00016738	154	182	125	145	–	–	8	9		Unknown
*PAB00030190**	333	361	397	317	14	57	17	10	AY473625	Longifolene synthase
PAB00014531	405	433	369	389	84	127	18	11		Unknown
PAB00028465	–	–	–	–	–	–	7	12		Unknown
PAB00029469	375	403	339	359	54	97	18	10		Unknown
PAB00051831	375	403	339	359	63	106	18	10		Unknown
*PAB00015508**	286	314	250	270	–	–	10	8	AY473623	(-)-linalool synthase
PAB00045518	–	–	–	–	–	–	8	7		Unknown
PAB00053419	486	514	450	470	186	229	17	11		Unknown
PAB00056821	154	182	118	138	–	–	12	6		Unknown
*PAB00023441**	367	395	331	351	50	93	17	10	AY473626	Myrcene synthase
*PAB00037353**	371	399	335	355	63	106	17	10	AY473622	(-)-α/β-pinene synthase
PAB00019578	174	202	138	158	–	–	7	10		Unknown
*PAB00001995**	368	396	332	352	61	104	18	10	AY473622	(-)-α/β-pinene synthase

*Gene functions could be proposed based on best matches by querying sequence entries from a previous study on cDNA cloned and functionally characterized Norway spruce TPSs (Martin et al. 2004) based on a > 80% similarity threshold for amino acid sequences; the NCBI accession no. of the matched TPS genes originating from the cloned cDNAs (from Martin et al. 2004) and their putative functions are provided. No other TPSs from Martin et al. (2004) could be matched here to our TPSs Norway spruce genomic database using our overall highly stringent criteria for sequence filtering conserved structure, and sequence characteristics, including conserved sequence motifs, intron number, and exon sizes (Trapp and Croteau 2001a, b). Therefore, knowing about the conserved motifs in gene sequence composition and uncovering other gene structural features are key factors to a first approach toward understanding gene functional differentiation of TPSs.

11.3.1 Analysis of Conserved Motifs Within TPSs

TPSs are classified into two types, Class I and Class II, on the basis of TPS structure and reaction mechanism in plants (Christianson 2006; Hayashi et al. 2006; Köksal et al. 2011). Almost all plants contain Class I, with "DDXXD" and "NSE/DTE" motifs being considered a feature of Class I, while the Class II TPS contains the signature sequence motif "DXDD" (Christianson 2006). We searched for the conserved amino acid residues within the 26 Norway spruce TPS genes, and the "DDXXD" motif was found at position 807 amino acids (from the N-terminus within the multiple sequence alignment (Fig. 11.2)). Noteworthy, a conserved "RxR" motif located upstream of the "DDXXD" motif at position 770 was found. The "RxR" motif also varied within the TPS-e subfamily occurring as "RXK", which is consistent with a previous report (Zhou et al. 2019). This "RXR" motif was found to be shared across all subfamilies except for TPS-c. Furthermore, we found "DDXXD" and "RXR" motifs (where present) spanning the same sequence distance between each other (34 amino acids) in each of the TPS genes of Norway spruce.

Subsequently, in order to further identify the motifs shared between the TPS genes from Norway spruce, we used the MEME online tools (Bailey et al. 2009) and analyzed the conserved

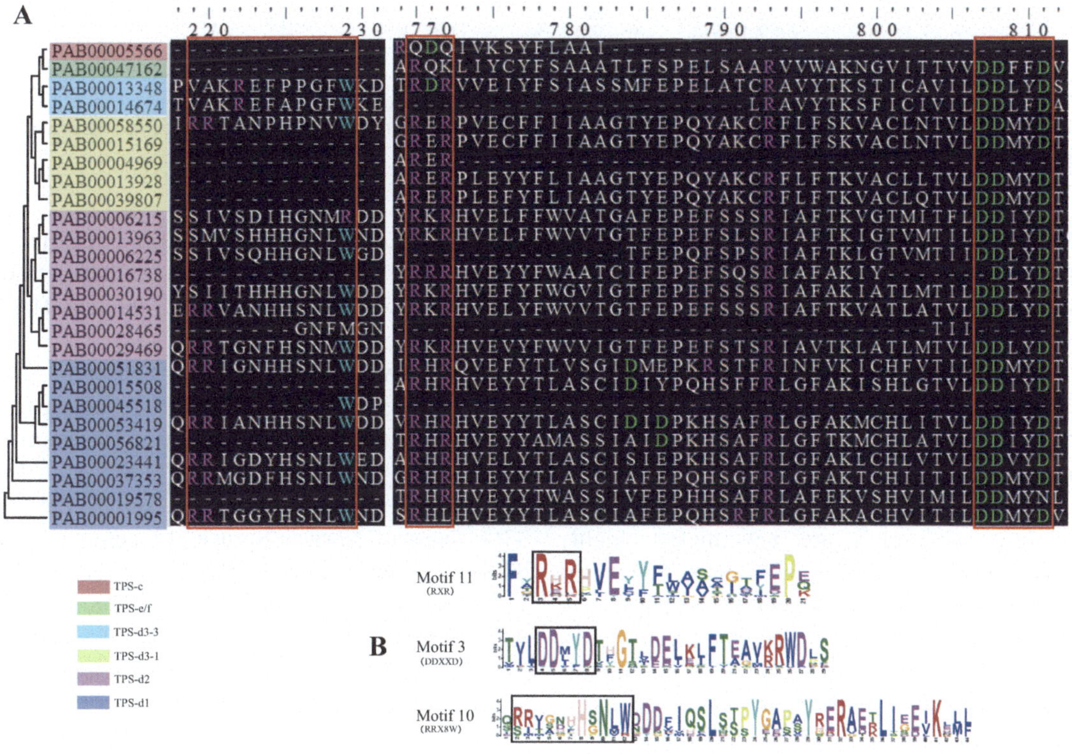

Fig. 11.2 A Conserved motifs ("DDXXD", "RXR", and "RRX8W") detected for the 26 Norway spruce TPS genes; B sequence motifs detected with MEME tool

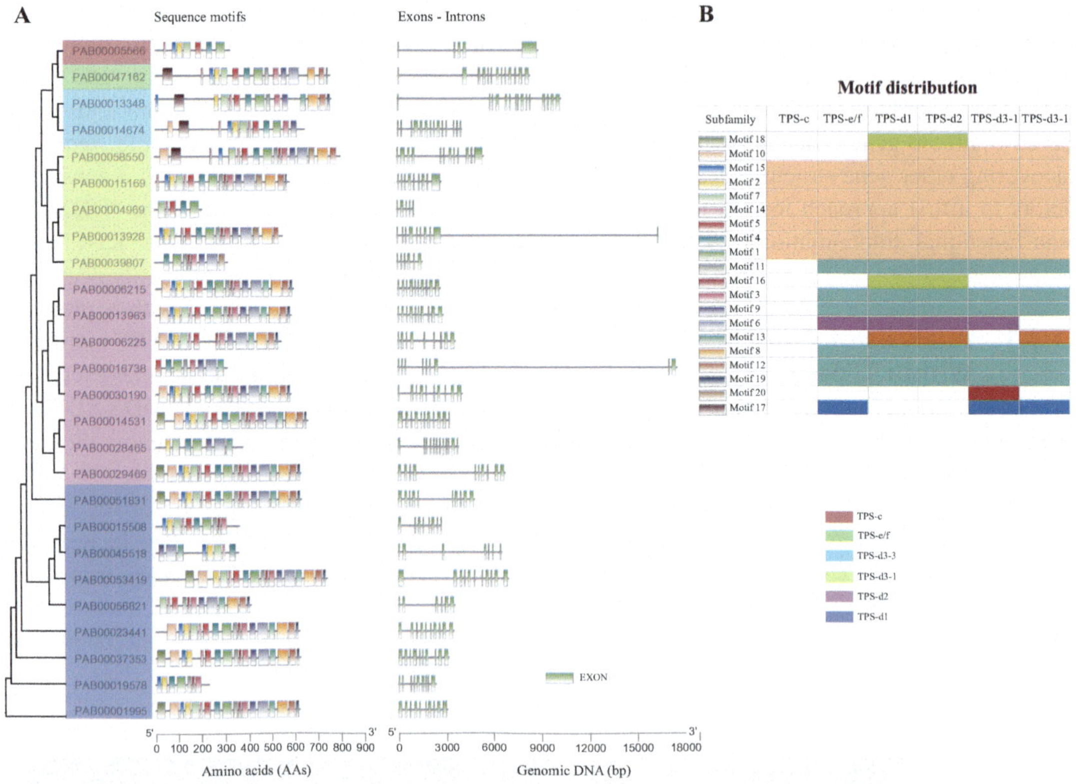

Fig. 11.3 **A** Structural variability among TPS genes from Norway spruce. Twenty putative conserved motifs have been labeled with different colors; exons are indicated by green boxes; introns are indicated by black lines. **B** Different colors denoting TPS gene subfamilies

domains of the TPSs. In total, 20 motifs were detected (Table 11.3, Figs. 11.3, and 11.4). Figure 11.4 clearly shows the detailed amino acid composition of each motif sequence. Among the 26 designated TPS genes from Norway spruce, 20 TPSs were found to contain both, Motif 3 ("DDXXD") and Motif 11 ("RXR"), and also other motifs with several highly conservative amino acids within each gene, indicating that different subfamilies are nevertheless highly conserved in some functional features. The results by MEME analysis are generally consistent with the abovementioned results of conserved motifs. Except for TPS-c, the "DDXXD" motif was found shared among all subfamilies. Furthermore, three genes from Norway spruce were found to lack the typical motif 3 ("DDXXD" motif). These genes are *PAB00045518*, *PAB00028465*, and *PAB00004969*, from TPS-d1, TPS-d2, and TPS-d3–1 clades, respectively.

The "RRX8W" motif is highly conserved and is located near the N-terminus of monoterpene synthases (mostly cyclases) from angiosperms' TPS-b clade and gymnosperms' TPS-d–1 clade, respectively (Chen et al. 2011). Among the 26 TPS genes of Norway spruce, 13 genes of the TPS-d subfamily contain this conserved motif, i.e., genes belonging to the TPS-d1, TPS-d2, and TPS-d3 subfamilies, respectively. The position of the "RRX8W" motif within the alignment is shown in Fig. 11.2.

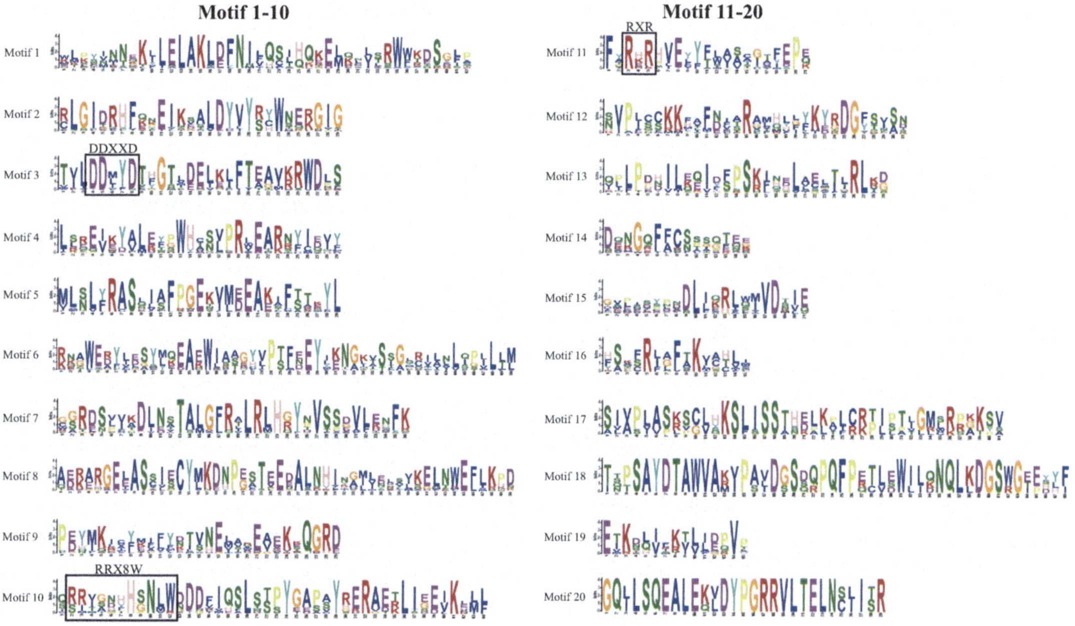

Fig. 11.4 Details of the 20 motifs detected by MEME

11.3.2 Gene Structural Characteristics

To better understand the structural sequence features of each individual gene, and, in fact, also the entirety of the evolutionary diversity between different subfamilies, we retrieved Norway spruce gene annotations from TreeGenes (Wegrzyn et al. 2008). Then, we examined the phylogeny, the composition of conservative domains and the exon-intron features for all Norway spruce TPS members using TBtools (Chen et al. 2018). Among these retrieved TPS genes, 57.7% contained 15 or more motifs, while 26.9% had less or equal 10 motifs. We found that TPSs containing > 15 motifs were structurally more conserved, while those with fewer motifs were highly diverse.

The analysis of exon-intron structures revealed that the number of exons for TPS genes can be found between 8 and 12. It is worth noting that the number of exons in these genes seems to be highly correlated with the number of motifs (Fig. 11.3). Nystedt et al. (2013) found that Norway spruce has accumulated long introns and also showed that some introns can be very long (exceeding 10 Kbp) compared to those in other plant species. In this study, we observed that two TPS genes, namely *PAB00016738* (unknown, TPS-d2) and *PAB00013928* (E-α-bisabolene synthase), both exhibited one abnormally long intron each (Fig. 11.3).

11.3.3 From Sequence to Function

By querying the 26 Norway spruce TPS genes identified here, we associated these sequences with the functionally characterized TPS enzymes set from a previous Norway spruce TPSs study (Martin et al. 2004) based on the threshold of > 80% similarity of amino acid sequence. According to the query results, we found 10 TPS genes potentially involved in five functionally distinct enzymatic processes, while all other 16 TPS genes' functions remain unknown (Table 11.3 and Fig. 11.1). Of note, we also found all genes from the TPS-d3–1 subfamily collapsed onto one particular enzyme function ("E-α-bisabolene synthase"); contrary to this result, the TPS-d1 subfamily genes were matched to individual enzyme targets and probably have indeed diverse functioning (Table 11.3 and Fig. 11.1).

11.4 TPSs' Gene Expression

Gene expression patterns can provide important clues on gene functional divergence (Meng et al. 2019). By analyzing the expression profiles of TPS subfamily members across different tissues and at different developmental stages, the functional differentiation of TPS members and the potential roles of TPS genes can be revealed. In general, members from the same subfamily, similar in gene structure and sequence composition, should display similar expression patterns with minor differences regarding expression levels (Meng et al. 2019). In order to reveal the potential relationship between gene structure and gene function, we analyzed TPS genes' expression profiles, the relationship between gene expression level and gene structure in Norway spruce.

11.4.1 Differential Expression Profiles of TPS Genes

For further exploration, the 26 TPS genes' IDs from the Norway spruce genome assembly were used to obtain each gene's and the overall expression pattern related heat map representation (Figs. 11.5 and 11.6) from Congenie.org (http://congenie.org/eximage). Details of Norway spruce samples from which TPS genes' expression data were collected are depicted in Fig. 11.6, including collection time, sample tissue specifics, such as identity, growth stage, and health condition.

Generally, TPS genes' expression from the same subfamily displayed the same pattern with

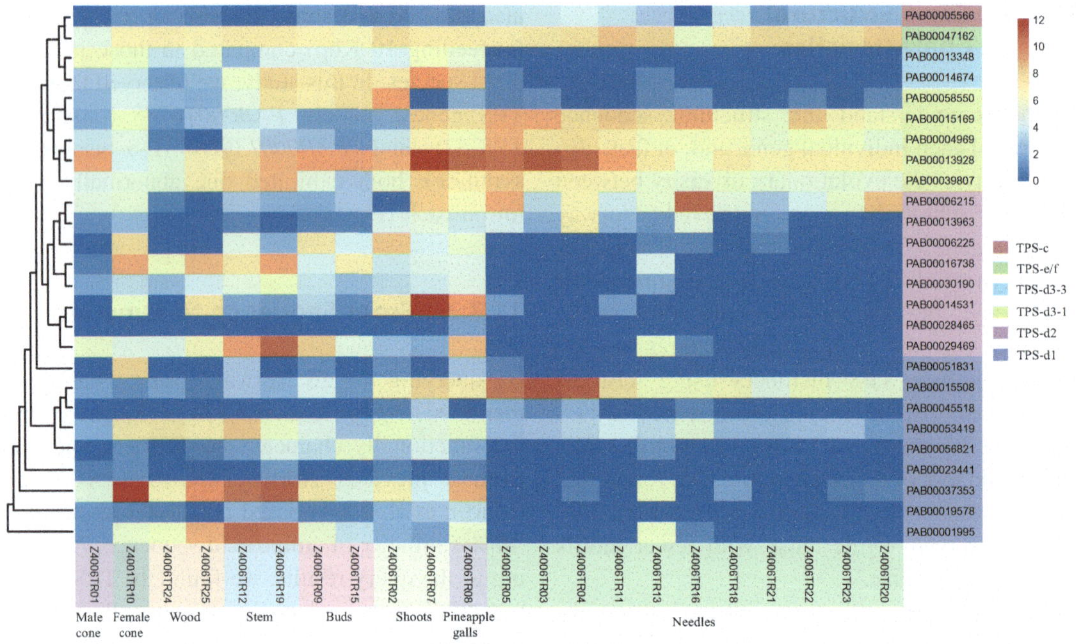

Fig. 11.5 Expression profiling of 26 TPS genes from Norway spruce. Different colors on the right separate the different TPS subfamilies across gene IDs, and the different colors on the bottom of the figure correspond to the different tissue types (Congenie.org: http://congenie.org/eximage). See Fig. 11.6 for further explanations on plant tissue and imposed stress specifics

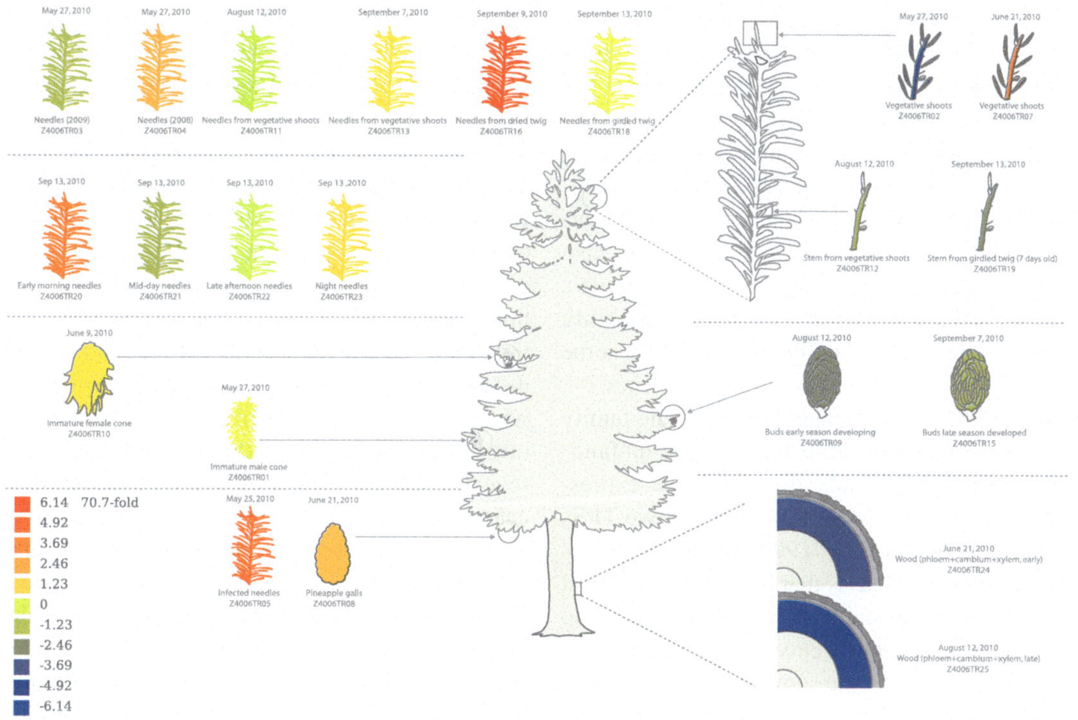

Fig. 11.6 Details on the plant tissue sources and the expression profiles of a Norway spruce TPS gene family member, *PAB0006215*, across these sampled tissues. (Congenie.org: http://congenie.org/eximage)

regard to their expression in the spruce needle tissue [TPS-d3–1, e.g., with the exception of only one among the five E-α-bisabolene synthases (*PAB00058550* not expressed in needles), and also with the exception of *PAB00015508*, a (-)-linalool synthase, among the large number of TPS-d1 members (not expressed in needles), here highly expressed in early spring needles and in infected needles; TPS-d2 members such as its representative longifolene synthase *PAB00030190* not expressed in needles with the exception of subfamily member *PAB0006215*, whose exceptional overall expression pattern is discussed in more details below], Fig. 11.5. However, certain more refined differences in expression patterns were obvious across the different development stages of needles as well as the different environmental conditions employed to this study of Norway spruce gene expression. Co-expression analysis revealed that genes having close sequence relationships might also have the same or similar functional properties. This is again based on the assumption that the shared structural features among TPS gene family members belonging to the same phylogenetic subfamily implies that such genes are indeed functionally more similar to each other. It is worth noting that the TPS gene expression levels in pathogen infested needle samples (Z40006TR05 coded in Figs. 11.5 and 11.6) are generally higher than in uninfected, healthy needles. The production of oleoresin, a complex blend of terpenes, is an important physical and chemical barrier against biotic stress imposed by insects and pathogens on conifer's actively growing tissues (Byun-McKay et al. 2006). Therefore, when plants are challenged, such as parasitized by diseases, insect pests, or mechanically damaged, they will initiate a stress response and secrete a substantial amount of terpenes to defend themselves against such stressors. In this respect, the observed trend in increased TPS gene expression in the infected needle sample indicates that the plant initiated the defense response within the needles to counteract the pathogen invasion. Not surprisingly, expression levels of certain TPS genes were also elevated in tissues that were gall infested (Z40006TR08). In addition, for needles sampled from dried twig (Z40006TR16), we found one TPS gene member from clade TPS-d2 (*PAB0006215*) whose expression level was substantially higher under these conditions than in any other tissue or condition sampled (Figs. 11.5 and 11.6). Differential TPS gene expression patterns were also observed that were related to a specific tissue's differential growth stages or seasonal variability and therefore may in fact be stimulated by some internal factors as well as environmental cues.

In other tissue types than needles, gene family members originating from within TPS subfamilies showed significantly different gene expression patterns (with the exception of the two TPSs *PAB00037353*, *PAB00001995*, and annotated as (-)-α/β-pinene synthases that both showed almost perfectly matched high expression in stem and wood, Fig. 11.5). Differences in the expression of genes from within the same subfamily in those tissues indicate that genes have functionally diverged. As an example for such a pattern, it is worth mentioning that in the female cone (Z3001TR10) the expression level of the TPS-d1 member *PAB00037353* was highest among all tested TPS genes, which may indicate its potential functional role also correlated to female cone development.

11.4.2 The Relationship Between Gene Expression and Gene Structure of TPSs

In order to verify a possible association between gene structure and gene expression, these two factors were further examined. We found no relationship between the exon number and gene expression overall. Among the TPS genes (4) that do not contain the conserved domains motifs 3 (DDXXD) and 11 (RXR) (Table 11.3), with the exception of *PAB00004969* (TPS-d3–1, E-α-bisabolene synthase, and high expression across all needle samples and within pineapple gall infested tissue, Fig. 11.5), Norway spruce genes *PAB00005566* (TPS-c), *PAB00028465* (TPS-d2), and *PAB000045518* (TPS-d1) expression were extremely low or even absent (Figs. 11.3 and 11.5). We speculate that these three genes are in fact pseudogenes, mainly due to their lack of these sequence motifs. For three other genes in clade TPS-d1 which do have the two motifs but for which we found only extremely low or absent expression across the wide range of tissues sampled and stresses imposed (e.g., the myrcene synthase gene *PAB00023441*), it is still possible that their expression is elicited by different cues or in different tissue types such as bark (+phloem) which was not sampled here. In addition, a previous study on white spruce (Stival Sena et al. 2014) revealed that the length of introns may be intricately related to gene function and expression level such that ubiquitously expressed genes generally contain more intronic sequence than tissue preferential genes. In our study, the expression of the Norway spruce TPS gene *PAB00013928* with an abnormally long intron displayed consistently higher expression than all other genes within the same subfamily. Within the TPS-d3–3 gene family, expression is high in needle tissue but *PAB00013928* even displayed the highest expression values among all members. The second spruce TPS gene for which extremely long introns were uncovered (*PAB00016738*), while being hardly expressed in needle tissue (consistent with gene expression patterns for TPS-d2 clade and to which this member belongs), showed overall (across all other sampled tissues than needle) the highest gene expression levels in this clade. Based on this observation, we can conclude that introns may—to a certain extent—regulate TPSs gene expression and that the variation in introns size may also cause gene functional divergence in Norway spruce.

11.5 Conclusion and Perspectives

Terpenes are important plant metabolites and play key roles in plant growth and development, and thus, are involved in various aspects of plant physiological activities. The characterization of the TPS gene family helps us to explore normal development but also plant's defense systems.

The phylogenetic reconstruction and classification of the TPS gene family of Norway spruce will provide a valuable basis for further studying terpenes' functional evolution in the future. Upcoming studies should focus on structural and gene expression variation among Norway spruce TPSs and uncover their functional divergence in a refined experimental context.

References

Bailey TL, Boden M, Buske FA, Frith M, Grant CE et al (2009) MEME SUITE: tools for motif discovery and searching. Nucl Acids Res 37(suppl_2):W202–W208

Bateman A, Coin L, Durbin R, Finn RD, Hollich V et al (2004) The Pfam protein families database. Nucl Acids Res 32(suppl_1): D138–D141

Bohlmann J, Keeling CI (2008) Terpenoid biomaterials. Plant J 54(4):656–669

Bohlmann J, Meyer-Gauen G, Croteau R (1998) Plant terpenoid synthases: Molecular biology and phylogenetic analysis. Proc Natl Acad Sci USA 95(8):4126–4133

Booth JK, Bohlmann J (2019) Terpenes in Cannabis sativa – from plant genome to humans. Plant Sci 284:67–72

Byun-McKay A, Godard K-A, Toudefallah M, Martin DM, Alfaro R et al (2006) Wound-induced terpene synthase gene expression in Sitka spruce that exhibit resistance or susceptibility to attack by the white pine weevil. Plant Physiol 140(3):1009–1021

Capella-Gutiérrez S, Silla-Martínez JM, Gabaldón T (2009) trimAl: a tool for automated alignment trimming in large-scale phylogenetic analyses. Bioinformatics 25(15):1972–1973

Chaw S-M, Liu Y-C, Wu Y-W, Wang H-Y, Lin C-YI et al (2019) Stout camphor tree genome fills gaps in understanding of flowering plant genome evolution. Nat Plants 5(1):63–73

Chen F, Tholl D, Bohlmann J, Pichersky E (2011) The family of terpene synthases in plants: a mid-size family of genes for specialized metabolism that is highly diversified throughout the kingdom. Plant J 66 (1):212–229

Chen C, Xia R, Chen H, He Y (2018) TBtools, a toolkit for biologists integrating various biological data handling tools with a user-friendly interface 289660

Christianson DW (2006) Structural biology and chemistry of the terpenoid cyclases. Chem Rev 106(8):3412–3442

Dudareva N, Martin D, Kish CM, Kolosova N, Gorenstein N et al (2003) (E)-beta-ocimene and myrcene synthase genes of floral scent biosynthesis in snapdragon: function and expression of three terpene synthase genes of a new terpene synthase subfamily. Plant Cell 15(5):1227–1241

Falara V, Akhtar TA, Nguyen TTH, Spyropoulou EA, Bleeker PM et al (2011) The tomato terpene synthase gene family. Plant Physiol 157(2):770–789

Finn RD, Clements J, Eddy SR (2011) HMMER web server: interactive sequence similarity searching. Nucl Acids Res 39(suppl_2): W29–W37

Gardner MF (1993) Pinaceae – drawings and descriptions of the genera Abies, Cedrus, Pseudolarix, Keteleeria, Nothotsuga, Tsuga, Cathaya, Pseudotsuga, Larix and Picea. A. Farjon. Koeltz Scientific Books, D-6240 Königstein, Federal Republic of Germany. 1990. pp. xii + 330; 117 illustrations (mostly line drawings); 124 maps. ISBN 3 87429 298 3. Forming volume 121 of Regnum Vegetabile; ISSN 0080-0694, DM 260. *Edinburgh Journal of Botany* 50(1): 121-122

Hayashi K-i, Kawaide H, Notomi M, Sakigi Y, Matsuo A et al (2006) Identification and functional analysis of bifunctional ent-kaurene synthase from the moss *Physcomitrella patens*. FEBS Lett 580(26):6175–6181

Hu X-G, Liu H, Jin Y, Sun Y-Q, Li Y et al (2016) *De novo* transcriptome assembly and characterization for the widespread and stress-tolerant conifer *Platycladus orientalis*. PLoS ONE 11(2):e0148985

Keeling CI, Weisshaar S, Lin RPC, Bohlmann J (2008) Functional plasticity of paralogous diterpene synthases involved in conifer defense. Proc Natl Acad Sci USA 105(3):1085–1090

Keeling CI, Weisshaar S, Ralph SG, Jancsik S, Hamberger B, Dullat HK, Bohlmann J (2011) Transcriptome mining, functional characterization, and phylogeny of a large terpene synthase gene family in spruce (*Picea* spp.). BMC Plant Biol 11:43

Köksal M, Jin Y, Coates RM, Croteau R, Christianson DW (2011) Taxadiene synthase structure and evolution of modular architecture in terpene biosynthesis. Nature 469(7328):116–120

Külheim C, Padovan A, Hefer C, Krause ST, Köllner TG et al (2015) The Eucalyptus terpene synthase gene family. BMC Genomics 16(1):450

Martin DM, Fäldt J, Bohlmann J (2004) Functional characterization of nine Norway spruce *TPS* genes and evolution of gymnosperm terpene synthases of the *TPS*-d subfamily. Plant Physiol 135(4):1908–1927

McGarvey DJ, Croteau R (1995) Terpenoid metabolism. Plant Cell 7(7):1015–1026

McKay SAB, Hunter WL, Godard K-A, Wang SX, Martin DM et al (2003) Insect attack and wounding induce traumatic resin duct development and gene expression of (-)-pinene synthase in Sitka spruce. Plant Physiol 133(1):368–378

Meng D, Cao Y, Chen T, Abdullah M, Jin Q et al (2019) Evolution and functional divergence of MADS-box genes in Pyrus. Sci Rep 9(1):1266

Mewalal R, Rai DK, Kainer D, Chen F, Külheim C et al (2017) Plant-derived terpenes: a feedstock for specialty biofuels. Trends Biotechnol 35(3):227–240

Nakamura T, Yamada KD, Tomii K, Katoh K (2018) Parallelization of MAFFT for large-scale multiple sequence alignments. Bioinformatics 34(14):2490–2492

Nguyen L-T, Schmidt HA, von Haeseler A, Minh BQ (2014) IQ-TREE: a fast and effective stochastic algorithm for estimating maximum-likelihood phylogenies. Mol Biol Evol 32(1):268–274

Nystedt B, Street NR, Wetterbom A, Zuccolo A, Lin Y-C et al (2013) The Norway spruce genome sequence and conifer genome evolution. Nature 497:579

Ohta T (2000) Evolution of gene families. Gene 259 (1):45–52

Phillips MA, Croteau RB (1999) Resin-based defenses in conifers. Trends Plant Sci 4(5):184–190

Savard L, Li P, Strauss SH, Chase MW, Michaud M, Bousquet J (1994) Chloroplast and nuclear gene sequences indicate late Pennsylvanian time for the last common ancestor of extant seed plants. Proc Natl Acad Sci USA 91(11):5163–5167

Shalev TJ, Yuen MMS, Gesell A, Yuen A, Russell JH et al (2018) An annotated transcriptome of highly inbred *Thuja plicata* (Cupressaceae) and its utility for gene discovery of terpenoid biosynthesis and conifer defense. Tree Genet Genomes 14(3):35

Smit SJ, Vivier MA, Young PR (2019) Linking terpene synthases to sesquiterpene metabolism in grapevine flowers. Front Plant Sci. https://doi.org/10.3389/fpls.2019.00177

Stival Sena J, Giguère I, Boyle B, Rigault P, Birol I et al (2014) Evolution of gene structure in the conifer *Picea glauca*: a comparative analysis of the impact of intron size. BMC Plant Biol 14(1):95

Trapp S, Croteau R (2001a) Defensive resin biosynthesis in conifers. Annu Rev Plant Physiol Plant Mol Biol 52 (1):689–724

Trapp SC, Croteau RB (2001b) Genomic organization of plant terpene synthases and molecular evolutionary implications. Genetics 158(2):811–832

Van Bel M, Diels T, Vancaester E, Kreft L, Botzki A et al (2017) PLAZA 4.0: an integrative resource for functional, evolutionary and comparative plant genomics. Nucl Acids Res 46(D1): D1190–D1196

Wegrzyn JL, Lee JM, Tearse BR, Neale DB (2008) TreeGenes: a forest tree genome database. Int J Plant Genomics 412875

Zhou S-S, Xing Z, Liu H, Hu X-G, Gao Q et al (2019) In-depth transcriptome characterization uncovers distinct gene family expansions for *Cupressus gigantea* important to this long-lived species' adaptability to environmental cues. BMC Genomics 20(1):213

Spruce Phenolics: Biosynthesis and Ecological Functions

12

Almuth Hammerbacher,
Louwrance P. Wright, and Jonathan Gershenzon

Abstract

Phenolics are organic compounds that play an important role in the physiology of plants and their ecological interactions with biotic as well as abiotic factors. The genus *Picea* (spruce) produces a large array of structurally and functionally diverse phenolic compounds, including stilbenes, flavonoids, lignin, and phenylpropanoids. In this chapter, we review the current knowledge on the biosynthesis and ecological roles of the major classes of phenolic compounds in spruce. We also elaborate on the methods used to study the biosynthesis and functions of these compounds in this genus. Finally, we touch on how climate change is expected to affect the biosynthesis of phenolics in spruce and identify the most important directions for future research.

12.1 Introduction

Phenolics are abundant, structurally diverse compounds which are naturally produced by all plant species. The core structures of these compounds are aromatic rings with at least one functional group, such as a carboxylic acid or a hydroxyl group. In plants, phenolic rings are coupled and decorated in thousands of different ways to form a large diversity of unique structures supporting an equally large array of functions (reviewed by Lattanzio et al. 2008).

The earliest phenolics are thought to have been the flavonoids, which functioned in the transition of plants from an aqueous environment to land as sunscreens by absorbing UV rays (Stafford 2000). Flavonoids consist of a three-ring core skeleton (C_6-C_3-C_6—see for more structural details further down) which is modified by hydroxylation, methylation, glycosylation, acetylation, and polymerization to form a large variety of compounds. These include the flavonols, flavones, isoflavonoids, anthocyanidins, and flavan-3-ols. Lignans evolved shortly after the flavonoids and also played an important role in the transition of plants from aquatic to terrestrial environments (Stafford 2000). Lignans are monolignol dimers (C_6-C_3-dimers—see discussion of their structural details in the appropriate chapter section) that function as defense compounds against pathogenic microbes and herbivores (Dar and Arumugam 2013). Lignin polymers $(C_6\text{-}C_2)_n$ evolved from these compounds,

A. Hammerbacher (✉)
Department of Zoology and Entomology, Forestry and Agricultural Biotechnollogy Institute, University of Pretoria, Pretoria, South Africa
e-mail: almuth.hammerbacher@fabi.up.ac.za

L. P. Wright · J. Gershenzon
Max Planck Institute for Chemical Ecology, Jena, Germany

I. M. Porth and A. R. De la Torre (eds.), *The Spruce Genome*, Compendium of Plant Genomes,
https://doi.org/10.1007/978-3-030-21001-4_12

and provide land plants with structural support for upright growth and water transport (Stafford 2000). Due to its importance in plant morphology and resilience, lignin is the second most abundant biopolymer in plants after cellulose (Stafford 2000). Apart from the ubiquitous flavonoids and lignin derivatives, a wide variety of other structurally unique phenolic compounds evolved in plants that are only synthesized by a few plant linages. These compounds, which are often antimicrobial or anti-herbivore defenses, also provide color, aroma and astringency to vegetable and fruit crops (Lattanzio et al. 2008).

The shikimate pathway (Fig. 12.1), in addition to providing the amino acid phenylalanine as a precursor for the biosynthesis of most phenolics, also produces salicylic acid, a plant hormone, as well as gallic acid which is the precursor of hydrolyzable tannins and ellagitannins (Lattanzio et al. 2008). Phenylalanine is the precursor of the phenylpropanoid pathway whose first step involves the elimination of the amine group by the enzyme phenylalanine ammonia-lyase (PAL) to form *trans*-cinnamic acid (Fig. 12.1). This step is thought to be the rate-limiting step in the biosynthesis of phenolic compounds in most plants (Howels et al. 1996). In the next step, *trans*-cinnamic acid is converted by cinnamate-4-hydroxylase (C4H) to *p*-coumaric acid. This molecule is then activated by coupling it to coenzyme A (CoA) via a 4-coumaroyl-CoA ligase (4CL) enzyme to form *p*-coumaroyl-CoA. The CoA activated form of *p*-coumaric acid serves as the substrate for most of the downstream reactions in the biosynthesis of phenolic compounds, including lignin and flavonoids (reviewed by Frazer and Chapple 2011). In addition, some plant species have modified the biosynthetic steps leading toward the formation of ubiquitous compounds to produce more unique phenolics. For example, modifications of the flavonoid pathway result in compounds without the typical 3-ring flavonoid backbone, such as the yellow aurones in snapdragon as well as stilbene phytoalexins (Vogt 2010). The polyketide pathway also produces phenolic compounds by coupling multiple malonyl- and acetyl-CoA moieties to each other or to starter phenylpropanoid units, as in flavonoid and stilbene biosynthesis (Vogt 2010).

Conifers from the family Pinaceae produce a wide variety of different phenolic compounds ranging from lignin for structural support to simple phenylpropanoid glycosides as anti-herbivore defenses. Apart from lignin, which is deposited into secondary cell walls, many phenolics in the Pinaceae are produced in parenchyma cells in the bark, foliage, wood, and roots. Interestingly, a specialized cell type was identified in the phloem of many different conifers which accumulates phenolic compounds upon wounding or pathogen infection (Franceschi et al. 2000; Hudgins et al. 2003, 2004, 2006; Li et al. 2012; Celedon et al. 2017). However, only in the last few years has research been conducted to characterize the contents of these phenolic-accumulating cells (Li et al. 2008, 2012) and to study the biosynthesis and function of the compounds identified (Hammerbacher et al. 2011, 2014, 2018, 2019).

Among the Pinaceae, the most extensive catalogue of phenolic compounds and the genes involved in their biosynthesis have been described for the genus *Picea* (spruce). In this chapter, we will review the current knowledge on the biosynthesis and ecological function of phenolics in this genus. We will also touch on new research trends and important gaps in knowledge. Finally, we will briefly discuss the effects of global warming, high atmospheric CO_2 levels, and abiotic stress on the biosynthesis of these compounds and what impact climate change will have on their ecological functions.

12.2 Studying the Biosynthesis and Biological Functions of Spruce Phenolics

Spruce is a long-lived, genetically diverse genus of slow-growing trees that is only rarely propagated clonally. For this reason, functional analysis of genes is more complex than in model species such as *Arabidopsis*, maize, rice, or poplar. Therefore, most studies on the functional characterization of genes in spruce were

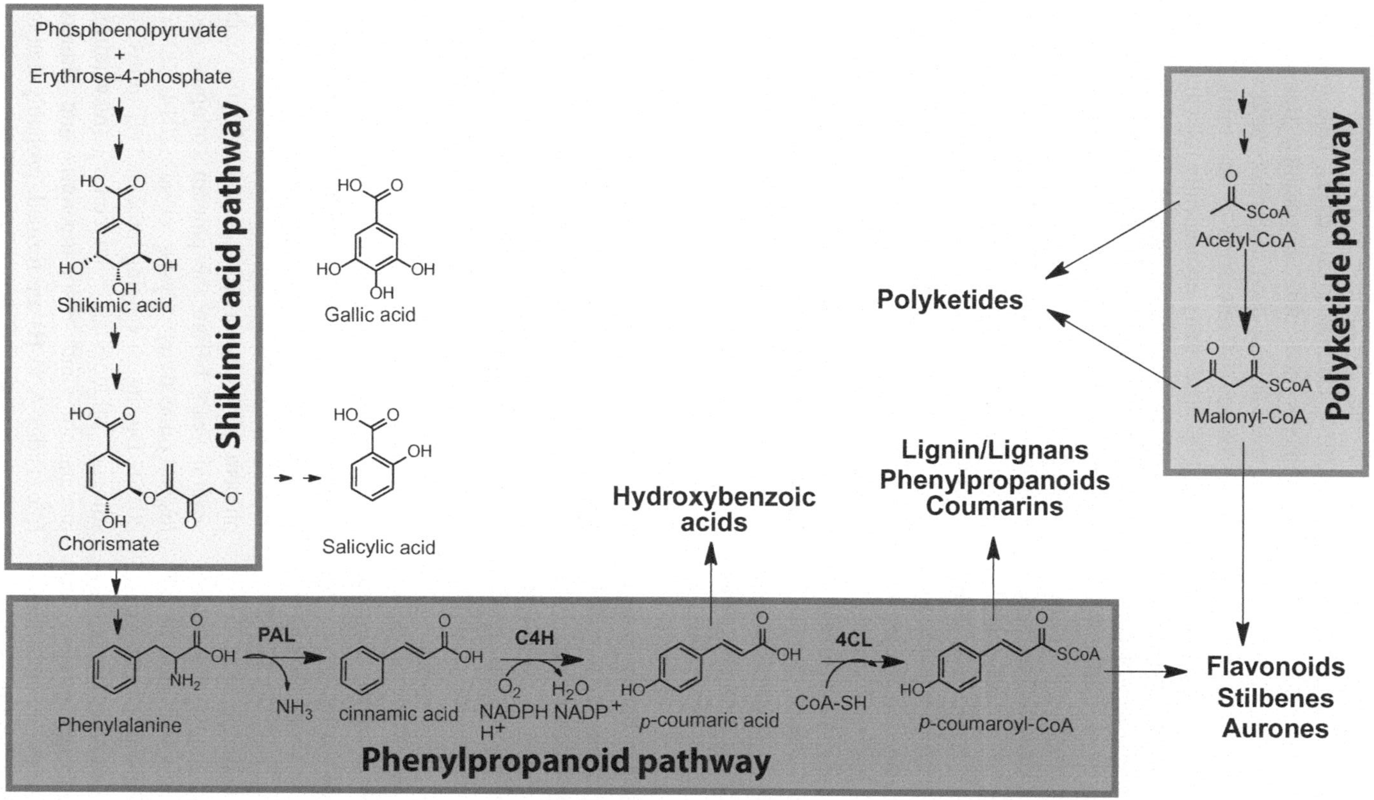

Fig. 12.1 An overview of the biosynthesis of phenolic compounds in plants. The shikimic acid pathway produces small aromatic compounds such as gallic acid and salicylic acid as well as the amino acid phenylalanine. Phenylalanine is transformed within the phenylpropanoid pathway to *para*-coumaroyl-Coenzyme A (*p*-coumaroyl-CoA) which is the precursor for the biosynthesis of many different phenolic compound classes in plants. For the biosynthesis of flavonoids, stilbenes and aurones, a phenylpropanoid moiety serves as the starter unit for several cycles of the polyketide pathway. PAL: phenylalanine ammonia lyase; C4H: cinnamate-4-hydroxylase; 4CL: 4-coumarate CoA ligase; CoA: Coenzyme A

conducted by heterologous expression of a cloned open reading frame (ORF) of the gene of interest and in vitro functional assays of the expressed enzymes (e.g., Hamberger et al. 2011). Heterologous hosts used to express spruce enzymes include *Escherichia coli* and *Saccharomyces cerevisiae* cell lines (e.g., Hamberger et al. 2011; Hammerbacher et al. 2011). Due to the codon bias of each expression system, the coding regions often have to be synthesized ex vivo with an optimal complement of codons for the selected heterologous host. Furthermore, in many cases, expression protocols require optimization in order to achieve sufficient protein yields for enzyme assays. For example, in expressing spruce stilbene synthases, active enzymes were only obtained from heterologous expression in *E. coli* when expressed at very low temperatures by specialized bacterial strains co-expressing cold-resistant chaperonins to aid in protein folding (Hammerbacher et al. 2011).

On the other hand, stable transformation of spruce to either overexpress biosynthetic genes or silence their expression using RNA interference (RNAi) was shown to be successful in characterizing genes involved in biosynthesis of specialized metabolites (Bedon et al. 2009; Schmidt et al. 2010; Hammerbacher et al. 2011, 2014, 2019; Mageroy et al. 2017a, b). Transgenic overexpression lines were shown to readily synthesize the expected metabolites at higher levels than the wild type or vector control lines with no detectable pleiotropic effects on growth or development (Schmidt et al. 2010; Hammerbacher et al. 2011). An added advantage of using transgenic spruce for enzyme characterization is that the contribution of the gene to the tree's phenotype can be evaluated (Bedon et al. 2009). For example, the use of stably transformed spruce in the functional characterization of stilbene synthases (STS) was critical to demonstrate that the STS enzymes, although only synthesizing resveratrol as an intermediate product, still contribute to the formation of the major stilbenes in the plant (Hammerbacher et al. 2011).

However, stable transformation of spruce is a labor intensive, time consuming, and costly process, as it necessitates a dedicated laboratory for spruce somatic embryogenesis and therein trained personnel. Also, many transformants do not show the desired phenotype in the sapling stage in the greenhouse (e.g., Hammerbacher et al. 2014; Dalman et al. 2017). Due to legal impediments and time constraints, transgenic spruce trees are not used to conduct field studies on the ecological relevance of the traits being studied because many ecologically relevant traits can only be studied in mature trees. For this reason, most functional studies related to phenolic metabolism are based on multiple lines of evidence, including metabolite accumulation profiles and transcript accumulation patterns of specific genes following a stimulus as well as in vitro enzyme characterization and in vitro bioassays (Hammerbacher et al. 2011, 2014, 2019).

12.3 Biosynthesis of Spruce Phenolics and Their Ecological Functions

12.3.1 Stilbenes

Stilbenes (C_6-C_2-C_6—Fig. 12.2) are the most abundant phenolic compounds in spruce bark, wood, and roots, but are less abundant in needles (Hammerbacher et al. 2011). The most abundant stilbene in spruce is the tetrahydroxystilbene glucoside *trans*-astringin. Other stilbene glucosides produced by spruce are *trans*-piceid (a trihydroxystilbene glucoside) and *trans*-isorhapontin (a trihydroxy-methoxystilbene glucoside) (Kieselev et al. 2016). These compounds are predominantly synthesized in the axial phloem parenchyma from June to August over several years before the phloem collapses and loses connectivity (Jyske et al. 2014, 2015, 2016). During this aging process, *cis*-forms of stilbene glucosides, aglycones (resveratrol, piceatannol, and rhapontigenin), and dimers (piceasides A to H) are formed, probably due to oxidative processes (Li et al. 2008; Jyske et al. 2016). Stilbenes are more abundant in older tree bark compared to younger bark (Jyske et al. 2015), but the levels of these compounds can

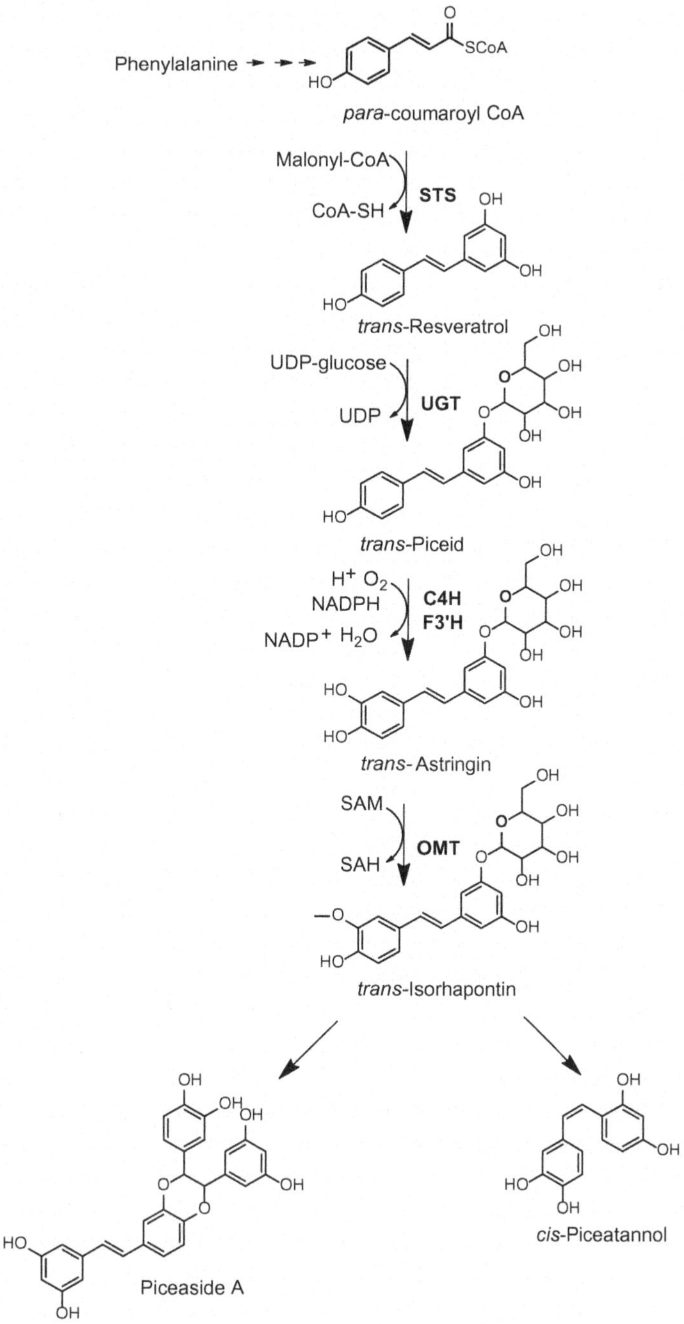

Fig. 12.2 Stilbene biosynthesis in spruce. STS: chalcone synthase; CoA: Coenzyme A; UGT: UDP-glucose:flavonoid *β*-D-glucosyltransferase; C4H: cinnamoyl-4-hydroxylase; F3'H: flavonoid 3'-hydroxylase; OMT: flavonoid O-methyltransferase; SAM: S-adenosyl-L-methionine; SAH: S-adenosyl-L-homocysteine. The most abundant stilbene in spruce is *trans*-astringin

increase more rapidly in younger trees following an external stimulus inducing their biosynthesis (Jyske et al. 2016).

In the first step of stilbene biosynthesis in spruce (Fig. 12.2), the condensation of *p*-coumaroyl-coenzyme A (CoA) with three molecules of malonyl-CoA is catalyzed by the enzyme stilbene synthase (STS) to form a tetraketide, which is cyclized via a Claisen condensation to form the trihydroxystilbene *trans*-resveratrol. In Norway spruce, we identified and characterized two enzymes which formed resveratrol in vitro as well as in transgenic saplings (Hammerbacher et al. 2011). Two full-length gene models for STS were also identified in the *P. glauca* genome (Warren et al. 2015). In *P. jezoensis*, four candidate genes were identified all of which are independently regulated at the transcriptional level and are thought to contribute to stilbene accumulation at different times of the year (Kieselev et al. 2016).

After cyclization, *trans*-resveratrol is glycosylated to form *trans*-piceid by a *p*-hydroxybenzoic acid-UDPG-glucosyltransferase in grapevine (Hall and de Luca 2007). A similar conversion is thought to take place in spruce (Fig. 12.2), but no enzyme with a confirmed stilbene-glucosyltransferase function has been identified in this genus yet. *Trans*-astringin is formed from piceid by an NADPH-dependent microsomal cytochrome P450 monooxygenase enzyme. Such transformations have been shown in bacteria (Le et al. 2017). It is thought that in plants these reactions are catalyzed either by a cinnamate 4-hydroxylase or a flavonoid-3'-hydroxylase (Hammerbacher et al. 2011). Similarly, the methylation of *trans*-astringin to form *trans*-isorhapontin has not been characterized in spruce yet. O-methyltransferases with broad substrate specificities were shown to methylate stilbenes to produce different products, including pterostilbene from resveratrol in grape (Rimando et al. 2012) and pinosylvin 3-O-monomethyl ether from the pine stilbene pinosylvin (Chiron et al. 2000). Apart from the three glucosides produced by enzymatic transformations, the other stilbenes detected in spruce are thought to be produced by spontaneous oxidative processes. It is well known that *trans*-stilbenes are readily isomerized to *cis*-forms by UV light, reactive oxygen species, and pH imbalances (Waldeck 1991). Similarly, dimerizations to form piceasides and deglucosylations of stilbenes are likely the product of natural oxidative processes occurring in plants during aging, biotic or abiotic stress (Fig. 12.2; Jyske et al. 2016).

Stilbenes are potent antifungal compounds in grapevine and other stilbene-producing plants, acting as both phytoanticipins (constitutive defense molecules) and phytoalexins (induced anti-microbials). However, in spruce their biological and ecological roles are not entirely clear. We showed that *PgSTS* genes were transcriptionally up-regulated in white spruce during infection by the European spruce bark beetle-associated fungus, *Endoconidiophora polonica* (Hammerbacher et al. 2011), and there was a corresponding increase in stilbene content in the bark of white spruce saplings. However, white spruce is not a natural host for this fungus. In a similar experiment using Norway spruce, *E. polonica's* natural host, increased transcript accumulation of *PaSTS* genes was also observed, but lower levels of stilbenes were detected in fungus-infected bark tissues than in uninfected control trees (Hammerbacher et al. 2013). Similar stilbene glucoside declines were also described in mature Norway spruce trees infected by *E. polonica* (Brignolas et al. 1995a; Virii et al. 2001; Li et al. 2012; Krajnc et al. 2014) or by a *Heterobasidon* spp. (Danielsson et al. 2011). Concomitant with stilbene declines, increased levels of aglucones (Virii et al. 2001) or piceasides (Danielsson et al. 2011; Li et al. 2012) were detected in mature bark tissue after fungal infection. The lower levels of stilbene glucosides and increased levels of aglucones and piceasides were explained by later studies, which showed that numerous bark beetle-associated fungi (Zhao et al. 2019), and most notably *E. polonica*, could degrade stilbene glucosides during spruce infection by oxidation, deglucosylation and dimerization (Hammerbacher et al. 2013). These fungi could even use stilbenes as carbon sources (Wadke et al. 2016). However, stilbenes also declined in experiments where only fungal elicitors were used without a living fungus, strongly supporting the hypothesis that intrinsic oxidative processes during tree defense against fungal invasion can also trigger dimer and aglucone accumulation (Cvikrová et al. 2008; Malá et al. 2011).

Although fungi adapted to growing on spruce may have effective detoxification mechanisms for stilbene glucosides, the compounds may still have a negative effect on their growth. For example, *E. polonica* showed a small but significant decrease in growth on media amended with high concentrations of astringin (Hammerbacher et al. 2011). Similarly, the weakly pathogenic cosmopolitan wood decay fungi *Phaeolus schweinitzii* and *Sparassis crispa* were shown to be sensitive to low stilbene concentrations in Sitka spruce (Woodward and Pearce 1988). On the other hand, *Heterobasidion* spp. and *Rhizina undulata* showed high tolerance to astringin and isorhapontin, and these compounds even stimulated the growth of *Armillaria ostoyae* and *Stereum sanguinolentum* (Woodward and Pearce 1988). Furthermore, genotypes with high susceptibility to the biotrophic spruce needle bladder rust, *Chrysomyxa rhododendri*, generally contained higher stilbene levels and significantly higher levels of *cis*-stilbenes (Ganthaler et al. 2017). Stilbenes might therefore be an effective defense against poorly adapted opportunistic tree pathogens, but have little effect on specialist spruce pathogens.

Although the roles of spruce stilbenes as defenses against herbivores and as allelochemicals have rarely been studied, there are indications that the compounds may play a more important role in these processes than previously thought. For example, it was shown that the great spruce bark beetle preferred to oviposit in bark with low stilbene levels (Storer and Speight 1996). Faccoli and Schlyter (2007) also showed that the European spruce bark beetle, *Ips typographus*, produced shorter tunnel lengths in artificial medium amended with high resveratrol concentrations compared to low concentrations. An STS gene was also identified as a strong molecular marker in spruce for resistance against feeding by the white pine weevil, *Pissodes strobi* (Porth et al. 2011). Circumstantial evidence also suggests that stilbenes have some allelopathic effects against neighboring plants (Chong et al. 2009). For example, stilbenes from Engelmann spruce were shown to inhibit germination and seedling growth of other conifer species (Taylor and Shaw 1983). However, more research is still required to unravel the roles of stilbenes against herbivory and as allelochemicals in spruce.

12.3.2 Flavonoids

12.3.2.1 The Early Flavonoid Pathway: Flavanones and Dihydroflavonols

Flavonoid and stilbene biosynthesis are similar in spruce as both processes require a CoA-activated *p*-coumaric acid which is coupled in the first enzymatic step to three malonyl-CoA molecules to form a linear tetraketide. But, while stilbenes are cyclized via a Claisen condensation to a C_6-C_2-C_6 molecule (Fig. 12.2), the flavonoid C_6-C_3-C_6 backbone is produced via an aldol condensation resulting in the formation of chalcone (Fig. 12.3). This reaction is carried out by chalcone synthase enzymes (CHS). In the white spruce genome, 15 candidate CHS genes were discovered (Warren et al. 2015) of which at least 8 are expressed in Norway spruce bark constitutively (Hammerbacher et al. 2011). Following its synthesis, chalcone is isomerized to the flavanone naringenin by chalcone isomerase (CHI; 5 full-length gene models exist in white spruce; Warren et al. 2015). Naringenin can then be further transformed by hydroxylation of the B-ring to form eryodictyol. Warren et al. (2015) described 3 full-length genes with a putative flavonol-3'hydroxylase (F3'H) function in white spruce based on sequence similarity. The B-ring hydroxylation step catalyzed by F3'H has not been biochemically characterized in spruce, but was shown in other plant species to be an essential step for producing the precursors for the synthesis of major downstream flavonoids (Fig. 12.3).

The flavanones naringenin and eryodictyol accumulate to high levels in some pine species, e.g., in *Pinus tabuliformis* (Cheng et al. 2016). In contrast, although multiple studies have shown increased transcriptional activity of chalcone synthease (CHS) in spruce after fungal infection, insect feeding, or elicitation with the defense hormone methyl jasmonate (Brignolas et al.

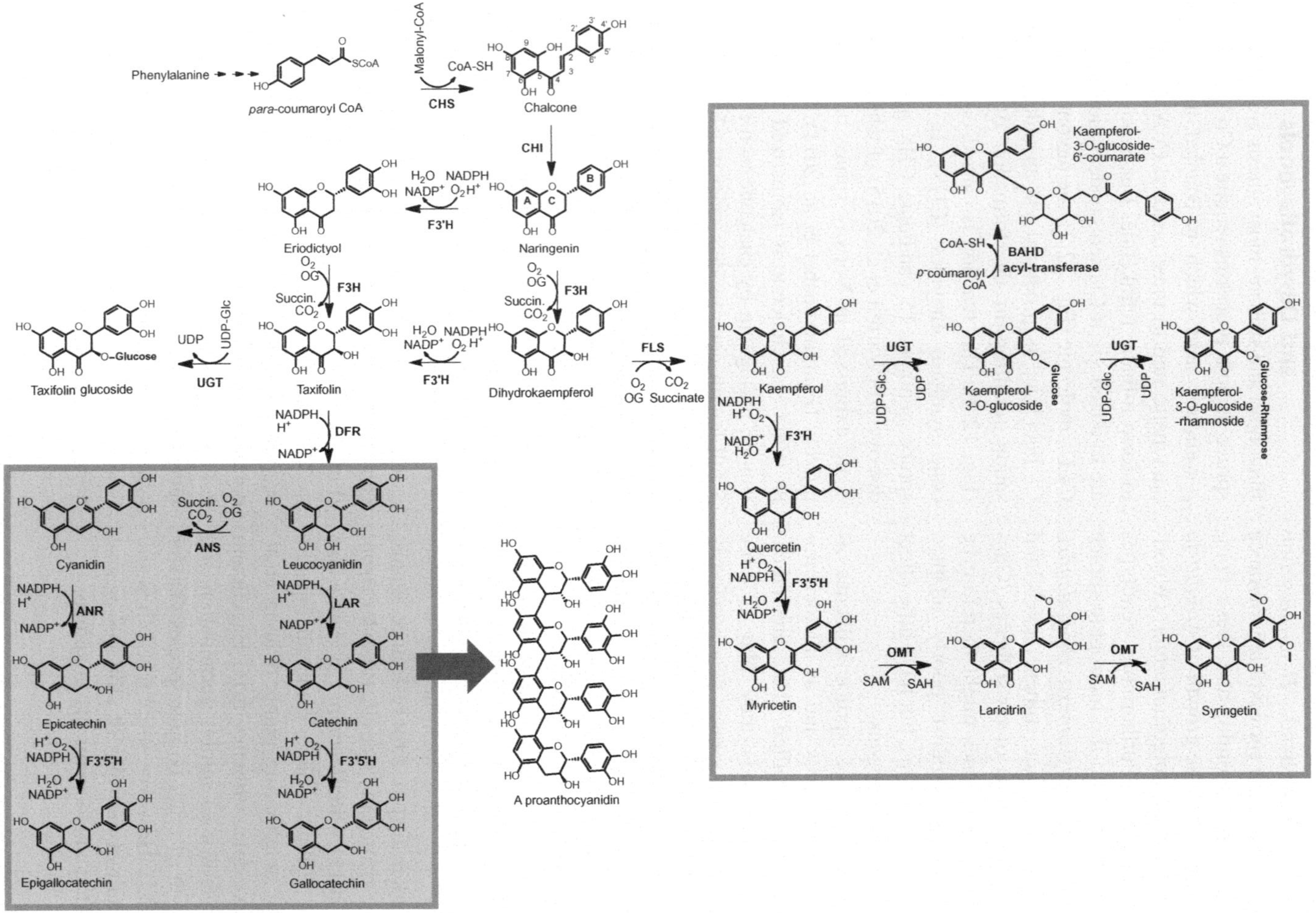

Fig. 12.3 Flavonoid biosynthesis in spruce. Light grey box: flavonol biosynthesis; dark grey box: late flavonoid pathway. CHS: chalcone synthase; CoA: coenzyme A; CHI: chalcone isomerase; F3'H: flavonoid 3'-hydroxylase; F3H: flavanol-3-hydroxylase; OG: oxoglutarate; Succin: succinate; UGT: UDP-glucose:flavonoid β-D-glucosyltransferase; UDP-glc: uridine diphosphate glucose; DFR: dihydroflavonol reductase; ANS: anthocyanidin synthase; ANR: anthocyanidin reductase; LAR: leucoanthocyanidin reductase, F3'5'H: flavonol-3'5'-hydroxylase; FLS: flavonol synthase; OMT: flavonoid O-methyltransferase; SAM: S-adenosyl-L-methionine; SAH: S-adenosyl-L-homocysteine

1995a, b; Richard et al. 2000; Nagy et al. 2004; Porth et al. 2011), chalcone itself only accumulates to low final concentrations in treated trees (Hammerbacher et al. 2019).

For the synthesis of downstream products, naringenin or eryodictyol is hydroxylated in the next step on the 3-position of the C-ring by a 2-oxoacid-dependent flavanone-3-hydroxylase (F3H) to form dihydrokaempferol or dihydroquercetin (taxifolin), respectively (Hammerbacher et al. 2019). This biosynthetic step, which is the rate-limiting step for the production of all downstream products of the flavonoid pathway in some herbaceous plants (Pelletier and Shirley 1996), was also shown to be a key step in defense metabolite production and a regulator of flavan-3-ol biosynthesis in spruce (Hammerbacher et al. 2019). In Norway spruce, only a single gene was identified and characterized encoding this important enzyme (Hammerbacher et al. 2019), although a number of other 2-oxoacid-dependent dioxygenases have been annotated in the white spruce genome as putative F3H genes (Warren et al. 2015).

Spruce has been shown to accumulate high concentrations of taxifolin in its bark, wood, and needles, mainly in the form of the 3-O-glucoside, but also as the aglucone (Brignolas et al. 1995a, b, Slimestad 2003). In response to fungal infection, the F3H enzyme was shown to be highly expressed in Norway spruce (Danielsson et al. 2011; Hammerbacher et al. 2019). The increased gene expression in response to infection by *E. polonica* resulted in the accumulation of taxifolin, taxifolin-glucoside as well as downstream metabolites in the pathway (Everson et al. 2000; Hammerbacher et al. 2019). Taxifolin accumulation was also strongly associated with Norway spruce resistance to infection by *E. polonica* (Brignolas et al. 1995a, b; Zhao et al. 2011). Taxifolin activity assays using *E. polonica* confirmed that this compound is toxic to the fungus and inhibited its growth by more than 50% (Hammerbacher et al. 2019). Evidence also suggests that taxifolin is an effective defense against herbivore damage. For example, in Scots pine needles, taxifolin inhibited feeding of the sawfly, *Diprion pini* (Auger et al. 1994). Taxifolin also inhibited tunneling of *I. typographus* in artificial medium (Faccoli and Schlyter 2007) and a significant decrease in weight gain was observed in beetles feeding on taxifolin amended medium compared to the un-amended control (Hammerbacher et al. 2019).

12.3.2.2 Flavonols

Dihydroflavonols are either converted to flavonols or alternatively to anthocyanidins and flavan-3-ols in the late flavonoid pathway (Fig. 12.3). A wide diversity of flavonols are produced in spruce needles but no knowledge is available about their biosynthesis (Slimestad 2003). From other plant species, it is known that the key enzyme in this branch point of the pathway is the oxogluterate dependent flavonol synthase enzyme (FLS). In petunia, for example, FLS can typically accept dihydrokaempferol and taxifolin as substrates to form kaempferol and quercetin, respectively (Holton et al. 1993). Once the flavonols are formed, they can undergo a multitude of modifications. Kaempferol can be hydroxylated by F3'H to form quercetin, and quercetin can be further converted to myricetin via an enzymatic conversion by flavonol-3'5'-hydroxylase (F3'5'H). However, in the only F3'5'H enzyme that has thus far been characterized in Norway spruce, this conversion was not demonstrated (Hammerbacher et al. 2018).

Quercetin and myricetin can be methylated by an S-adenosyl-L-methionine-dependent methyltransferase on the 3' hydroxyl group to form isorhamnetin or laricitrin, respectively. A second methylation of laricitrin on the 5' hydroxyl group yields syringetin. A group of four putative quercetin methyltransferases were identified in a white spruce genetic mapping study (Pavy et al. 2012). Interestingly, this gene family arose by two duplication events, one predating the split between gymnosperms and angiosperms and a second duplication after the two lineages split. Two of these methyltransferases from Norway spruce were characterized and shown to convert 3',4' and 3',4',5' hydroxylated flavonols and flavones as well as 3,4 and 3,4,5-hydroxylated phenolic acids to their respective methoxylated forms (Kim et al. 2010).

The next steps of flavonol synthesis involve the addition of glucose or rhamnose units to the flavonol to form the typical glucosides and rutinosides that have been discovered in spruce needles (Slimestad and Hostettmann 1996; Slimestad 2003). These reactions are catalyzed by UDP-glucose: flavonol 3-O-glycosyltransferase enzymes. Although the family of genes encoding putative enzymes with this function is extensive (Hammerbacher, unpublished), only a single enzyme has been characterized in Norway spruce to date. This glucosyltransferase can accept the substrates kaempferol, quercetin, and isorhamnetin and uses UDP-glucose as the sugar donor (Heilemann and Strack 1991). Finally, in spruce, 6″-acetylated glucosides for all six known flavonol glucosides from spruce have been described, for example, quercetin 3-(6″-acetyl-glucoside) or kaempferol 3-(6″-coumaroyl-glucoside) (Slimestad and Hostettmann 1996; Slimestad 2003). The biosynthesis of these compounds typically involves a CoA-activated carboxylic acid, such as acetyl-CoA or *p*-coumaroyl-CoA which is coupled by an acyltransferase enzyme (BAHD-family enzyme) to the C_6 position of the glucose moiety of the flavonol glucoside. Such reactions have been shown to occur in *Arabidopsis* (Luo et al. 2007), but very little is known about BAHD-family enzymes in spruce (Fig. 12.3).

Flavonols play an important role in spruce resistance to abiotic stress. They have been shown to accumulate in needles in response to damage by UV radiation (Fischbach et al. 1999) and are deposited into the cell walls of the outer epidermis during oxidative stress (Hoque and Remus 1999) where they scavenge reactive oxygen species (Pukaki 2004). These compounds are regulated by light (OuYang et al. 2015) and accumulate in summer and autumn in developing buds to assist them in resisting cold winter temperatures (Pukacki and Kamińska-Rożek 2013; Lee et al. 2014). Spruce trees from high elevations have been shown to accumulate more of these compounds, probably due to a higher cold-tolerance requirement (Pukacki and Modrzyński 1998). In addition, methylated and acetylated flavonols from spruce have been implicated in defense against fungi (Song et al. 2011) and myricetin and laricitrin aglucones correlated with unsuccessful attacks by the European spruce bark beetle on Norway spruce indicating that they may play a role in anti-herbivore defense (Schiebe et al. 2012).

12.3.2.3 Late Flavonoid Pathway: Anthocyanidins, Flavan-3-ols, and Proanthocyanidins

Dihydroflavonols, instead of serving as substrates for flavonol biosynthesis, can enter the late flavonoid pathway (Fig. 12.3). In contrast to flavonol biosynthesis, this branch point has been shown to be more active in the woody tissues than in the needles of spruce (Hammerbacher et al. 2019). In the first step of this pathway, dihydroflavonols are reduced by dihydroflavonol reductase enzymes (DFR) to leucoanthocyanidins (Peters and Constabel 2002). These highly unstable compounds are then either converted to anthocyanidins or to flavan-3-ols.

The biosynthesis and ecological roles of anthocyanidins and flavan-3-ols have been often studied in spruce. In other plant species it is known that anthocyanidins are synthesized by oxidation of leucoanthocyanidins via 2-oxoacid-dependent dioxygenase enzymes, known as anthocyanidin synthases (ANS; Wilmouth et al. 2002). Following their synthesis, they are modified by glycosylation and acetylation in a similar manner as the flavonols. Alternatively, anthocyanidins can be reduced by anthocyanidin reductase enzymes (ANR) in a reaction requiring NADPH as a co-factor to form the flavan-3-ol, (−)-epicatechin (Xie et al. 2003). Two ANR enzymes have been annotated in Norway spruce, but their enzymatic activities have not been confirmed yet (Hammerbacher et al. 2014). Interestingly, free (−)-epicatechin accumulates in spruce only to very low levels, but they represent the major subunit of proanthocyanidins in spruce (Hammerbacher et al. 2014, 2018).

Leucocyanidin, instead of functioning as a substrate for anthocyanidin biosynthesis, can be directly converted to the flavan-3-ol (+)-catechin, the major starter unit for proanthocyanidin biosynthesis in spruce. This conversion is carried out in spruce by four NADPH-dependent reductase enzymes known as leucoanthocyanidin reductases (LAR; Hammerbacher et al. 2014). Furthermore, (+)-catechin can be hydroxylated on the 5' position by the CYP450 monooxygenase, flavonol-3'5'-hydroxylase (F3'5'H), to produce (+)-gallocatechin. A single enzyme with this function was described in Norway spruce (Hammerbacher et al. 2018), but a second non-transcribed gene encoding a putative F3'5'H enzyme was annotated in the white spruce genome (Warren et al. 2015).

In the last step of the flavonoid pathway in spruce, flavan-3-ols are coupled to each other by $4 \rightarrow 8$ bonds to form dimers. The (+)-catechin usually is the starter unit and is coupled to (−)-epicatechin to form procyanidin B1 or (−)-epigallocatechin to form prodelphinidin B1. Other dimers are also formed in spruce, but at much lower concentrations than the B1 type. Further polymerizations can result in the formation of long proanthocyanidin chains with very high molecular weights (Fig. 12.3). The molecular processes governing polymerization are not well understood in plants yet. In Norway spruce bark, (+)-catechin is the major free flavan-3-ol monomer and occurs at a ratio of 10:1 to (+)-gallocatechin with almost no detectable levels of (−)-epicatechin or (−)-epigallocatechin (Hammerbacher et al. 2014). Structures of proanthocyanidin polymers are far less well known. We (Hammerbacher et al. 2014) could separate and quantify polymers up to 8-mers and showed an exponential decrease in quantities of individual proanthocyanidins with an increase in the degree of polymerization. Hydrolysis of proanthocyanidins into their monomeric subunits resulted in a ratio of 7:2:0.1:1 for the subunits (−)-epicatechin, (+)-catechin, (+)-gallocatechin, (−)-epigallocatechin. This clearly shows that (−)-epicatechin is the major subunit in spruce proanthocyanidins, although it does not accumulate as a monomer (Hammerbacher et al. 2014, 2018).

The response of the last steps in the flavonoid pathway to fungal infection has been studied thoroughly in Norway spruce. The transcription of all four LAR enzymes catalyzing the conversion of leucocyanidin to (+)-catechin was up-regulated in response to infection by the bark beetle-associated fungus, *E. polonica* (Hammerbacher et al. 2014). Transcript abundances of *PaLAR1*, *PaLAR3,* and *PaLAR4* were shown to increase early after fungal infection (2–7 days), whereas *PaLAR2* transcripts accumulate 14–28 days after fungal infection as well as after wounding. The transcriptional activity of *LAR* genes during fungal infection correlated well with (+)-catechin and proanthocyanidin accumulation, which increased exponentially during the first 7 days after fungal infection (Hammerbacher et al. 2014, 2018). Furthermore, *PaLAR2* and *PaLAR3* transcription were also up-regulated in response to *Heterobasidion* spp. infection (Danielsson et al. 2011). A variant of *PaLAR3* was identified as a high confidence marker for resistance against this fungus (Lind et al. 2014) and was later shown to occur naturally in two allelic forms in Norway spruce populations, of which one form is more transcriptionally active during infection by the heart-rot fungus *Heterobasidion parviporum*, resulting in higher accumulation of (+)-catechin in the bark of infected trees and increased resistance to the fungus (Nemesio-Gorritz et al. 2016). Furthermore, transcription of *PaF3'5'H*, catalyzing the conversion of (+)-catechin to (+)-gallocatechin was also significantly up-regulated following infection by *E. polonica*, and resulted in a significant accumulation of (+)-gallocatechin and proanthocyanidins containing this compound (Hammerbacher et al. 2018). Transcription of *PaF3'5'H* also increased in response to *H. parviporum* infection (Kovalchuk et al. 2019).

Li et al. (2012) showed that (+)-catechin biosynthesis is localized in phloem parenchyma cells and that the metabolite significantly increased in concentration following fungal attack. Flavan-3-ols, especially (+)-catechin, were also shown to accumulate in spruce needles to high levels in response to other plant pathogens, for example, in response to the spruce needle blight fungus, *Sirococcus conigenus*

(Bahnweg et al. 2000). These compounds might therefore be effective defenses against fungal pathogens in spruce. (+)-Catechin was shown to retard growth of both *E. polonica* (Hammerbacher et al. 2014) as well as *H. parviporum* (Nemesio-Gorriz et al. 2016). Proanthocyanidins also retard in vitro growth of *E. polonica*, but are not as effective as pure (+)-catechin (unpublished data; Hammerbacher, Wright and Gershenzon). Interestingly, (+)-gallocatechin is significantly less effective in retarding the growth of *E. polonica*, but results in lower levels of melanin accumulation in fungal cell walls, making the fungus potentially more susceptible to environmental stresses (Hammerbacher et al. 2018).

In comparison to other phenolics, catechin was shown to be less toxic than the dihydroflavonol taxifolin (Hammerbacher et al. 2019). However, this study also showed that a mixture of (+)-catechin and taxifolin was more toxic to *E. polonica*, than each compound alone, demonstrating that mixtures of chemical defense compounds can have synergistic effects. Flavan-3-ols are also not effective against all fungal pathogens. For example, Norway spruce genotypes with high susceptibility to the biotrophic spruce needle bladder rust, *Chrysomyxa rhododendri*, contained higher (+)-gallocatechin levels and similar (+)-catechin levels compared to resistant genotypes (Ganthaler et al. 2017). Furthermore, although (+)-catechin inhibits the growth of *E. polonica*, this fungus as well as other bark beetle-associated ophiostomatoid fungi are capable of degrading this compound and of using it as a carbon source (Wadke et al. 2016; Zhao et al. 2019).

Levels of (+)-Catechin and proanthocyanidin also increase in response to insect attack. For example, Norway spruce showed sustained, elevated levels of these compounds following attack by *I. typographus*, the European spruce bark beetle (Rohde et al. 1996). This beetle is sensitive to (+)-catechin and has been shown to make shorter tunnels in bark medium amended with catechin in no-choice assays (Faccoli and Schlyter 2007) and gained less weight when feeding on this medium (Hammerbacher et al. 2019). Furthermore, in choice assays it was demonstrated, that *I. typographus* avoided medium amended with (+)-catechin unless a detoxifying ophiostomatoid fungus was present (Zhao et al. 2019).

Concentrations of (+)-catechin in spruce were correlated to tree tolerance to abiotic stresses. Levels of catechin increased late in the season in Sitka spruce during cold acclimatization and trees with higher cold-hardiness contained higher levels of catechin in their foliage (Dauwe et al. 2012). Catechin is also important in UV B tolerance of spruce trees (Fischbach et al. 1999). Furthermore, experiments with spruce cell cultures showed that flavan-3-ol monomers as well as polymers are produced during apoplastic oxidative stress (Laitinen et al. 2017).

12.3.3 Lignin, Lignans, and Neolignans

12.3.3.1 Lignin

Lignin is the second most abundant polymer made by plants. This macromolecule is laid down in the secondary cell walls and serves to strengthen the cell wall matrix. Similar to the biosynthesis of flavonoids and stilbenes, the major precursor for the biosynthesis of lignin is *p*-coumaroyl-CoA (Fig. 12.4). In spruce, the biosynthesis of these molecules has not been characterized on a biochemical level, but from other plant species it is known that *p*-coumaroyl-CoA is coupled by a shikimate: hydroxycinnamoyl-CoA transferase (HCT) enzyme to shikimic acid. In white spruce, Warren et al. (2015) identified seven full-length gene models for this enzyme. In the next step, the *p*-coumaroyl-shikimate ester is hydroxylated on the 3-position by a *p*-coumarate-3-hydroxylase (C3H), of which three full-length gene models have been discovered in white spruce (Warren et al. 2015) and in Norway spruce (Koutaniemi et al. 2007). After hydroxylation, HCT removes the shikimate moiety from the ester and releases caffeoyl-CoA. Methylation at the 3-position of caffeoyl-CoA is catalyzed by a caffeoyl-CoA O-methyltransferase enzyme (CCOMT—Fig. 12.4). Spruce has a large gene family

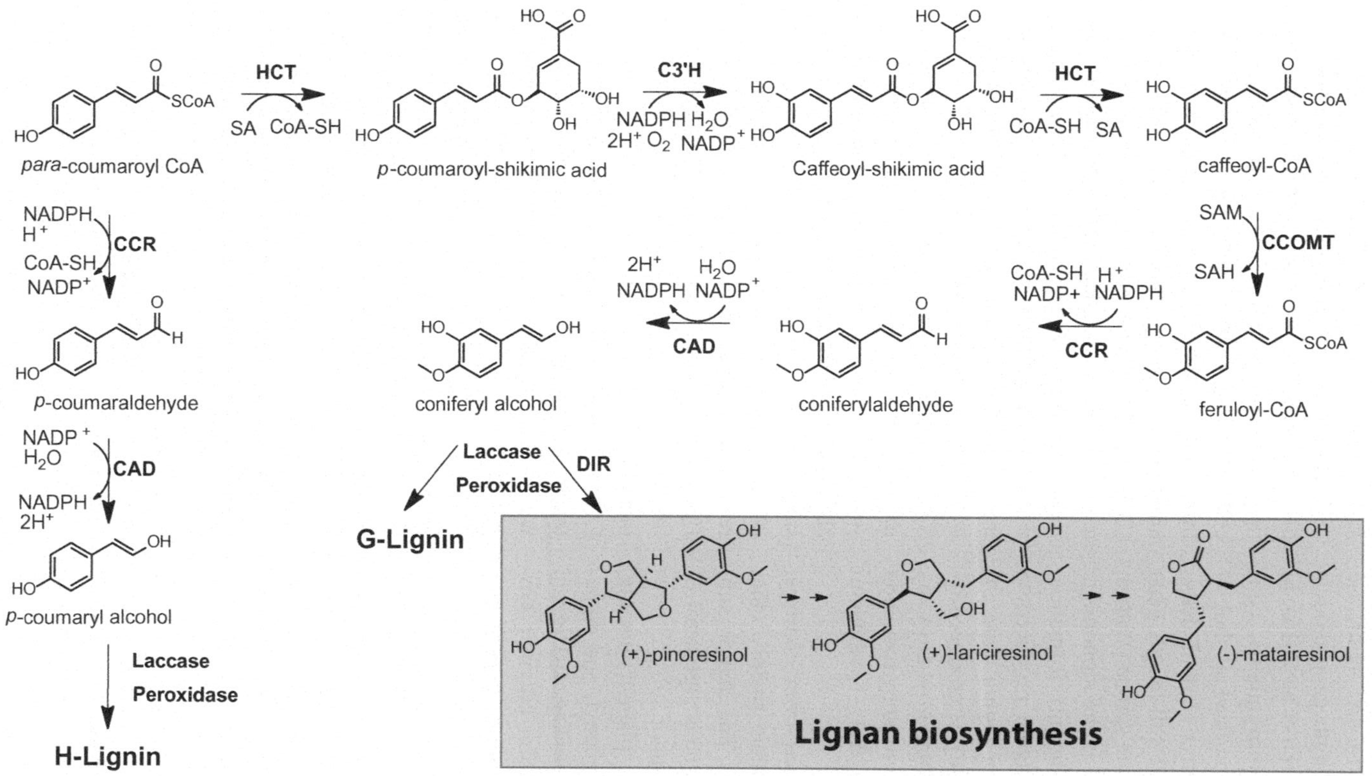

Fig. 12.4 Lignin biosynthesis pathway and schematic representation of lignan biosynthesis (grey box). HCT: hydroxycinnamoyl-CoA shikimate/quinate hydroxycinnamoyl transferase; SA: shikimic acid; C3'H: *p*-coumaroyl 3'-hydroxylase; CCOMT: caffeoyl-CoA O-methyltransferase; SAM: S-adenosyl-L-methionine; SAH: S-adenosyl-L-homocysteine; CCR: Cinnamoyl-CoA reductase; CAD: coniferyl alcohol dehydrogenase; H-lignin: hydroxyphenyl lignin; G-lignin: guaiacyl lignin; DIR, dirigent protein

encoding this enzyme with 13 full-length gene models (Warren et al. 2015). Reduction and hydrolysis of caffeoyl-CoA to coniferyl aldehyde is catalyzed by cinnamoyl-CoA reductase (CCR) with at least 25 full-length gene models recorded in white spruce (Warren et al. 2015). In the last step of monolignol biosynthesis, coniferyl aldehyde is reduced to coniferyl alcohol by an NADPH-dependent cinnamyl alcohol dehydrogenase (CAD) for which 3 full-length gene models were discovered in white spruce (Warren et al. 2015). Coniferyl alcohol is the monomeric unit for guaiacyl lignin (G-lignin) formation, which is the major type of lignin produced in conifer softwoods. The minor lignin type, *p*-hydroxyphenyl lignin (H-lignin), is produced by direct transformation of *p*-coumaroyl-CoA by CCR to *p*-coumaraldehyde followed by reduction to *p*-coumaryl alcohol by CAD (Koutaniemi et al. 2007).

The lignin polymers formed from coniferyl and *p*-coumaryl alcohols contain mainly ether and carbon–carbon linkages. The major substructure in spruce lignin is guaiacylglycerol-β-aryl ether (40–60%). A few minor substructures of spruce lignin include phenylcoumaran, diarylpropane, pinoresinol, biphenyl, and diphenyl ether interphenylpropane linkages (Higuchi 1990). The mechanism of lignin polymer synthesis is still under debate, but empirical evidence suggests that the polymerization occurs through a random radical-coupling process involving peroxidases (POX) and laccases (LAC) with hydrogen peroxide or molecular oxygen as electron donors (Boerjan et al. 2003). Genes for these enzymes occur in large families in spruce, for example, in Norway spruce 21 *PaPOX* unigenes and 6 *PaLAC* genes were annotated (Koutaniemi et al. 2007).

Lignification is an important process in wood formation and secondary cell wall synthesis. Transcriptional profiles of genes involved in wood formation are available for Sitka as well as Norway spruce (Friedmann et al. 2007; Koutaniemi et al. 2007). In both studies, high expression levels were observed for the majority of genes encoding the enzymes involved in monolignol biosynthesis as well as laccases and peroxidases in young developing xylem. The genes with the highest expression in developing wood of both Sitka and Norway spruce were the *LAC*, *POX*, *CCR*, and *CCOMT* (Fig. 12.4).

Lignin is an important defense against herbivory because it increases cell toughness and indigestibility. Notably, spruce contains stone cells in its bark tissue. These sclerified cells have thickened, lignified cell walls that are often arranged in small bundles. Whitehill et al. (2016a, b, 2019) showed that these stone cells are a very effective defense against the white pine weevil, *P. strobi*. Stone cells with their lignified cell walls contributed in these studies to mandible wear of the insects and could not be digested by neonate larvae. In consequence, resistant Sitka spruce varieties contained significantly more stone cells in cortex tissue compared to susceptible varieties. Stone cells of resistant trees also contained higher levels of G-lignin than stone cells of susceptible trees (Whitehill et al. 2016a). Transcript profiling of susceptible and resistant varieties revealed an increased level of *PgC3H* transcription in the resistant trees. Furthermore, Porth et al. (2011) reported an increased number of eQTLs for laccase, CCOMT, and peroxidases associated with resistance to *P. strobi* in white spruce, and white pine weevil feeding also activated transcription of *PsLAC* genes in Sitka spruce 48 h after herbivore damage in a microarray study (Ralph et al. 2006a). However, Ralph et al. (2006a) reported that the overall induced transcriptional response of the lignin biosynthesis pathway to weevil feeding was not significantly higher than that of mechanically wounded trees. Interestingly, feeding by the western spruce budworm on Sitka spruce activated many more genes, including *PsLAC* and *PsCCR*, compared to mechanical wounding or weevil feeding (Ralph et al. 2006a). However, stone cells have not yet directly been linked to spruce budworm resistance, as these insects mostly feed on the foliage of spruce trees.

Lignin was also shown to be an effective defense against infection by phytopathogenic microbes, especially fungal infection. In Norway spruce, an increase in lignin content after inoculation with *Heterobasidion* spp. was recorded in resistant tree genotypes (Fossdal et al. 2012). Genes that were transcriptionally up-regulated in wood during infection by *Heterobasidion* spp. included *PaLAC* and *PaPOX* genes (Koutaniemi et al. 2007; Yaqoob et al. 2012). Interestingly, although this fungus mainly attacks wood, the reaction toward this fungus was much stronger in bark, with an up-regulation of *PaLAC*, *PaC3H*, *PaC4H*, *PaCCOMT,* and *PaCAD* transcription after infection (Deflorio et al. 2011; Fossdal et al. 2012). On the other hand, Kovalchuk et al. (2019) also showed increased expression of *PaLAC*, *PaPOX,* and *PaCCOMT* genes in wood of *H. annosum* infected Norway spruce trees.

12.3.3.2 Lignans and Neolignans

Instead of complete polymerization, monolignols can be coupled into shorter chains. In spruce, a number of monolignol dimers and oligomers have been described. Dimeric forms are known as lignans (Fig. 12.4) and neolignans. Lignans in spruce are formed by the stereospecific coupling of two coniferyl alcohol units linked by a β-β (8-8′) bond to form (+) pinoresinol (Fig. 12.4; Fedorova et al. 2016). The stereospecific coupling is mediated by dirigent proteins (Davin and Lewis 2000). The dimer can then undergo multiple reductions and dehydrogenations to form the major lignan in spruce, (−) matairesinol (Fig. 12.4; Piispanen et al. 2008). Other lignans reported in spruce include (+) hydroxymatairesinol, (+) lariciresinol, (−) isolariciresinol, (−) secoisolariciresinol, (−) todolactol A, (−) isoliovil, and (−) conidenrin (Piispanen et al. 2008). Neolignans are dimeric compounds derived from coniferyl alcohol, *p*-coumaryl alcohol, or any of the phenolic acid precursors of the monolignols. They differ from lignans in that the monomers are not β-β linked. A few neolignans have been described in spruce, such as *(2S,3S)*-2,3-dihydro-2-(4-hydroxy-3-methoxyphenyl)-3-(hydroxymethyl)-1,4-benzodioxin-6-propanol, but their biosynthesis and biological functions have not been studied in spruce yet (Strack et al. 1989).

Lignans have been shown to be potent antifungal compounds in many plant species (Dar and Arumugam 2013) and circumstantial evidence suggests that they fulfill a similar role in spruce. A number of studies have shown that lignans accumulate in wood infected with *Heterobasidion* spp. (Stenlid and Johansson 1987, Kovalchuk et al. 2019). A concomitant increase in the expression of genes encoding dirigent proteins in trees showing symptoms of *Heterobasidium* spp. infection also supports the role of lignans in antifungal defense in spruce (Kovalchuk et al. 2019). Interestingly, dirigent protein-encoding genes were also highly expressed in wounded and herbivore-damaged spruce shoots (Ralph et al. 2006b, 2007), and were highly up-regulated in response to spruce budworm feeding. These compounds might therefore also have a function in spruce as insect deterrents in addition to their antifungal properties, but more research on this topic is required.

12.3.4 Phenylpropanoids

Apart from the ubiquitous phenolic acids, cinnamic acid, coumaric acid, and ferulic acid, a number of different phenylpropanoid glycosides have been described in spruce. These compounds include the coumarin glucoside, skimmin, hydoxybenzoic acid glucosides, and hydroxyacetophenone glucosides (e.g., Strack et al. 1989). All of these compounds are derived from coumaric acid (Neish 1959; Loscher and Heide 1994; Vialart et al. 2012), but details on their biosynthesis in spruce are still lacking (Fig. 12.1).

Among the phenylpropanoids, only the hydroxyacetophenone glucosides, picein, and pungenin, accumulate to high concentrations in spruce needles. Two UDP-dependent glucosyltransferases were described in white spruce that transfer glucose to the aglucones during their biosynthesis rendering them non-toxic to plant cells (Mageroy et al. 2017a). However, upon herbivory, the needles express a β-glucosidase enzyme to release the aglucones piceol and pungenol (Mageroy et al. 2015, 2017b) which are toxic to insects. High expression of the β-glucosidase (Mageroy et al. 2015; Parent et al.

2018) as well as high accumulation of acetophenone glucosides in spruce needles (Delvas et al. 2011) are both markers for resistance against the spruce budworm in white spruce. Apart from their important role in anti-herbivore defense, the aglucones also accumulate in spruce needles in response to environmental stress (Hoque 1985), but their role during abiotic stress is still under investigation (Zhang et al. 2018).

12.4 Biosynthesis of Spruce Phenolics in a Changing Climate

Global climate change has complex effects on plants. On the one hand, elevated temperatures and higher CO_2 increase growth and carbon fixation (Robinson et al. 2012). As nitrogen and phosphorus are still limiting factors for plant productivity, it is predicted, that the additional fixed carbon will accumulate mainly in the form of carbohydrates and non-nitrogenous secondary metabolites (Lindroth 2010). On the other hand, higher temperatures and lower levels of precipitation will limit this productivity (Dale et al. 2001). Spruce is adapted to colder climates and is sensitive to drought stress (Prunier et al. 2011). The massive bark beetle attacks, killing millions of spruce trees worldwide, are known to be the result of increased temperatures and drought, clearly illustrating the disastrous effects that climate change already has on this tree genus (Bentz et al. 2010). As phenolics are an important component of spruce defense against biotic and abiotic stress, the effects of increased levels of CO_2 and temperature, as well as the effects of lower precipitation on the biosynthesis of phenolics are of great interest.

In angiosperms, it was shown that increased CO_2 levels increase the flux through the phenylpropanoid pathway, resulting in higher accumulation of phenolics (Robinson et al. 2012 and references therein). An opposite effect is seen in plants grown at warmer temperatures at ambient CO_2 levels, where a significant decline in phenolics was observed (e.g., Kuokkanen et al. 2001). In agreement with observations made in angiosperms, higher temperatures were shown to result in a decline in phenolics in spruce (Sallas et al. 2003; Riikonen et al. 2012; Zhang et al. 2018). The effect of higher CO_2 levels was less clear, with both increased as well as reduced phenolic levels reported for two different experiments (Sallas et al. 2003; Zhang et al. 2018). In pine, on the other hand, higher CO_2 levels resulted in an increased accumulation of phenolics in the needles (Gebauer et al. 1997; Zhang et al. 2018). Interestingly, in a study on spruce where drought was simulated by reducing CO_2 levels, a significant decrease in phenolics was observed compared to ambient CO_2 levels (Huang et al. 2019). This supports the hypothesis that higher CO_2 levels could result in higher accumulation of phenolic compounds in spruce. Taken together, these results indicate that both drought as well as higher temperatures will result in lower accumulation of phenolics in spruce and that these two effects will cancel out the positive effect of high CO_2 on growth as well as biosynthesis of phenolics. In this scenario, lower levels of phenolics could decrease spruce resistance to both abiotic as well as biotic stresses, which could result in a decline of the genus. Unfortunately, manifestations of such declines can already be observed worldwide in spruce (Bentz et al. 2010). However, this genus also possesses an impressive arsenal of other chemical defenses, such as terpenoid oleoresin and piperidine alkaloids. Experimental evidence suggests that these compounds will accumulate to much higher levels under elevated temperatures (Virjamo et al. 2014). Therefore, due to the plasticity of its chemical defenses as well as its high genetic diversity in wild populations, definitive predictions on how this genus will exactly adapt to future climate conditions cannot be made.

12.5 Future Research Directions

Phenolics are compounds which play important roles in growth and defense. Over the last decade, progress has been made in elucidating the biosynthesis of some ecologically important branches of the phenylpropanoid pathway in spruce, for example, the stilbene and late flavonoid pathways (Hammerbacher et al. 2011, 2014). A few exceptionally interesting studies were published recently, placing spruce phenolics into an ecological or applied context. For example, the roles of acetophenones and stone cells as anti-herbivore defenses in spruce and the mechanism by which these defenses are deployed have recently been elucidated (Mageroy et al. 2015; Whitehill et al. 2016a). This information can now be used by tree breeders to select for more resistant genotypes. In another series of studies, flavan-3-ol defenses against fungal infection were studied in Norway spruce, resulting in the discovery of a genetic marker linked to partial resistance against *Heterobasidion* spp. infection (Nemesio-Gorriz et al. 2016).

With the recent development of a set of tools to manipulate bark beetles in the laboratory, the effects of phenolic compounds against this most devastating pest complex can be evaluated, opening up new avenues of research (Hammerbacher et al. 2019). As genome sequence information for white spruce and Norway spruce is becoming more extensive and complete, more tools are becoming available to study phenolic metabolism in greater detail and at a larger scale. Furthermore, curated annotations of coding genes in the genomes now allow for more targeted candidate gene selection and functional characterization (Nystedt et al. 2013; Warren et al. 2015). However, in order to conduct large-scale studies on the ecological importance of phenolics, more genes encoding enzymes in the biosynthetic pathways of these compounds should be functionally annotated to support and complement bioinformatics approaches.

Climatic conditions in the regions where spruce is endemic are predicted to become warmer and dryer (Serreze et al. 2000). For this reason, spruce will experience more abiotic stress which will favor pest and disease outbreaks. For this reason, research should focus on mechanisms by which abiotic stress can be mitigated. The flavonoids are known protectants against abiotic stress. These anti-oxidant compounds could potentially counteract the effects of drought stress and oxidative cell damage and thus are important candidates for further studies. The lignans, which are known anti-microbial compounds have not been studied extensively in spruce either. Since these compounds could be important in defense against pathogen attack as well as abiotic stress, research to elucidate their ecological roles should be encouraged to close this important knowledge gap.

References

Auger MA, Jay-Allemand C, Bastien C, Geri C (1994) Quantitative variations of taxifolin and its glucoside in *Pinus sylvestris* needles consumed by *Diprion pini* larvae. Ann des sci for 51(2):135–146

Bahnweg G, Schubert R, Kehr RD, Müller-Starck G, Heller W, Langebartels C, Sandermann H Jr (2000) Controlled inoculation of Norway spruce (*Picea abies*) with *Sirococcus conigenus*: PCR-based quantification of the pathogen in host tissue and infection-related increase of phenolic metabolites. Trees 14 (8):435–441

Bedon F, Levasseur C, Grima-Pettenati J, Séguin A, MacKay J (2009) Sequence analysis and functional characterization of the promoter of the *Picea glauca* cinnamyl alcohol dehydrogenase gene in transgenic white spruce plants. Plant Cell Rep 28(5):787–800

Bentz BJ, Régnière J, Fettig CJ, Hansen EM, Hayes JL, Hicke JA, Kelsey RG, Negrón JF, Seybold SJ (2010) Climate change and bark beetles of the western United States and Canada: direct and indirect effects. BioScience 60(8):602–613

Boerjan W, Ralph J, Baucher M (2003) Lignin biosynthesis. Annu Rev Plant Biol 54(1):519–546

Brignolas F, Lieutier F, Sauvard D, Yart A, Drouet A, Claudot AC (1995a) Changes in soluble-phenol content of Norway spruce (Picea abies) phloem in response to wounding and inoculation with Ophiostoma polonicum. Eur J For Path 25(5):253–265

Brignolas F, Lacroix B, Lieutier F, Sauvard D, Drouet A, Claudot AC, Yart A, Berryman AA, Christiansen E (1995b) Induced responses in phenolic metabolism in two Norway spruce clones after wounding and inoculations with *Ophiostoma polonicum*, a bark beetle-associated fungus. Plant Physiol 109(3):821–827

Celedon JM, Yuen MM, Chiang A, Henderson H, Reid KE, Bohlmann J (2017) Cell-type-and tissue-specific transcriptomes of the white spruce (Picea glauca) bark unmask fine-scale spatial patterns of constitutive and induced conifer defense. Plant J 92 (4):710–726

Cheng C, Xu L, Xu D, Lou Q, Lu M, Sun J (2016) Does cryptic microbiota mitigate pine resistance to an invasive beetle-fungus complex? Implications for invasion potential. Sci Rep 6:e33110

Chiron H, Drouet A, Claudot AC, Eckerskorn C, Trost M, Heller W, Ernst D, Sandermann H (2000) Molecular cloning and functional expression of a stress-induced multifunctional O-methyltransferase with pinosylvin methyltransferase activity from Scots pine (*Pinus sylvestris* L.). Plant Mol Biol 44(6):733–745

Chong J, Poutaraud A, Hugueney P (2009) Metabolism and roles of stilbenes in plants. Plant Sci 177(3):143–155

Cvikrová M, Malá J, Hrubcová M, Eder J, Foretová S (2008) Induced changes in phenolic acids and stilbenes in embryogenic cell cultures of Norway spruce by culture filtrate of *Ascocalyx abietina*. J Plant Dis Prot 115(2):57–62

Dale VH, Joyce LA, McNulty S, Neilson RP, Ayres MP, Flannigan MD, Hanson PJ, Irland LC, Lugo AE, Peterson CJ, Simberloff D (2001) Climate change and forest disturbances: climate change can affect forests by altering the frequency, intensity, duration, and timing of fire, drought, introduced species, insect and pathogen outbreaks, hurricanes, windstorms, ice storms, or landslides. BioScience 51(9):723–734

Dalman K, Wind JJ, Nemesio-Gorriz M, Hammerbacher A, Lundén K, Ezcurra I, Elfstrand M (2017) Overexpression of PaNAC03, a stress induced NAC gene family transcription factor in Norway spruce leads to reduced flavonol biosynthesis and aberrant embryo development. BMC Plant Biol 17(1):e6

Danielsson M, Lundén K, Elfstrand M, Hu J, Zhao T, Arnerup J, Ihrmark K, Swedjemark G, Borg-Karlson AK, Stenlid J (2011) Chemical and transcriptional responses of Norway spruce genotypes with different susceptibility to *Heterobasidion* spp. infection. BMC Plant Biol. 11(1):e154

Dar AA, Arumugam N (2013) Lignans of sesame: purification methods, biological activities and biosynthesis–a review. Bioorg Chem 50:1–10

Dauwe R, Holliday JA, Aitken SN, Mansfield SD (2012) Metabolic dynamics during autumn cold acclimation within and among populations of Sitka spruce (*Picea sitchensis*). New Phytol 194(1):192–205

Davin LB, Lewis NG (2000) Dirigent proteins and dirigent sites explain the mystery of specificity of radical precursor coupling in lignan and lignin biosynthesis. Plant Physiol 123(2):453–462

Deflorio G, Horgan G, Woodward S, Fossdal CG (2011) Gene expression profiles, phenolics and lignin of Sitka spruce bark and sapwood before and after wounding and inoculation with *Heterobasidion annosum*. Phys. Mol. Plant Path. 75(4):180–187

Delvas N, Bauce É, Labbé C, Ollevier T, Bélanger R (2011) Phenolic compounds that confer resistance to spruce budworm. Entomol Exp Appl 141(1):35–44

Evensen PC, Solheim H, Høiland K, Stenersen J (2000) Induced resistance of Norway spruce, variation of phenolic compounds and their effects on fungal pathogens. For Pathol 30(2):97–108

Faccoli M, Schlyter F (2007) Conifer phenolic resistance markers are bark beetle antifeedant semiochemicals. Agric. For Entomol 9(3):237–245

Fedorova TE, Fedorov SV, Babkin VA (2016) Oligolignans in the wood of *Picea obovata* Ledeb. Russ J Bioorg Chem 42(7):712–715

Fischbach RJ, Kossmann B, Panten H, Steinbrecher R, Heller W, Seidlitz HK, Sandermann H, Hertkorn N, Schnitzler JP (1999) Seasonal accumulation of ultraviolet-B screening pigments in needles of Norway spruce (*Picea abies* (L.) Karst.). Plant Cell Environ 22 (1):27–37

Fossdal CG, Nagy NE, Hietala AM, Kvaalen H, Slimestad R, Woodward S, Solheim H (2012) Indications of heightened constitutive or primed host response affecting the lignin pathway transcripts and phenolics in mature Norway spruce clones. Tree Phys 32 (9):1137–1147

Franceschi VR, Krokene P, Krekling T, Christiansen E (2000) Phloem parenchyma cells are involved in local and distant defense responses to fungal inoculation or bark-beetle attack in Norway spruce (Pinaceae). Am J Bot 87(3):314–326

Fraser CM, Chapple C (2011) The phenylpropanoid pathway in *Arabidopsis*. Arab Book 9:e152

Friedmann M, Ralph SG, Aeschliman D, Zhuang J, Ritland K, Ellis BE, Bohlmann J, Douglas CJ (2007) Microarray gene expression profiling of developmental transitions in Sitka spruce (*Picea sitchensis*) apical shoots. J Exp Bot 58(3):593–614

Ganthaler A, Stöggl W, Mayr S, Kranner I, Schüler S, Wischnitzki E, Sehr EM, Fluch S, Trujillo-Moya C (2017) Association genetics of phenolic needle compounds in Norway spruce with variable susceptibility to needle bladder rust. Plant Mol Biol 94(3):229–251

Gebauer RL, Strain BR, Reynolds JF (1997) The effect of elevated CO_2 and N availability on tissue concentrations and whole plant pools of carbon-based secondary compounds in loblolly pine (*Pinus taeda*). Oecologia 113(1):29–36

Hall D, De Luca V (2007) Mesocarp localization of a bifunctional resveratrol/hydroxycinnamic acid glucosyltransferase of Concord grape (Vitis labrusca). Plant J 49(4):579–591

Hamberger B, Ohnishi T, Hamberger B, Séguin A, Bohlmann J (2011) Evolution of diterpene metabolism: Sitka spruce CYP720B4 catalyzes multiple oxidations in resin acid biosynthesis of conifer defense against insects. Plant Physiol 157(4):1677–1695

Hammerbacher A, Ralph SG, Bohlmann J, Fenning TM, Gershenzon J, Schmidt A (2011) Biosynthesis of the major tetrahydroxystilbenes in spruce, astringin and isorhapontin, proceeds via resveratrol and is enhanced by fungal infection. Plant Physiol 157(2):876–890

Hammerbacher A, Schmidt A, Wadke N, Wright LP, Schneider B, Bohlmann J, Brand WA, Fenning TM, Gershenzon J, Paetz C (2013) A common fungal associate of the spruce bark beetle metabolizes the stilbene defenses of Norway spruce. Plant Physiol 162 (3):1324–1336

Hammerbacher A, Paetz C, Wright LP, Fischer TC, Bohlmann J, Davis AJ, Fenning TM, Gershenzon J, Schmidt A (2014) Flavan-3-ols in Norway spruce: biosynthesis, accumulation, and function in response to attack by the bark beetle-associated fungus *Ceratocystis polonica*. Plant Physiol 164(4):2107–2122

Hammerbacher A, Raguschke B, Wright LP, Gershenzon J (2018) Gallocatechin biosynthesis via a flavonoid 3′, 5′-hydroxylase is a defense response in Norway spruce against infection by the bark beetle-associated sap-staining fungus *Endoconidiophora polonica*. Phytochemistry 148:78–86

Hammerbacher A, Kandasamy D, Ullah C, Schmidt A, Wright LP, Gershenzon J (2019) Flavanone-3-hydroxylase plays an important role in the biosynthesis of spruce phenolic defenses against bark beetles and their fungal associates. Front Plant Sci 10 (in press)

Heilemann J, Strack D (1991) Flavonol glucosyltransferase from Norway spruce needles. Phytochemistry 30(6):1773–1776

Higuchi T (1990) Lignin biochemistry: biosynthesis and biodegradation. Wood Sci Technol 24(1):23–63

Holton TA, Brugliera F, Tanaka Y (1993) Cloning and expression of flavonol synthase from *Petunia hybrida*. Plant J 4(6):1003–1010

Hoque E (1985) Norway spruce dieback: occurrence, isolation and biological activity of p-hydroxy acetophenone and p-hydroxy acetophenone-O-glucoside and their possible roles during stress phenomena. Eur J For Pathol 15(3):129–145

Hoque E, Remus G (1999) Natural UV-screening mechanisms of Norway spruce (*Picea abies* [L.] Karst.) needles. Photochem Photobiol 69(2):177–192

Howles PA, Sewalt VJ, Paiva NL, Elkind Y, Bate NJ, Lamb C, Dixon RA (1996) Overexpression of L-phenylalanine ammonia-lyase in transgenic tobacco plants reveals control points for flux into phenylpropanoid biosynthesis. Plant Physiol 112(4):1617–1624

Huang J, Hammerbacher A, Weinhold A, Reichelt M, Gleixner G, Behrendt T, van Dam NM, Sala A, Gershenzon J, Trumbore S, Hartmann H (2019) Eyes on the future–evidence for trade-offs between growth, storage and defense in Norway spruce. New Phytol 222(1):144–158

Hudgins JW, Christiansen E, Franceschi VR (2003) Methyl jasmonate induces changes mimicking anatomical defenses in diverse members of the Pinaceae. Tree Physiol 23(6):361–371

Hudgins JW, Christiansen E, Franceschi VR (2004) Induction of anatomically based defense responses in stems of diverse conifers by methyl jasmonate: a phylogenetic perspective. Tree Physiol 24(3):251–264

Hudgins JW, Ralph SG, Franceschi VR, Bohlmann J (2006) Ethylene in induced conifer defense: cDNA cloning, protein expression, and cellular and subcellular localization of 1-aminocyclopropane-1-carboxylate oxidase in resin duct and phenolic parenchyma cells. Planta 224(4):e865

Jyske T, Laakso T, Latva-Mäenpää H, Tapanila T, Saranpää P (2014) Yield of stilbene glucosides from the bark of young and old Norway spruce stems. Biomass Bioenergy 71:216–227

Jyske TM, Suuronen JP, Pranovich AV, Laakso T, Watanabe U, Kuroda K, Abe H (2015) Seasonal variation in formation, structure, and chemical properties of phloem in *Picea abies* as studied by novel microtechniques. Planta 242(3):613–629

Jyske T, Kuroda K, Suuronen JP, Pranovich A, Roig-Juan S, Aoki D, Fukushima K (2016) *In planta* localization of stilbenes within *Picea abies* phloem. Plant Physiol 172(2):913–928

Kim BG, Kim DH, Sung SH, Kim DE, Chong Y, Ahn JH (2010) Two O-methyltransferases from *Picea abies*: characterization and molecular basis of different reactivity. Planta 232(4):837–844

Kiselev KV, Grigorchuk VP, Ogneva ZV, Suprun AR, Dubrovina AS (2016) Stilbene biosynthesis in the needles of spruce *Picea jezoensis*. Phytochemistry 131:57–67

Koutaniemi S, Warinowski T, Kärkönen A, Alatalo E, Fossdal CG, Saranpää P, Laakso T, Fagerstedt KV, Simola LK, Paulin L, Rudd S (2007) Expression profiling of the lignin biosynthetic pathway in Norway spruce using EST sequencing and real-time RT-PCR. Plant Mol Biol 65(3):311–328

Kovalchuk A, Zeng Z, Ghimire RP, Kivimäenpää M, Raffaello T, Liu M, Mukrimin M, Kasanen R, Sun H, Julkunen-Tiitto R, Holopainen JK (2019) Dual RNA-seq analysis provides new insights into interactions between Norway spruce and necrotrophic pathogen *Heterobasidion annosum* sl. BMC Plant Biol 19(1):e2

Krajnc AU, Novak M, Felicijan M, Kraševec N, Lešnik M, Zupanec N, Komel R (2014) Antioxidative response patterns of Norway spruce bark to low-density *Ceratocystis polonica* inoculation. Trees 28 (4):1145–1160

Kuokkanen K, Julkunen-Tiitto R, Keinänen M, Niemelä P, Tahvanainen J (2001) The effect of elevated CO 2 and temperature on the secondary chemistry of Betula pendula seedlings. Trees 15(6):378–384

Laitinen T, Morreel K, Delhomme N, Gauthier A, Schiffthaler B, Nickolov K, Brader G, Lim KJ, Teeri TH, Street NR, Boerjan W (2017) A key role for apoplastic H_2O_2 in Norway spruce phenolic metabolism. Plant Physiol 174(3):1449–1475

Lattanzio V, Kroon PA, Quideau S, Treutter D (2008) Plant phenolics—secondary metabolites with diverse functions. Recent Adv Polyphen Res 1:1–35

Le TK, Jang HH, Nguyen HT, Doan TT, Lee GY, Park KD, Ahn T, Joung YH, Kang HS, Yun CH (2017) Highly regioselective hydroxylation of polydatin, a resveratrol glucoside, for one-step synthesis of astringin, a piceatannol glucoside, by P450 BM3. Enz Microb Technol 97:34–42

Lee YK, Alexander D, Wulff J, Olsen JE (2014) Changes in metabolite profiles in Norway spruce shoot tips during short-day induced winter bud development and long-day induced bud flush. Metabolomics 10(5):842–858

Li SH, Niu XM, Zahn S, Gershenzon J, Weston J, Schneider B (2008) Diastereomeric stilbene glucoside dimers from the bark of Norway spruce (*Picea abies*). Phytochemistry 69(3):772–782

Li SH, Nagy NE, Hammerbacher A, Krokene P, Niu XM, Gershenzon J, Schneider B (2012) Localization of phenolics in phloem parenchyma cells of Norway spruce (*Picea abies*). ChemBioChem 13(18):2707–2713

Lind M, Källman T, Chen J, Ma XF, Bousquet J, Morgante M, Zaina G, Karlsson B, Elfstrand M, Lascoux M, Stenlid JA (2014) *Picea abies* linkage map based on SNP markers identifies QTLs for four aspects of resistance to *Heterobasidion parviporum* infection. PLoS ONE 9(7):e101049

Lindroth RL (2010) Impacts of elevated atmospheric CO_2 and O_3 on forests: phytochemistry, trophic interactions, and ecosystem dynamics. J Chem Ecol 36(1):2–21

Loscher R, Heide L (1994) Biosynthesis of *p*-hydroxybenzoate from *p*-coumarate and p-coumaroyl-coenzyme A in cell-free extracts of *Lithospermum erythrorhizon* cell cultures. Plant Physiol 106(1):271–279

Luo J, Nishiyama Y, Fuell C, Taguchi G, Elliott K, Hill L, Tanaka Y, Kitayama M, Yamazaki M, Bailey P, Parr A (2007) Convergent evolution in the BAHD family of acyl transferases: identification and characterization of anthocyanin acyl transferases from *Arabidopsis thaliana*. Plant J 50(4):678–695

Mageroy MH, Parent G, Germanos G, Giguère I, Delvas N, Maaroufi H, Bauce É, Bohlmann J, Mackay JJ (2015) Expression of the β-glucosidase gene Pgβglu-1 underpins natural resistance of white spruce against spruce budworm. Plant J 81(1):68–80

Mageroy MH, Jancsik S, Saint Yuen MM, Fischer M, Withers SG, Paetz C, Schneider B, Mackay J, Bohlmann J (2017a) A conifer UDP-sugar dependent glycosyltransferase contributes to acetophenone metabolism and defense against insects. Plant Physiol 175(2):641–651

Mageroy MH, Lachance D, Jancsik S, Parent G, Séguin A, Mackay J, Bohlmann J (2017b) *In vivo* function of Pgβglu-1 in the release of acetophenones in white spruce. PeerJ 5:e3535

Malá J, Hrubcová M, Máchová P, Cvrčková H, Martincová O, Cvikrová M (2011) Changes in phenolic acids and stilbenes induced in embryogenic cell cultures of Norway spruce by two fractions of *Sirococcus strobilinus* mycelian. J For Sci 57(1):1–7

Nagy NE, Fossdal CG, Krokene P, Krekling T, Lönneborg A, Solheim H (2004) Induced responses to pathogen infection in Norway spruce phloem: changes in polyphenolic parenchyma cells, chalcone synthase transcript levels and peroxidase activity. Tree Physiol 24(5):505–515

Neish AC (1959) Biosynthesis of pungenin from C14-labelled compounds by Colorado spruce. Can J Bot 37(5):1085–1100

Nemesio-Gorriz M, Hammerbacher A, Ihrmark K, Källman T, Olson Å, Lascoux M, Stenlid J, Gershenzon J, Elfstrand M (2016) Different alleles of a gene encoding leucoanthocyanidin reductase (PaLAR3) influence resistance against the fungus *Heterobasidion parviporum* in *Picea abies*. Plant Physiol 171(4):2671–2681

Nystedt B, Street NR, Wetterbom A, Zuccolo A, Lin YC, Scofield DG, Vezzi F, Delhomme N, Giacomello S, Alexeyenko A, Vicedomini R (2013) The Norway spruce genome sequence and conifer genome evolution. Nature 497(7451):579

OuYang F, Mao JF, Wang J, Zhang S, Li Y (2015) Transcriptome analysis reveals that red and blue light regulate growth and phytohormone metabolism in Norway spruce [*Picea abies* (L.) Karst.]. PloS One 10(8):e0127896

Parent GJ, Giguère I, Mageroy M, Bohlmann J, MacKay JJ (2018) Evolution of the biosynthesis of two hydroxyacetophenones in plants. Plant Cell Environ 41(3):620–629

Pavy N, Pelgas B, Laroche J, Rigault P, Isabel N, Bousquet J (2012) A spruce gene map infers ancient plant genome reshuffling and subsequent slow evolution in the gymnosperm lineage leading to extant conifers. BMC Biol 10(1):e84

Pelletier MK, Shirley BW (1996) Analysis of flavanone 3-hydroxylase in *Arabidopsis* seedlings (Coordinate regulation with chalcone synthase and chalcone isomerase). Plant Physiol 111(1):339–345

Peters DJ, Constabel CP (2002) Molecular analysis of herbivore-induced condensed tannin synthesis: cloning and expression of dihydroflavonol reductase from trembling aspen (Populus tremuloides). Plant J 32(5):701–712

Piispanen R, Willför S, Saranpää P, Holmbom B (2008) Variation of lignans in Norway spruce (*Picea abies* [L.] Karst.) knotwood: within-stem variation and the effect of fertilisation at two experimental sites in Finland. Trees 22(3):317–328

Porth I, Hamberger B, White R, Ritland K (2011) Defense mechanisms against herbivory in *Picea*: sequence evolution and expression regulation of gene family members in the phenylpropanoid pathway. BMC Genomics 12(1):e608

Prunier J, Laroche J, Beaulieu J, Bousquet J (2011) Scanning the genome for gene SNPs related to climate adaptation and estimating selection at the molecular level in boreal black spruce. Mol Ecol 20(8):1702–1716

Pukacki PM (2004) Effect of water deficit on oxidative stress and degradation of cell membranes in needles of Norway spruce (*Picea abies*). Acta Physiol Plant 26(4):431–442

Pukacki PM, Modrzyński J (1998) The influence of ultraviolet-B radiation on the growth, pigment production and chlorophyll fluorescence of Norway spruce seedlings. Acta Physiol Plant 20(3):245–250

Pukacki PM, Kamińska-Rożek E (2013) Reactive species, antioxidants and cold tolerance during deacclimation of *Picea abies* populations. Acta Physiol Plant 35(1):129–138

Ralph SG, Yueh H, Friedmann M, Aeschliman D, Zeznik JA, Nelson CC, Butterfield YS, Kirkpatrick R, Liu J, Jones SJ, Marra MA (2006a) Conifer defence against insects: microarray gene expression profiling of Sitka spruce (Picea sitchensis) induced by mechanical wounding or feeding by spruce budworms (Choristoneura occidentalis) or white pine weevils (Pissodes strobi) reveals large-scale changes of the host transcriptome. Plant Cell Environ 29(8):1545–1570

Ralph S, Park JY, Bohlmann J, Mansfield SD (2006b) Dirigent proteins in conifer defense: gene discovery, phylogeny, and differential wound-and insect-induced expression of a family of DIR and DIR-like genes in spruce (Picea spp.). Plant Mol Biol 60(1):e21

Ralph SG, Jancsik S, Bohlmann J (2007) Dirigent proteins in conifer defense II: Extended gene discovery, phylogeny, and constitutive and stress-induced gene expression in spruce (*Picea* spp.). Phytochemistry 68(14):1975–1991

Richard S, Lapointe G, Rutledge RG, Séguin A (2000) Induction of chalcone synthase expression in white spruce by wounding and jasmonate. Plant Cell Physiol 41(8):982–987

Riikonen J, Kontunen-Soppela S, Ossipov V, Tervahauta A, Tuomainen M, Oksanen E, Vapaavuori E, Heinonen J, Kivimäenpää M (2012) Needle metabolome, freezing tolerance and gas exchange in Norway spruce seedlings exposed to elevated temperature and ozone concentration. Tree Physiol 32(9):1102–1112

Rimando AM, Pan Z, Polashock JJ, Dayan FE, Mizuno CS, Snook ME, Liu CJ, Baerson SR (2012) In planta production of the highly potent resveratrol analogue pterostilbene via stilbene synthase and O-methyltransferase co-expression. Plant Biotech J 10 (3):269–283

Robinson EA, Ryan GD, Newman JA (2012) A meta-analytical review of the effects of elevated CO2 on plant–arthropod interactions highlights the importance of interacting environmental and biological variables. New Phytol 194(2):321–336

Rohde M, Waldmann R, Lunderstädt J (1996) Induced defence reaction in the phloem of spruce (*Picea abies*) and larch (*Larix decidua*) after attack by *Ips typographus* and *Ips cembrae*. For Ecol Manag 86(1–3): 51–59

Sallas L, Luomala EM, Utriainen J, Kainulainen P, Holopainen JK (2003) Contrasting effects of elevated carbon dioxide concentration and temperature on Rubisco activity, chlorophyll fluorescence, needle ultrastructure and secondary metabolites in conifer seedlings. Tree Physiol 23(2):97–108

Schiebe C, Hammerbacher A, Birgersson G, Witzell J, Brodelius PE, Gershenzon J, Hansson BS, Krokene P, Schlyter F (2012) Inducibility of chemical defenses in Norway spruce bark is correlated with unsuccessful mass attacks by the spruce bark beetle. Oecologia 170 (1):183–198

Schmidt A, Wächtler B, Temp U, Krekling T, Séguin A, Gershenzon J (2010) A bifunctional geranyl and geranylgeranyl diphosphate synthase is involved in terpene oleoresin formation in *Picea abies*. Plant Physiol 152(2):639–655

Serreze MC, Walsh JE, Chapin FS, Osterkamp T, Dyurgerov M, Romanovsky V, Oechel WC, Morison J, Zhang T, Barry RG (2000) Observational evidence of recent change in the northern high-latitude environment. Climat. Change 46(1–2):159–207

Slimestad R (2003) Flavonoids in buds and young needles of *Picea, Pinus and Abies*. Biochem Syst Ecol 31 (11):1247–1255

Slimestad R, Hostettmann K (1996) Characterisation of phenolic constituents from juvenile and mature needles of Norway spruce by means of high performance liquid chromatography–mass spectrometry. Phytochem Anal 7(1):42–48

Song F, Song G, Dong A, Kong X (2011) Regulatory mechanisms of host plant defense responses to arbuscular mycorrhiza. Acta Ecol Sin 31(6):322–327

Stafford HA (2000) The evolution of phenolics in plants. Recent Adv Phytochem 34:25–54

Stenlid J, Johansson M (1987) Infection of roots of Norway spruce (*Picea abies*) by *Heterobasidion annosum*: II. Early changes in phenolic content and toxicity. Eur J Forest Pathol 17(4–5):217–226

Storer AJ, Speight MR (1996) Relationships between *Dendroctonus micans* Kug. (Coleoptera: Scolytidae) survival and development and biochemical changes in Norway spruce, *Picea abies* (L.) Karst., phloem caused by mechanical wounding. J Chem Ecol 22 (3):559–573

Strack D, Heilemann J, Wray V, Dirks H (1989) Structures and accumulation patterns of soluble and insoluble phenolics from Norway spruce needles. Phytochemistry 28(8):2071–2078

Taylor RJ, Shaw DC (1983) Allelopathic effects of Engelmann spruce bark stilbenes and tannin–stilbene combinations on seed germination and seedling growth of selected conifers. Can J Bot 61(1):279–289

Vialart G, Hehn A, Olry A, Ito K, Krieger C, Larbat R, Paris C, Shimizu BI, Sugimoto Y, Mizutani M, Bourgaud F (2012) A 2-oxoglutarate-dependent dioxygenase from *Ruta graveolens* L. exhibits p-coumaroyl CoA 2′-hydroxylase activity (C2′ H): a missing step in the synthesis of umbelliferone in plants. Plant J 70(3):460–770

Viiri H, Annila E, Kitunen V, Niemelä P (2001) Induced responses in stilbenes and terpenes in fertilized Norway spruce after inoculation with blue-stain fungus, *Ceratocystis polonica*. Trees 15(2):112–122

Virjamo V, Sutinen S, Julkunen-Tiitto R (2014) Combined effect of elevated UVB, elevated temperature and fertilization on growth, needle structure and phytochemistry of young Norway spruce (*Picea abies*) seedlings. Glob Chang Biol 20(7):2252–2260

Vogt T (2010) Phenylpropanoid biosynthesis. Mol Plant 3 (1):2–10

Wadke N, Kandasamy D, Vogel H, Lah L, Wingfield BD, Paetz C, Wright LP, Gershenzon J, Hammerbacher A (2016) The bark-beetle-associated fungus,

Endoconidiophora polonica, utilizes the phenolic defense compounds of its host as a carbon source. Plant Physiol 171(2):914–931
Waldeck DH (1991) Photoisomerization dynamics of stilbenes. Chem Rev 91(3):415–436
Warren RL, Keeling CI, Yuen MM, Raymond A, Taylor GA, Vandervalk BP, Mohamadi H, Paulino D, Chiu R, Jackman SD, Robertson G (2015) Improved white spruce (*Picea glauca*) genome assemblies and annotation of large gene families of conifer terpenoid and phenolic defense metabolism. Plant J 83(2):189–212
Whitehill JG, Henderson H, Strong W, Jaquish B, Bohlmann J (2016a) Function of Sitka spruce stone cells as a physical defence against white pine weevil. Plant Cell Environ 39(11):2545–2556
Whitehill JG, Henderson H, Schuetz M, Skyba O, Yuen MM, King J, Samuels AL, Mansfield SD, Bohlmann J (2016b) Histology and cell wall biochemistry of stone cells in the physical defence of conifers against insects. Plant Cell Environ 39 (8):1646–1661
Whitehill JG, Yuen MM, Henderson H, Madilao L, Kshatriya K, Bryan J, Jaquish B, Bohlmann J (2019) Functions of stone cells and oleoresin terpenes in the conifer defense syndrome. New Phytol 221(3):1503–1517
Wilmouth RC, Turnbull JJ, Welford RW, Clifton IJ, Prescott AG, Schofield CJ (2002) Structure and mechanism of anthocyanidin synthase from *Arabidopsis thaliana*. Structure 10(1):93–103
Woodward S, Pearce RB (1988) The role of stilbenes in resistance of Sitka spruce (*Picea sitchensis* (Bong.) Carr.) to entry of fungal pathogens. Physiol Mol Plant Pathol 33(1):127–149
Xie DY, Sharma SB, Paiva NL, Ferreira D, Dixon RA (2003) Role of anthocyanidin reductase, encoded by BANYULS in plant flavonoid biosynthesis. Science 299(5605):396–399
Yaqoob N, Yakovlev IA, Krokene P, Kvaalen H, Solheim H, Fossdal CG (2012) Defence-related gene expression in bark and sapwood of Norway spruce in response to *Heterobasidion parviporum* and methyl jasmonate. Physiol Mol Plant Pathol 77(1):10–16
Zhang Y, Virjamo V, Du W, Yin Y, Nissinen K, Nybakken L, Guo H (2018) Julkunen-Tiitto R (2018) Effects of soil pyrene contamination on growth and phenolics in Norway spruce (*Picea abies*) are modified by elevated temperature and CO_2. Environ Sci Pollut Res 1:1–12
Zhao T, Solheim H, Långström B, Borg-Karlson AK (2011) Storm-induced tree resistance and chemical differences in Norway spruce (*Picea abies*). Ann For Sci 8(3):657–665
Zhao T, Kandasamy D, Krokene P, Chen J, Gershenzon J, Hammerbacher A (2019) Fungal associates of the tree-killing bark beetle, Ips typographus, vary in virulence, ability to degrade conifer phenolics and influence bark beetle tunneling behavior. Fungal Ecol 38:71–79

Prospects: The Spruce Genome, a Model for Understanding Gymnosperm Evolution and Supporting Tree Improvement Efforts

13

Ilga M. Porth, Amanda R. De la Torre, and Yousry A. El-Kassaby

Abstract

The spruce reference genomes (*Picea abies*, *P. glauca*, *P. glauca* x *engelmannii*) were published eight years after the release of *Populus trichocarpa's* genome, the first forest tree species ever sequenced (Science 313:1596–1604 (2006)). Although sequencing these highly complex 20 Gigabases spruce genomes represented a major achievement, the current assemblies present many limitations and a chromosome-level assembly reference in this ecologically and economically important genus is still far from reach (Chaps. 2, 7, 8). Nevertheless, improvements in next-generation sequencing technologies and bioinformatic methods have led to important advancements as outlined in this book (Chap. 3). Increasingly complete and more accurate spruce genome sequences will serve the research community in many aspects. Foremost, they will enable improved annotations in the gene space, new discoveries in plant secondary metabolism, and support new breeding strategies for physiologically active compounds and beyond (Chaps. 5, 9, 12, and 13). An even more uncharted territory is the elucidation of epigenetic memory (Chaps. 4 and 6). In the context of adaptation to future climates, spruces are prime study subjects, as important representatives of boreal ecosystems (Chaps. 10 and 11). Following the example of plant species with less complex genomes, whole-genome resequencing will become state-of-the-art in population genomics of spruces and thus invaluable for finding the most important genetic/genomic variants underlying traits of interest. This volume details the progress of the spruce genome sequencing efforts as a cornerstone for exciting discoveries in gymnosperm genomics.

13.1 Main Text

Almost 15 years ago from this day, a book chapter on "Spruce" (Bousquet et al. 2007) appeared in the Volume of "Forest Trees" as part of the Compendium on "Genome Mapping and Molecular Breeding in Plants" (Springer ed.). The major focus at the time was on the

I. M. Porth (✉)
Départment des Sciences du Bois et de la Forêt, Faculté de Foresterie, de Géographie et Géomatique, Université Laval, 1030 Avenue de la Médecine, Quebec City, QC G1V 0A6, Canada
e-mail: ilga.porth@sbf.ulaval.ca

A. R. De la Torre
School of Forestry, Northern Arizona University, 200 E. Pine Knoll, Flagstaff, AZ 86011, USA
e-mail: Amanda.de-la-torre@nau.edu

Y. A. El-Kassaby
Department of Forest and Conservation Sciences, Faculty of Forestry, The University of British Columbia, 2424 Main Mall, Vancouver, BC V6T 1Z4, Canada
e-mail: y.el-kassaby@ubc.ca

I. M. Porth and A. R. De la Torre (eds.), *The Spruce Genome*, Compendium of Plant Genomes,
https://doi.org/10.1007/978-3-030-21001-4_13

construction of species-specific genetic linkage maps and performing comparative mapping among related species to study species evolution in the Pinaceae family. Gene space was sparsely represented, but a project on identifying and mapping conserved orthologous sets (COS) of markers for Pinaceae was well underway (Liewlaksaneeyanawin et al. 2009). The latest spruce maps have now reached a saturation with thousands or tens of thousands of genes (Bernhardsson et al. 2019; Pavy et al. 2017), only comparable to the ultra-dense maps in *Pinus taeda* (26k, De La Torre et al. 2019a) and *Ginkgo biloba* (Liu et al. 2017). Nowadays, such ultra-dense genetic linkage maps mainly serve to improve spruce genome annotation. In fact, high-resolution linkage maps have become invaluable to uncover errors in the genome assembly and guide anchoring and ordering scaffolds into putative chromosomes (Bernhardsson et al. 2019).

As mentioned on several occasions throughout this plant genome series volume, reduced representation sequencing of the massive spruce genome has become very popular. Restriction site-associated DNA (RAD) sequencing benefits substantially from the existence of an annotated reference genome as it helps to interpret results. Furthermore, an annotated spruce reference genome helps to identify the targets of selection that stem from population/landscape genomics studies, though it is clear that further experiments need to be conducted that support the role of the targets in spruce adaptation. An improved spruce reference genome annotation can, therefore, identify the biological processes that are generating these outliers in population/landscape screens. Nevertheless, field experiments such as reciprocal transplants are indispensable along with functional molecular studies that can provide the ultimate proof for a potentially adaptive phenotypic change.

Spruce, conifers, and gymnosperms in general, represent interesting study subjects to discover specialized plant metabolism. Several compounds that are bioactive against pests and pathogens could be isolated from resin and structurally characterized. These natural products can also be interesting for biotechnological applications such as "green" biofuels and bioproducts or in pharmacological uses (De la Torre et al. 2019b). Peter (2018) from the University of Florida, Gainsville, proposed that targeted terpene breeding and engineering has the potential to boost the industry. Substantially increased tissue content for terpenes will make planted and managed forest trees, such as spruces, much more valuable. The extraction of valuable secondary compounds can be done already *in forest* on live trees. However, so far, the genetic mechanisms implicated in terpene secretory tissues determination (resin duct identity) and development have not been deciphered but might likely be polygenic. Research on this subject is currently underway in pine (Peter 2018) but has not been done for spruces yet. Certainly, a better knowledge of the spruce genes and gene families implicated in important phenotypic variation is the first step to generate such spruce plantations with improved varieties through marker-assisted breeding, genomic selection, or genetic modifications (De la Torre et al. 2019b). Another important group of secondary compounds of interest are the phenolic metabolites (Witzell and Martin 2008), also highly diverse and abundant in spruces. But as with the terpenes, environmental and developmental factors affect the metabolic pathway flux, and spruces may show genetic trade-offs between allocation to constitutive and induced metabolic responses (Witzell and Martin 2008; Porth et al. 2018). Therefore, we need to better understand the complexity of the metabolic pathway networks in spruces. Their regulation remains a major challenge in research but deciphering the underlying genetics will facilitate future molecular breeding of agronomically useful traits (De la Torre et al. 2019b).

A relatively underdeveloped genomics research field in spruces is the study of reproductive development; therefore, we have no dedicated chapter in this volume. However, it is anticipated that a fully annotated spruce reference genome will facilitate efforts in this regard. Gene families of varying sizes seem to be implicated in the process of reproductive and seed development in spruce, and for a synopsis we refer the reader to De la Torre et al. (2019b) and the references of original work cited therein. It is interesting to note, however, that certain reproductive development gene family members in spruce appear to be intertwined in regulatory genetic networks for biotic stress responses in the species. Perhaps this is due to the involvement of some early members of these important molecular modules in seasonal growth regulation (phenology). Further research on this topic is warranted.

In 2014, the first genomic selection (GS) project for (white) spruce was published by Beaulieu et al. Several other Canadian spruces GS studies followed including for black spruce (also led by Dr. Jean Bousquet at Laval University, Quebec) and for Interior spruce (led by Dr. Yousry El-Kassaby at The University of British Columbia), as reviewed in the appropriate chapter of this volume. Recently, a genomic selection study for Norway spruce was published in 2018 (led by Dr. Harry Wu from the Umeå Plant Science Centre). In the most recent of these studies, the Norway spruce reference genome was used to annotate genetic variants. This provided insights into the location of the individual SNPs used in the analyses, i.e., in coding versus non-coding regions (Chen et al. 2018). Establishing strong genetic linkage (linkage disequilibrium in a population context) between the surveyed genetic variants and the QTN(s) [quantitative trait nucleotides, directly controlling trait variability] would be important to confidently predict a trait of interest in a population. However, in most cases, we do not have the knowledge about the genetic architecture of a trait we want to predict, except that the trait is polygenic and not of simple Mendelian inheritance. Because spruces are wind pollinated and outcrossing and spruce breeding populations have "shallow pedigree" (spruces have not yet undergone advanced breeding in contrast to many agricultural crops), the number of variants that need to be surveyed in order to capture this LD is enormous. Therefore, successful genomic prediction in spruce relies heavily on accurate genetic/genomic relatedness estimation between individuals of a population. Nevertheless, due to the fragmented condition of the white and Interior spruce genome assemblies (not anchored and not ordered along the chromosomes), there is currently not a significant advantage to use the genome information instead of de novo SNP markers obtainable from RAD sequencing, for example, in GS of spruce (Gamal El-Dien et al. 2015).

It is interesting to note here that so far only single nucleotide polymorphisms (since SNPs are highly abundant throughout the spruce genome) have exclusively been considered as genetic markers. An emerging important new marker system has recently been identified for spruce: variation in DNA copy numbers (CNVs) (Prunier et al. 2017a). The authors used a comparative hybridization array (aCGH) method because of the lack of a contiguous spruce reference genome in order to detect CNVs. Especially genic CNVs and that are inherited from one generation to the next are interesting, because gene CNVs could indeed result in dosage effects on phenotypic variation. Therefore, they represent potent new markers to be considered in quantitative trait variations for growth and phenology in spruce. In fact, structural variation such as CNVs should be implemented in future GS projects for spruces, especially in cases where heterosis is invoked for hybrid performance (Prunier et al. 2017b).

References

Beaulieu J, Doerksen T, Clément S, MacKay J, Bousquet J (2014) Accuracy of genomic selection models in a large population of open-pollinated families in white spruce. Heredity 113(4):343

Bernhardsson C, Vidalis A, Wang X, Scofield DG, Schiffthaler B, Baison J, Street NR, García-Gil MR, Ingvarsson PK (2019) An Ultra-dense haploid genetic map for evaluating the highly fragmented genome assembly of Norway spruce (*Picea abies*). Genes| Genomes|Genet 9(5):1623

Bousquet J, Isabel N, Pelgas B, Cottrell J, Rungis D, Ritland K (2007) Spruce. In: Kole C (eds) Forest trees. Genome mapping and molecular breeding in plants, vol 7. Springer, Berlin. https://doi.org/10.1007/978-3-540-34541-1_3

Chen Z, Baison J, Pan J et al (2018) Accuracy of genomic selection for growth and wood quality traits in two control-pollinated progeny trials using exome capture as the genotyping platform in Norway spruce. BMC Genom 19:946

De La Torre AR, Wilhite B, Neale DB (2019a) Environmental genome-wide association reveals climate adaptation is shaped by subtle to moderate allele frequency shifts in loblolly pine. Genome Biol Evol 11 (10):2976–2989

De La Torre AR, Piot A, Liu B, Wilhite B, Weiss M, Porth I (2019b) Functional and morphological evolution in gymnosperms: a portrait of implicated gene families. Evol Appl. https://doi.org/10.1111/eva.12839

Gamal El-Dien O, Ratcliffe B, Klápště J et al (2015) Prediction accuracies for growth and wood attributes of interior spruce in space using genotyping-by-sequencing. BMC Genom 16:370

Liewlaksaneeyanawin C, Zhuang J, Tang M, Farzaneh N, Lueng G, Cullis C, Findlay S, Ritland CE, Bohlmann J, Ritland K (2009) Identification of COS markers in the Pinaceae. Tree Genet Genomes 5 (1):247–255

Liu H, Cao F, Yin T, Chen Y (2017) A highly dense genetic map for *Ginkgo biloba* constructed using sequence-based markers. Front Plant Sci 8:1041

Pavy N, Lamothe M, Pelgas B, Gagnon F, Birol I, Bohlmann J, Mackay J, Isabel N, Bousquet J (2017) A high-resolution reference genetic map positioning 8.8 K genes for the conifer white spruce: structural genomics implications and correspondence with physical distance. Plant J 90(1):189–203

Peter GF (2018) Breeding and engineering trees to accumulate high levels of terpene metabolites for plant defense and renewable chemicals. Front Plant Sci 9:1672

Porth I, White R, Jaquish B, Ritland K (2018) Partial correlation analysis of transcriptomes helps detangle the growth and defense network in spruce. New Phytol 218(4):1349–1359

Prunier J, Caron S, Lamothe M et al (2017a) Gene copy number variations in adaptive evolution: The genomic distribution of gene copy number variations revealed by genetic mapping and their adaptive role in an undomesticated species, white spruce (*Picea glauca*). Mol Ecol 26:5989–6001

Prunier J, Caron S, MacKay J (2017b) CNVs into the wild: screening the genomes of conifer trees (Picea spp.) reveals fewer gene copy number variations in hybrids and links to adaptation. BMC Genomics 18:97

Witzell J, Martín JA (2008) Phenolic metabolites in the resistance of northern forest trees to pathogens—past experiences and future prospects. Can J Forest Res 38 (11):2711–2727